9/06

The Complex Forest

*Communities, Uncertainty, and
Adaptive Collaborative Management*

Carol J. Pierce Colfer

RESOURCES FOR THE FUTURE
  Washington, DC, USA

CENTER FOR INTERNATIONAL
FORESTRY RESEARCH
  Bogor, Indonesia

An RFF Press book
Published by Resources for the Future
1616 P Street NW
Washington, DC 20036–1400
USA
www.rffpress.org

A copublication of Resources for the Future (www.rff.org) and the Center for International Forestry Research (www.cifor.org)

Library of Congress Cataloging-in-Publication Data

The complex forest : communities, uncertainty, and adaptive collaborative management / by
   Carol J. Pierce Colfer.
   p. cm.
   "An RFF Press book"—T.p. verso.
   ISBN 1-933115-12-2 (cloth : alk. paper) — ISBN 1-933115-13-0 (pbk. : alk. paper)
   1. Forest management—Tropics—Citizen participation. 2. Sustainable forestry
   —Tropics—Citizen participation.   I. Colfer, Carol J. Pierce.

The paper in this book meets the guidelines for permanence and durability of the Committee on Production Guidelines for Book Longevity of the Council on Library Resources. This book was designed and typeset in Bembo and Gill Sans by Chrysalis Editorial. It was copyedited by Sally Atwater. Cover inset photo by Carol Colfer titled "Forest users and managers from Loa Loa on the Ivindo River, Gabon." Cover background photo by Carol Colfer titled "Mulut Mountain forest area, Pasir, East Kalimantan, Indonesia." The cover was designed by Rosenbohm Graphic Design.

ISBN 1-933115-12-2 (cloth)   ISBN 1-933115-13-0 (paper)

About Resources for the Future *and* RFF Press

RESOURCES FOR THE FUTURE (RFF) improves environmental and natural resource policymaking worldwide through independent social science research of the highest caliber. Founded in 1952, RFF pioneered the application of economics as a tool for developing more effective policy about the use and conservation of natural resources. Its scholars continue to employ social science methods to analyze critical issues concerning pollution control, energy policy, land and water use, hazardous waste, climate change, biodiversity, and the environmental challenges of developing countries.

RFF Press supports the mission of RFF by publishing book-length works that present a broad range of approaches to the study of natural resources and the environment. Its authors and editors include RFF staff, researchers from the larger academic and policy communities, and journalists. Audiences for publications by RFF Press include all of the participants in the policymaking process—scholars, the media, advocacy groups, NGOs, professionals in business and government, and the public.

About the Center for International Forestry Research

THE CENTER FOR INTERNATIONAL FORESTRY RESEARCH (CIFOR) was established in 1993 as part of the Consultative Group on International Agricultural Research (CGIAR) in response to global concerns about the social, environmental, and economic consequences of forest loss and degradation. CIFOR research produces knowledge and methods needed to improve the well-being of forest-dependent people and to help tropical countries manage their forests wisely for sustained benefits. This research is done in more than two dozen countries, in cooperation with numerous partners. Since it was founded, CIFOR has also played a central role in influencing global and national forestry policies.

Contents

Foreword

Global Uncertainty and Certitude in Sustainable Forest Management

SINCE SCHOLARS, PRACTITIONERS, AND policymakers first started to link human development and environmental conservation agendas two decades ago, there has been a wealth of study and practice in this new, hybrid field. But there has often been little institutional learning from individual projects, even less comparative study across nations and regions, and little if any success in discerning principles that can be applied to the globe as a whole. This impressive study by Carol J. Pierce Colfer helps to rectify this lacuna in the field of conservation and development, focusing on the sustainable interaction of people with forests.

This volume is based on a project that was launched in 1998 by the Center for International Research in Forestry (CIFOR), headquartered in Bogor, Indonesia, to try to figure out how to develop systems of adaptive collaborative management (ACM) for forest-dwelling communities worldwide. To achieve global coverage, the CIFOR project worked with other donors and their respective research programs, enabling the collaboration of some 90 scholars working at 30 different sites in 11 different countries. Across all 30 sites the project examined seven distinct variables or "analytic dimensions" that were hypothesized to affect forest sustainability: the extent of devolution of forest rights to local communities; the type of forest; the extent of population pressure on the forest; the nature of management goals for the forest; the extent of diversity among forest stakeholders; the level of conflict over forest resources; and the amount of social capital available for collective action by forest communities.

Reflexivity

The Complex Forest, which reports the results of this ambitious global project, is divided into two parts: the second half (Appendix) presents the details of the 30 sites from around the globe, and the first half presents the overall findings that Dr. Colfer has synthesized from these studies. This first part of the book is not merely reflective, however, it is also self-reflexive (which is quite uncommon in policy-related works like this one). It reads, in part, as a sort of "how-to" manual for the conduct of global projects on peoples and forests. An entire chapter (Chapter 2) contains an archaeology of the ideas that underlie the project. This includes details regarding how the project was run, what sorts of problems were encountered, and how they were overcome (or not).

This foregrounding of the operational reality of the CIFOR project reflects its explicit, guiding belief that the same principles of adaptive collaborative management that the project scientists were attempting to apply to local forest communities should also be applied to their own activities as scholars. This represents a very rare but salutary eliding of the typically formidable barrier between scientific subject and object. Altogether, this volume offers a highly unusual but potentially powerful look "backstage" at a major conservation and development program. For this reason alone, anyone engaged in academic or policy-related comparative work on forests and people should read this book.

Anti-Determinism

Consistent with the reflexive tone of this book is its broad questioning of the scientific presumptions and authority that underlie much of contemporary conservation and development interventions. Drawing on the broad, comparative sweep of the study, Dr. Colfer observes that the forests of the world are characterized by highly complex and diverse social and biological systems, for which monistic, deterministic planning is simply inappropriate.

Dr. Colfer begins her discussion of this subject by acknowledging, in company with many astute observers today, that local communities demonstrate far more "agency" in their management of forest resources than was once credited to them. Given the presence of local agency in highly complex settings, Dr. Colfer suggests that conservation and development planning will inevitably encounter the unexpected; there will be surprises. As a result, she says, projects like CIFOR's must contain mechanisms for gathering feedback, and there must be capacity within the project itself for adaptation based on this feedback. But even if we succeed in learning while doing, Dr. Colfer says that our expectations must still be modest: she writes that "We need, in short, to let go of our illusion that we can plan and control our–and especially anyone else's–destiny." She suggests that the most we can expect is to help to "enhance the conditions" for the emergence of adaptive collaborative management.

Given her meticulous care in identifying the limits to our capacity for both understanding and affecting forest sustainability, one might expect to see few if any global generalizations from this project. And indeed, Colfer says that the project scientists abandoned at the outset the insistence on standardized, cross-site data sets that are favored by most such projects. One of the volume's surprises, therefore, is its convincing demonstration that the important principles regarding forest sustainability and the variables that affect it are in fact generalizable across most if not all of the study sites.

The assertion of seemingly universal principles regarding forest sustainability is all the more impressive, given the study's explicit renunciation of a standardized, monistic, deterministic approach. It suggests, perhaps counter-intuitively, that such deterministic approaches are actually inimical to generalization, and that it is only through an embrace of complexity, difference, and uncertainty that we can successfully arrive at generalizable principles. Furthermore, these findings suggest that we need to relinquish part of the authoritative stance of science in order, paradoxically, to be more effective. This is a powerful message and a rich book for anyone interested in the contemporary global challenge of how relations between people and forests are best governed.

MICHAEL R. DOVE
YALE UNIVERSITY

Dedication

This book is dedicated to my well-loved step children, Amy Hope Dudley Sweitzer and Brian Chapman Dudley. The information in this book is far removed from their passion for seventeenth century literature and history, respectively, and my worldview—my vision of a better world—differs, I think, from theirs. Perhaps this book can stimulate further constructive dialogue and even greater mutual under-standing—to complement the love that already binds us.

Acknowledgments

THE RESEARCH REPORTED HERE HAS been produced with the generous financial assistance of the following donors, whose contributions I gratefully acknowledge: The Asian Development Bank (ADB, RETA 5812: Planning for Sustainability of Forests through Adaptive Comanagement); the U.K. Department for International Development (DFID); the European Union (EU, B7-6201 Tropical Forestry Budget Line); the Research Fellow Partnership Programme for Agriculture, Forestry and Environment of the Swiss Agency for Development and Co-operation (SDC) and the Centre for International Agriculture (ZIL) at the Swiss Federal Institute of Technology Zurich; Intercooperation; the Ford Foundation; the German Agency for Technical Cooperation (GTZ); the International Fund for Agricultural Development (IFAD, Making Space for Local Forest Management in Asia Project); the International Tropical Timber Organization (ITTO project pd 12/97 rev.1(f): Forest, Science and Sustainability: Bulungan Model Forest); the United States Agency for International Development (US-AID); and of course, the Center for International Forestry Research (CIFOR) and Cornell University's Department of Natural Resources. The views expressed herein are those of the author and do not reflect any official opinion of the donors.

The Adaptive Collaborative Management (ACM) team and I have also had strong support from our main partner institutions (listed in Table 11-2). In each site, we have also had close and beneficial interaction with other collaborators too numerous to mention but whose assistance—routine or sporadic—we have very much appreciated.

At CIFOR, I could not have done without the able administrative assistance of both Rahayu Koesnadi and Dina Hubudin. Both routinely perform tasks above and beyond the call of duty—for which I shall be eternally grateful. I am also thankful for the support of CIFOR's administrators, including at various times Jeff Sayer and David Kaimowitz (previous and current Directors General), and Dennis Dykstra and Ken MacDicken (previous Directors of Research), as well as CIFOR's Board of Directors. We have also benefited throughout this research process from the congenial and cooperative attitude of our colleagues in other CIFOR programs. They have given unstintingly of their expertise and very limited time, virtually whenever asked.

I would like to thank those outside CIFOR also who reviewed our ideas and our manuscripts, including Norman Uphoff, Jesse Ribot, Dianne Russell, Katherine

Hudson, and Jack Ruitenbeek; and those connected with this work who checked the manuscript for accuracy, including Chimere Diaw, Herlina Hartanto, Trikurnianti Kusumanto, Cynthia McDougall, Benno Pokorny, Kaspar Schmidt, and Samantha Stone. None of them bears any responsibility for any remaining errors.

I thank Resources for the Future—Don Reisman, John Deever, Grace Hill, Meg Keller—and especially Sally Atwater, for her systematic approach, her watchful eye, and her ability to reduce my verbosity to manageable levels.

The most fundamental thanks are due, however, to the communities and other stakeholders who have engaged with us in participatory action research. Their roles in this endeavor have been as central as our own—hopefully more so. They have used their time, creativity, networks, and institutions to bring about what we report here. I hope that the confidence they placed in us, in agreeing to experiment with us on their own social and environmental systems, has been well placed; and that they will see the continuing benefit—through ongoing social learning and collaboration—that we have all been trying to achieve.

CAROL J. PIERCE COLFER

CHAPTER ONE

Introduction

*H*OW ON EARTH CAN THE GLOBAL community create the conditions necessary to sustain the world's forests and forest communities? This was the question that troubled us, a group of researchers at the Center for International Forestry Research (CIFOR) in Bogor, Indonesia, in early 1998. We passionately wanted to help bring about forest sustainability—including its human dimension—in tropical forests and with forest peoples. In this book, we describe an approach we call adaptive collaborative management (ACM). We had just completed a successful four-year project developing and testing criteria and indicators (C&I) for sustainable forest management; another research effort was under way on devolution processes in forested environments. We felt that we had moved the global sustainability debate forward, helping create a more precise meaning for the term *sustainable*—a meaning that could be examined in the field. Yet we were not satisfied. We knew that we could fine-tune criteria and indicators for a very long time without changing the conditions in a single forest or among a single group of forest dwellers. We could not expect conditions in the forests to change just because we had developed more precise C&I to assess their condition. There needed to be some way to *create* these conditions that the C&I sets identified as critical for sustainability, in forests and forest communities, and on a broad scale.

Having spent years in humid tropical forests in Asia, supplemented by time in African and South American forests, we were very aware of the urgency of solving the environmental and human problems related to sustainability. It is tiresome to repeat the distressing figures for global forest destruction and degradation (Terborgh 1999) and the appalling losses—economic, cultural, environmental—that accrue daily to forest communities (e.g., Wickramasinghe 1994). *Seeing* the same processes at work in country after country is even more dismaying than reading about them. It was clear that something needed to be done quickly if we had any hope of reversing or even slowing the dramatic and destructive processes that were under way.

1

We chose to develop a research program to address the problems and simultaneously contribute to our scientific understanding of the processes involved. By 2002, we were working with a team of more than 90 researchers in Asia, Africa, and South America. Together, we have taken our initial ideas forward in 30 field settings. This book builds on the site-by-site analyses of the team members; without their contribution it could not have been written, and thus we are all authors. Yet it also provides conclusions based on my own systematic comparison of the results at each site. I distinguish between my own analyses and conclusions (which may or may not be shared by my colleagues) and the work of the group by shifting between first person singular and first person plural. Such an approach is not usual in scientific writing, but accuracy requires it.

In this book, I examine a range of forest community contexts from the perspectives of seven important issues and analyze the processes and outcomes. Fuller discussion of the cases and the seven issues is provided in the Appendix of Cases at the back of this book (additional, in-depth country-by-country and regional analyses are in Diaw et al. forthcoming; Fisher et al. forthcoming; Matose and Prabhu forthcoming; Hartanto et al. 2003; Kusumanto et al. 2005). This research demonstrates the fundamental importance of expanding the scientific repertoire of analytical approaches to include local variation, systems, interactions and connections among systems, and change. It also shows that scientists, policymakers, and other decisionmakers must recognize (a) the inherent unpredictability of complex adaptive systems; (b) the importance of *learning* in our attempts to deal with this complexity; and (c) the necessity and potential for working closely with the people who act within those systems—at all levels.

A Starting Point

As we sought to act on our understanding of sustainability, we were guided by several insights that form the intellectual underpinnings of ACM (see Chapter 2). The striking complexity and diversity of forest and human systems were central in our minds (Box 1-1).[1]

Box 1-1. Simple, Complicated, and Complex Systems

Simple systems are those with very few elements, which behave in a readily understood and predictable manner. Complicated systems are those with many elements that, once understood, still behave in a predictable manner. Complex systems, because of internal interactions and feedback mechanisms, tend to generate "surprises" (Ruitenbeek and Cartier 2001).

These systems presented variation within a particular locale as well as incredible diversity from place to place and from time to time. We concluded that solutions to local problems would best begin at the local level. The record of decades of attempts to structure standardized solutions to local forest problems from afar is dismal (cf. Scott 1998). We concluded that complexity and diversity must be taken as givens and that recurring surprises would accompany any management effort.

Related to such complexity is the fact that in many areas, multiple management systems are at work simultaneously. A logging company may have a contract to log a large area where several ethnic groups have lived and managed the forest for centuries for multiple uses, adjacent to or even overlapping a conservation area, all overseen in some fashion by the national government. Typically, the communities' traditional uses and management practices are ignored or criminalized. Sustainable management of the forests will, we argue, require a process of negotiation wherein the stakeholders work together to resolve local forest management problems equitably. This will require at least communication, cooperation, negotiation, and conflict management—the collaborative elements of ACM.

That local communities often have functioning or partially functioning management systems of their own is an oft-neglected aspect. In some tropical forests, the only element of sustainability in forest management is the indigenous system (e.g., Colfer et al. 2001; cf. Clay 1988; Redford and Padoch 1992). In others, the indigenous system has been systematically undermined by the state (Fairhead and Leach 1996; Leach and Mearns 1996; Peluso 1994; Bennett 2002; Sarin et al. 2003; Oyono forthcoming) but still maintains the kernels from which more sustainable practices might grow. In still other places, settlers with little knowledge of forest management (transmigrants in Indonesia, settlers in Brazil, Porro 2001) have been moved into forested areas. The differences in peoples' forest involvement have implications for how forests should be managed. Our conclusion is that local communities represent an underutilized resource in forest management, and that they have the biggest stake in managing the forests well. Governments and industry have done badly managing many forests in the recent past; communities have been managing their forests for millennia. We were convinced that combining indigenous knowledge and motivation, on the one hand, with some outside information and some balance of power in management, on the other, could be a worthwhile approach.

We also noted the lack of feedback mechanisms in the policy context. Policymakers often made apparently reasonable national-level plans that, when applied in particular contexts, had unanticipated effects. Granting timber concessions to foreign companies or allocating areas for national parks, for instance, has often resulted in the unintentional loss of local resources to local communities (cf. Brechin et al. 2003). Effective avenues for providing feedback about such failures (or successes) were rare. Critical opportunities to learn from failures were missed, and malfunctioning management systems continued to hobble along, either ineffective or damaging. Our experience suggested that plans rarely work as originally conceived (cf. Stacey et al. 2000) and that successful management requires regular feedback. We thought that simplified adaptations of criteria and indicators might serve at the local level as effective monitoring tools that could provide useful feedback to improve management. C&I or some other monitoring approach could form the basis for the process of social learning required for management to be responsive—the adaptive element of ACM.

Negotiated resolution of problems at local levels must be accepted at other levels. If, for instance, a community without legal rights to its traditional forest worked with a logging company in Indonesia to allocate rights to particular areas among them, the agreement would not likely be honored unless the government of Indonesia were involved. Our research, therefore, had to find links between local communities and actors at other scales—whether governments, companies, or actors like the World Bank and international nongovernmental organizations (NGOs). We also had to find avenues and mechanisms for effective, two-way communication between individuals and groups with vastly differing power. ACM would not work unless local communities were empowered. Indeed, empowerment has been a concern from the beginning. Disempowered groups needed a place at the negotiating table; the uneven playing field had to be leveled somehow.

CIFOR's ACM Strategy

With those issues in mind, we went to work on developing an approach to managing forests that would be consistent with those insights, address the sustainability problems of forests and forest communities, and provide results of use to others. The approach we have crafted involves three components:

1. *Horizontal component.* Stakeholders in a particular forest begin working together to solve the problems of that forest and the people who live in and around it.
2. *Vertical component.* Local communities and actors at other scales develop effective mechanisms for two-way communication, cooperation, and conflict resolution.
3. *Diachronic component.* The stakeholders learn together, over time, in an iterative fashion about the management of their resources and their communities.

This led to our definition of adaptive collaborative management (Box 1-2):

Box 1-2. Definition of Adaptive Collaborative Management

Adaptive collaborative management (ACM), in our usage, is a value-adding approach whereby people who have interests in a forest agree to act together to plan, observe, and learn from the implementation of their plans while recognizing that plans often fail to achieve their stated objectives. ACM is characterized by conscious efforts among such groups to communicate, collaborate, negotiate, and seek out opportunities to learn collectively about the impacts of their actions.[2]

Figure 1-1 provides a graphic representation of some of the changes we anticipate as we move from one stage of adaptiveness and collaboration to another.

Our research strategy has two streams, one focused on catalyzing the ACM process in the field (the participatory action research process), and the other focused on analyzing what happens (the assessment process and strategy).

1. The Participatory Action Research Process

In a given site,[3] the ACM facilitator[4] has as a primary, ultimate responsibility to catalyze the process described in the ACM definition. The first task, used in both streams, is to conduct five context studies, focused on

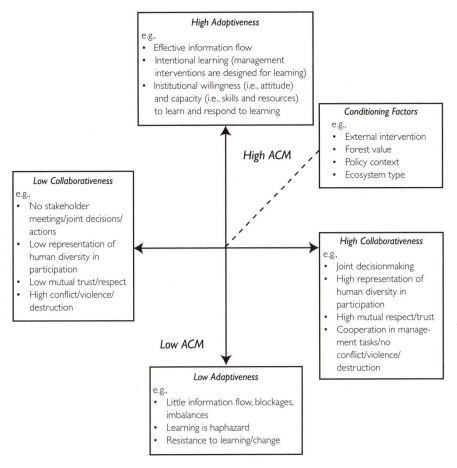

High Adaptiveness
e.g.,
• Effective information flow
• Intentional learning (management interventions are designed for learning)
• Institutional willingness (i.e., attitude) and capacity (i.e., skills and resources) to learn and respond to learning

Conditioning Factors
e.g.,
• External intervention
• Forest value
• Policy context
• Ecosystem type

High ACM

Low Collaborativeness
e.g.,
• No stakeholder meetings/joint decisions/actions
• Low representation of human diversity in participation
• Low mutual trust/respect
• High conflict/violence/destruction

High Collaborativeness
e.g.,
• Joint decisionmaking
• High representation of human diversity in participation
• High mutual respect/trust
• Cooperation in management tasks/no conflict/violence/destruction

Low ACM

Low Adaptiveness
e.g.,
• Little information flow, blockages, imbalances
• Learning is haphazard
• Resistance to learning/change

Figure 1-1: *Dimensions of ACM and Elements of High and Low Collaborativeness and Adaptiveness (McDougall 2000)*

• stakeholder identification;
• historical trends;
• policy context;
• biophysical and social characteristics (based on criteria and indicators); and
• level of adaptiveness and collaboration among stakeholders.

These studies inform the ACM facilitator and protect against egregious errors based on ignorance of the local context. During this process, the facilitator establishes rapport with the community or communities involved, and prepares to conduct participatory action research. These studies are also vital to the assessment process described next.

Participatory action research drives the ACM process. The point is to catalyze a process within the local community in which the people—or some subgroup thereof—can jointly plan improvements in local conditions (sometimes in collaboration with other stakeholders), gain power and skills in dealing with others, and develop a self-monitoring system to enhance sustainability. Designed to "enhance the capacity of individuals to improve or change their own lives" (Cleaver 2002), participatory action research proceeds in a systematic, iterative mode, as shown in Figure 1-2.

This approach inevitably results in considerable variation from site to site, as will become clear in the ensuing chapters. There are many routes forward. As McDougall (2002a, 90) put it, the field teams "had the ACM elements as 'beacons' but were 'making their own paths.'"

We have also had a strong interest in enhancing equity. In many ACM sites, women, particular ethnic groups, and low-ranking castes tend to be marginalized in any efforts to improve conditions. People in these categories tend to be left out of decisionmaking that directly affects their subsistence base, to have their traditionally defined ownership of and use rights to land and other resources ignored, and to be neglected when benefits are distributed to communities. One route to achieving more equitable treatment is for such disenfranchised people to develop a stronger power base in dealing with outsiders. This has meant an explicit emphasis in our work on power and its distribution and functioning. Cooke and Kothari (2002) point out that many advocates of participation have been naïve about power, and we have tried to ensure that we were not. We have expressed our interest and willingness to work with comparatively powerless groups,[5] organized separate meetings for them, tutored those who will deal with powerful outsiders, provided training for transformation (based on Paulo Freire's approach), ensured equitable access to public platforms in meetings we facilitated, worked with locals (elite and nonelite) to make local political processes more transparent, required equitable representation in visits and training we funded, contributed to the public view that increased equity was desirable, and so on. Considerable effort has gone into ensuring that marginalized groups—who often have very different interests in and knowledge about forests—are included in the ACM process (Nemarundwe 2005; Mutimukuru et al. 2005; Sharma 2002; Siagian 2001; Tiani et al. 2005; Diaw and Kusumanto 2005; Pokorny et al. 2005; Sitaula 2001; Pandey 2002; Dangol 2005).

In most countries where an ACM site exists, the teams have set up national steering committees whose purpose is to ensure that the research will be useful to the country and its people. Our intention was also that high-profile members from the government, academia, and NGOs could serve as channels for disseminating our results.

The ultimate test of the ACM process will be whether the process continues after the ACM facilitator withdraws. We hoped to have several years to initiate this process on all of our sites, but most sites began with only three-year funding windows.

2. The Assessment Process and Strategy

The second stream, the attempt to understand what works and does not work in our attempts to catalyze an ACM process, is a primary focus of this book. Three broad research questions apply to all of our work:

1. Under what conditions does the ACM process work?
2. What are the tools, methods, and concepts that are useful in catalyzing an ACM process?
3. What are the impacts of this process on the forests and on the people involved?

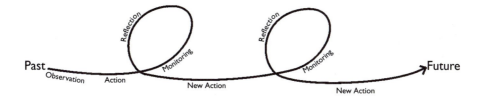

Figure 1-2. *The Worm*

The analysis in this book began as an attempt to answer the first question—what are the conditions under which ACM works (and is needed)? The findings move us away from this reductionist, analytical interpretation to a conclusion more consistent with Ruitenbeek and Cartier's (2001) ideas about the emergent properties of complex systems (discussed below). I retain the reductionist analytical structure (a) as a mechanism to test the adequacy of the reductionist approach itself for understanding complex adaptive systems, and (b) as an organizing framework for dealing with issues widely considered important in forest management.

The ACM teams have produced useful tools, methods, and concepts,[6] pertaining to question 2 above, and more are anticipated. We have also observed impacts on local social capital, institutions, and processes; levels of activity; management decisions made; and activities undertaken (see Chapter 3). But this analysis is based on at most two and a half years in the field, so it is too early to predict major or long-term changes in well-being or forests. Our evolving theoretical perspective acknowledges the difficulty of tracing specific impacts to specific interventions; here we focus on plausible causal links between actions and outcomes.

Early on, we identified variables that we considered likely to be important in analyzing the effectiveness of ACM. Our site selection reflects a range of values on each of these variables. In Nepal, for example, a great deal of authority has devolved to local communities; in Zimbabwe, authority is centralized. Chimaliro, in Malawi, has high population densities; Rantau Buta, in Indonesia, is sparsely populated.

We determined that all sites would have one or more communities involved to some degree in forest management (broadly defined), and more than one stakeholder. Community involvement in forest use and management is central to the problems we were trying to address and the solutions we hoped would be forthcoming with ACM; we needed more than one stakeholder because of our interest in collaboration. We then identified the following seven analytical dimensions (described in more detail in the Appendix of Cases) that would guide our research:

1. *Devolution status (formal and informal).* We expected that devolution in the form of official laws or policies that devolve authority over natural resources to local communities might make the ACM process we were trying to catalyze easier and more sustainable.

2. *Forest type.* Forest quality (whether a location was generally "forest rich" or "forest poor," as well as whether the forest contained valuable commercial species) often affects the willingness of governments to allow communities to be-

come formally involved in forest management; the general perception was that the more degraded the forest, the more involvement governments permit.

3. *Population pressure.* Many have noted the adverse effects of growing population on forest quality, and some suggest that higher population densities might lead to greater eagerness on the part of communities to solve natural resource problems.

4. *Management goals.* We anticipated that the different management goals of timber concessionaires, plantation managers, conservation agencies, subsistence, and other interests would influence our efforts to spark ACM. The power of the timber industry seemed likely to affect the process, compared with situations in which communities and governments were the only actors. Similarly, we suspected interesting and different implications for ACM in the juxtaposition of conservation projects and communities.

5. *Human diversity.* The number and diversity of stakeholders seemed likely to be important, both in their connection to management goals, and for the theoretical question of whether diversity is good or bad for collective action.

6. *Level of conflict.* We sought areas characterized by a significant level of conflict because of the ubiquity of conflict-ridden forest settings and the urgency of forest-related problems.

7. *Social capital.* We anticipated that social capital would be a precursor to collective action (Uphoff 2000; Krishna 2002b). At the outset, we had no reasonable methods for measuring social capital, so our estimates have been less systematic than we would have preferred. Measurement ideas have since become available in the literature (e.g., Krishna 2002a) and our own work (e.g., Akwah 2002; McDougall 2001a; see Appendix 10-1), which we expect to use as we move forward.

With the analytical dimensions clear, we selected 30 sites in 11 countries, seeking those that would give us a range of values for each dimension and represent the major tropical continents of Asia, Africa, and Latin America (Figure 1-3). We then sought potential donors and research and community partners in the host countries. The sites that we selected, with their rough positions on these various dimensions, are listed in Appendix 1-1. The geographic spread makes this book a global survey, with special emphasis on the tropics.

The ACM facilitators we selected for each site varied by discipline and level of expertise. Some were closely affiliated with and integrated into the work of local NGOs (e.g., in Brazil, Cameroon, Indonesia, and Nepal); others had a more dominant role themselves (e.g., in the Bolivia, Philippines, and Zimbabwe). The number of researchers involved also varied.

The first task of the facilitators was to undertake the context studies for the participatory action research process. Besides providing essential information to the ACM facilitator before beginning the participatory action research process, these studies were important for our attempt to evaluate the impact of our efforts on the forests and communities. Repeating these context studies yearly proved onerous; accordingly, we have conducted such studies twice only in the longest running sites. The studies double as impact studies and are supplemented by the ongoing recordkeeping of the ACM facilitators—daily field diaries, periodic reports on interactions among stakeholders and effects on the site, and accounts of the iterative learning cycles in the communities, as new things are tried and succeed or fail.

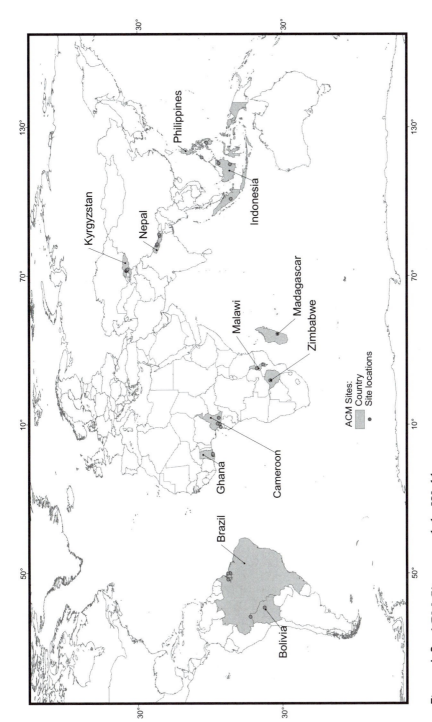

Figure 1-3. *ACM Sites around the World*

Close communication is essential for a multicountry study. We held a planning and training workshop in February 2000, followed by an International Steering Committee meeting.[7] A core group, resident in Bogor, Indonesia (CIFOR headquarters), met every two weeks for updates. Each Bogor-based person met regularly with his or her site facilitators, and every year several facilitators and partners joined us for meetings of the International Steering Committee. These meetings took place in Indonesia, Zimbabwe, Nepal, and the Philippines so that team members could visit sites in the different host countries. Various team members have visited other sites as well. In addition, we instituted a program newsletter for interchange (*ACM News*),[8] conducted a writing workshop for 35 team members in November 2001, and held regular conference calls among sites. Cross-site communication nevertheless remains difficult, given the distances and the vagaries of technology in remote areas.

Besides those core activities, we have followed up with specialized studies in areas of potential value. A subgroup within the research team has focused on modeling, including system dynamics models, multi-actor models, participatory systems models, and models of decisionmaking. Some of these models foster a shared view of a system among stakeholders (e.g., participatory systems analysis, Cwebe Working Group 2001); others enhance the negotiating potential of community members (CORMAS, Common Pool Resources and Multi-Agent System; Purnomo et al. 2001; Purnomo et al. 2003); still others make a system's functioning obvious to policymakers (Vanclay et al. 2003). Work thus far has focused on Indonesia, Zimbabwe, and Cameroon; in 2003 a special issue of *Small Scale Forestry Economics, Management and Policy* (Vol. 2, No. 2) was devoted primarily to this work.

Action research with communities in the Bulungan Research Forest in Indonesia, which began in the early 1990s, has produced several useful tools and concepts. Wollenberg et al.'s (2000) future scenario ideas initially came from this work, as have improvements in our understanding of how best to conduct community mapping (Anau et al. 2005). Such policy work is relevant to devolution, potential two-way communication links between communities and actors at other scales, and the ongoing political problems facing Indonesians (cf. Barr et al. 2001; Casson 2001; Moeliono and Djogo 2001).

A special study was conducted in Nepal, where government officials were concerned because we were implementing our action research in only four communities. They wanted us to add a survey covering some of the same issues on a broader scale (Ojha et al. 2002). In the Philippines, a group argued against collaborative management, stressing the need for groups to defend their own exclusive rights to traditional resources. We commissioned a study of this group and how they had dealt with the government from this perspective (Rice 2001).

Theoretical Orientation

Any theoretical orientation begins with some assumptions. Assumptions always seem obvious if you share them, surprising or dismaying if you don't.

In our work, we began by accepting the idea that forest people are part of complex and dynamic forest and human systems. These systems are constantly

---

Box 1-3. The Inevitability of Surprise

A complex adaptive system is capable of reorganizing and self-organizing itself, in response to random external shocks. Large economic systems, living organisms, neurological networks, and ecosystems are all examples of complex systems. Any forest ecosystem, coupled in turn to a variety of political, institutional, and social structures, is such a system. A common feature of complex systems is that they generate "surprises" (Ruitenbeek and Cartier 1998, 3).

Post-Newtonian social science starts by viewing the social world more in the way physicists now regard the world of quantum mechanics—as inherently and inescapably uncertain (Uphoff 1996, 399).

See Waldrop (1992) for pertinent ideas from complexity and chaos theory, written in lay language.

---

changing, nondeterministic, and interacting. We do not believe that prediction, at this level of complexity, is possible. And we do believe that surprise is inevitable (see Box 1-3).[9]

We also accept the idea that forest people (and others) have the capacity to act: they have "agency." And they have invaluable knowledge about how their own systems work. If conscious, adaptive management of forests can be catalyzed—as we hope it can—their involvement will be central.

We are also convinced that attempts to manage forests (and other natural resources) that ignore equity are doomed. Injustice breeds conflict and violence. And it is virtually always possible to find both a pragmatic and an ethical reason for strengthening equity.

With those ideas in mind, we initiated our program. Early on, Ruitenbeek and Cartier (2001) asked whether adaptive collaborative management was a package approach to be delivered by governments or others, or whether it was an emergent property of complex, multiscale, interacting systems. Could we make ACM "happen," or merely protect the conditions under which ACM might emerge? My own conclusion, after looking at the processes in the research sites and our results to date, is that we have strengthened the probability that adaptive collaborative management would emerge. *Enhancing* the conditions for the emergence of ACM is perhaps more proactive than *protecting* the conditions for its emergence. Although I have listed a series of steps that can be followed by others who would like to emulate the process, success depends on some parameters that we have found to be unusual in many forest management contexts—willingness to see failure as an opportunity to learn, recognizing that others play central roles in determining joint goals, looking at things from a nondeterministic systems perspective, and supporting and strengthening social capital and self-confidence in community members.

Three figures depict different ways of looking at what we have been doing. The first was created by David Kaimowitz, an economist, when we explained our research plans to him (Figure 1-4).

This simple model captures some critical elements in our approach:

- *Box A.* We are concerned about initial conditions (context studies and Chapters 6 through 10).

- *Box B.* We have proposed an approach to transforming the social processes involved in forest management and human governance (the 'steps' in ACM described above).
- *Box C.* We recognize that differences in our own behavior may affect outcomes (Chapter 11).
- *Box D.* We seek to learn what effects our efforts have (impact assessments, Chapter 5).

Still, this model seems too unidirectional for our purposes. It ignores, for instance, the important role of feedback in transformative processes and doesn't capture the progressive, iterative process exemplified by Figure 1-2, which incorporates new information, surprises emanating from the user's involvement in multiple systems operating at diverse scales, and human purpose (see Box 1-4).

Box 1-4. Of Knowledge, Action, and Ignorance

The rationality of sustainable management has to do with more than just what we know. It also extends to managing for what we do *not* know. Many systems we deal with are complex and often seem influenced by arbitrary events. As systems become more complex and larger, our ignorance of what might happen escalates as well. Acknowledging this ignorance is the first step in addressing it. The second step is acting on it: acting in such a way that the systems we steward—whether entire economies or single forest stands—are resilient to the shocks that our ignorance prevented us from foreseeing (Ruitenbeek and Cartier 1998, 21).

A unique property of human systems in response to uncertainty is the generation of new types of social structures (Gunderson et al. 2002, 325).

The third figure comes from *Panarchy: Understanding Transformations in Human and Natural Systems* (Gunderson and Holling 2002), whose framework may help us in the forthcoming analysis of our own data. In explaining the new term, *panarchy,* they say,

> … Our purpose is to develop an integrative theory to help us understand the changes occurring globally … The theory that we develop must of necessity transcend boundaries of scale and discipline. It must be capable of organizing our understanding of economic, ecological, and institutional systems. And it must explain situations where all three types of systems interact. The cross-scale, interdisciplinary, and dynamic nature of the theory has led us to coin the term panarchy for it. Its essential focus is to rationalize the interplay between change and persistence, between the predictable and unpredictable. Thus, we drew upon the Greek god Pan to capture an image of unpredictable change and upon notions of hierarchies across scales to represent structures that sustain experiments, test results, and allow adaptive evolution (Gunderson and Holling 2002, 5).

In their discussions of these transformations in human and natural systems, they postulate a recursive looping process (in the shape of a figure 8, Figure 1-5) in which ecological events and structures go through phases of exploitation, conservation, release, and reorganization (Holling and Gunderson 2002). An alternative

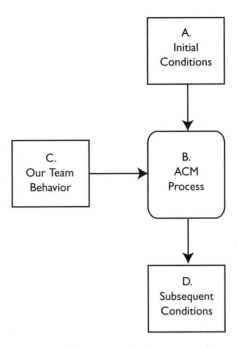

Figure 1-4. *A Unidirectional Model of ACM*

way of phrasing this process is policy plan, policy implementation, policy failure, and policy alternatives (Gunderson et al. 2002).

Holling and Gunderson (2002, 33–35) see *exploitation* (r) as the "rapid colonization of recently disturbed areas." *Conservation* (K) is a phase in which "slow accumulation and storage of energy and material are emphasized" (33). *Release* (Ω) refers to "creative destruction"—a phase in which "the tightly bound accumulation of biomass and nutrients becomes increasingly fragile (overconnected, in systems terms) until suddenly released by agents such as forest fires, drought, insect pests, or intense pulses of grazing." *Reorganization* (α) occurs when "soil processes minimize nutrient loss and reorganize nutrients so that they become available for the next phase of exploitation"(34–35). The ideas of *connectedness* (most extreme for the conservation phase, less so during the release phase) and *potential* (at its maximum in the conservation phase, less so in the reorganization phase) are also important, with both at their nadir during the exploitation phase (see Figure 1-5).

Other authors in the same collection build on these ideas in more social realms. Westley (2002), for instance, uses the model to describe a colleague's experience of trying to manage adaptively as a U.S. government bureaucrat. Gallopín (2002) imagines different scenarios for the earth's future, using this same conceptual framework.

Another idea that is important for the work reported here is the nested nature of a panarchy—the idea that systems function at different scales, and that elements of these interacting systems change at different rates, yielding extremely complex interactions that from the human perspective result in randomness (see Richardson 1991 for a fuller discussion of this idea).

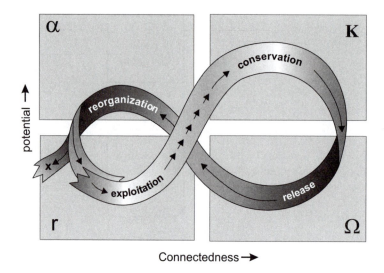

Figure 1-5. *Stylized Representation of the Four Important Functions and the Flows among Them in Transformations of Human and Natural Systems (from Holling and Gunderson 2002). Gunderson et al. (2002) show linked sets of these recursive loops, with successive, linked iterations of the adaptive cycle they describe (see also Chapter 12)*

In my view, each of these figures—the unidirectional model in Figure 1-4, the recursive, iterative loops in Figure 1-2, and the figure 8 in Figure 1-5—is a useful window into our work, comparable to Uphoff's (1996) call for "both-and" in addition to "either-or" thinking. These images represent different segments of the same processes. The unidirectional model (Figure 1-4) can be seen as a segment of the curving lines in either Figure 1-2 or Figure 1-5. Figure 1-4 represents a delimited time frame, a snapshot in an ongoing process. Sometimes we humans need to examine snapshots to help us deal with the complexity that defies human understanding.

Figure 1-2, which has been our mantra in ACM, captures the time element—things change, time passes.[10] And it captures, or at least recognizes, the notion of feedback. It also assumes that there can be a goal (something that is not entirely clear in the representation in Figure 1-5, perhaps purposely). Figure 1-2 implies the passage of time and is useful to reflect processes like those we have been trying to bring about, both in communities with participatory action research and in our research strategy to assess that process. This is a more general improvement over Figure 1-4, in that the latter only describes the research strategy undertaken by the researchers.

Figure 1-5 (the figure 8) is designed to deal with processes at all scales (geographical, conceptual, time). It represents processes that may be universal, that apply to a wide range of events, structures, and functions, including both streams of ACM (the catalyzing of adaptive and collaborative management processes in the field and the research strategy for assessing them). Just as the simple, unidirectional model is a segment of this iterative process, so the figure 8 captures the larger scale dynamic and repetitive processes. It seems to me that the kind of iteration exemplified in Figure 1-2 is a central part of people's roles as agents, as active influencers if not determinants of social change. This is a discussion I will return to in Chapter 12.

In the subsequent chapters, I describe the impacts that we have achieved and discuss the macro- and microlevel systemic features that we initially thought might determine success or failure. In discussing our impacts, I am looking for the "short-term, finer grained criteria of success that can usefully stand in for longer run broader goals" advocated by Ruitenbeek and Cartier (2001, 13) and the "plausible causal connections" suggested by Robert Fisher in our International Steering Committee meetings.

The issues affecting forests and forest communities represent a broad range of scales, pace of change, and geographic variation. We will see, as the analysis proceeds, whether the framework provided by Gunderson et al. is helpful. We may find that the use of participatory action research to make management and community decisionmaking more effective, sustainable, and equitable significantly affects the processes described in *Panarchy*. Or we may find, as Ruitenbeek and Cartier (2001) suggest, that adaptation of this kind is an emergent property that cannot be reliably instigated by human endeavor. If we follow the transformational teleology of Stacey et al. (2000), we may find a similar emergent result, based on the relations among the various actors involved. We agree with van Eijk (2000, 323):

> … it is evident that the dynamic and emergent nature of interactions taking place between farmers [or forest communities] and nature, including forces which lie beyond the interface situation itself, and between numerous other actors and their networks, puts the process of rural development [or forest management] beyond full human control.

Our Team's Management Approach

We were proposing a collaborative and adaptive approach for forest communities, but could we implement such a process among ourselves? I felt strongly that our team would be more successful if we could be collaborative and adaptive in our management style. There were few objections to this approach—though a few team members longed for more conventional leadership and direction.

We first developed a written conceptual framework. This document, drafted by Ravi Prabhu, Eva Wollenberg, and me, was critiqued by all team members, various external colleagues, and CIFOR administrators and went through 17 revisions. This process was important in getting buy-in from both the team members and the administrators who would need to trust us. We knew the task ahead would be difficult and demanding, and we wanted to make sure everyone was really on board. Although I had not read Senge's (1990) *The Fifth Discipline* at the time, I was looking for the kind of total commitment he describes, leading to outstanding synergies among team members.

As program leader, I was strongly committed to the idea of democratic, or horizontal, management (rather than a hierarchical organization), with each team member playing a vital role. This meant that I had to have high levels of trust and tolerance of uncertainty and risk. It also meant accepting, even encouraging, shifts in informal leadership at different times and in different contexts. This was doubly important for the field researchers doing the most critical work of the group. Those of us in more bureaucratic positions saw our task as one of support for them.

We all cultivated an attitude that problems or "failures" represented opportunities to learn and improve. Team members' formal and informal sharing of successes and failures in addressing particular problems was valuable for making improvements. I had to maintain this attitude as well and defend team members while such "corrections" played out. I improved my skills at public defense of our somewhat unconventional ideas.

We developed mechanisms to monitor our progress and alter our course as needed (reflection and correction). These included routine but short meetings, critiquing of each other's writings, joint site visits, even formal surveys. One technique we used was "before, during, and after" meetings in which we planned, monitored, and evaluated particular activities, sharing our experience, perceptions, and suggestions for future modification (proposed to us by Michael Hailu). Individual team members also sporadically convened meetings to discuss particularly thorny conceptual or tactical problems.

We agreed that attitudes of cooperation would be more effective than competition in this kind of work. Open sharing of information and insights among team members was crucial, so the atmosphere among team members had to be supportive. We tried to discuss and resolve any interpersonal problems. We recognized that strain and unresolved conflict could sap team members' energies, which were needed for more creative work.

The group was diverse in gender, citizenship, ethnicity, position, discipline, and religion. An explicit agreement to strive for intercultural understanding and equity was important. The Bogor-based team, for instance, comprised talkative senior people (including myself) who had to struggle to keep quiet, and quiet junior staff whose cultural predilections included overt respect for elders and people in formal positions of authority. One effective technique we used was a "talking stick," which allowed only the person holding it to talk; it could be handed to reticent teammates to encourage their input. Team members, including leaders, had to be attuned to such potential problems and ready to deal with them. Skilled facilitation was critical, but some problems—staff differentiations within the international research system,[11] entrenched ethnic and gender stereotypes—proved intransigent.

One of the most perplexing problems was effective communication among team members over long distances. Efforts to overcome these problems continue, via e-mail, an interactive Web site, joint publications, periodic meetings and workshops, site visits, and ACM News. The trust that is needed for such cooperative endeavors depends on effective communication.

Before I went on sabbatical in February 2002, I tried to persuade management to allow us to share responsibilities and have a junior coordinator, with backstopping from several senior team members, rather than a program leader. Although this proposal was not accepted, the new formal team leader came from within our team and was based in Africa at a branch office. A good portion of our plan was ultimately implemented de facto. I believe that the smooth transition when I left was due basically to the buy-in we had been able to engender among ourselves as a team—though this in no way minimizes the excellent leadership qualities of Ravi Prabhu, who succeeded me.

I recount our approach and experience because it demonstrates the degree to which we have been able to implement Stacey et al.'s (2000) admonition to develop a single theory of causation for management and for the behavior of those "managed." We catalyzed creative, proactive efforts on the part of most team members in difficult contexts. I would recommend our approach to those in other bureaucracies who seek to accomplish similar goals. I suspect something like this approach to management will be necessary for others who may want to emulate our work in communities.

Structure of the Book

This book begins with this introduction to the issues we have tried to address and the method we have used to do so.

Chapters 2 and 3 briefly describe the intellectual underpinnings of our approach. The proliferation of literature, theory, and evidence on the subjects addressed in this book is daunting, but those interested only in the pragmatics of ACM may skip this chapter. Chapter 4 describes the theoretical questions that pertain to the seven dimensions (devolution status, forest type, population pressure, management goals, human diversity, level of conflict, and social capital) by which we characterized our sites; the book's analysis is based on these facets.

Chapter 5 documents the impacts that we have had to date. The research has been conducted within the framework of participatory action research, so credit for these impacts actually goes to the communities more than to the research teams. The process of identifying these impacts has been inductive, achieved by visiting the sites, reading the pertinent written materials, and interviewing the field researchers. The impacts are closely related to the methods used to bring them about, so both impact and method are dealt with here. I have categorized this information in two ways. First, I looked for impacts that coincided with our three components: adaptiveness and horizontal and vertical collaboration. Then I sliced the results another way, looking for the more conventional impacts on livelihoods, environment, and empowerment. Once the early impacts were clear, I examined each site from the perspective of the seven variables we had identified as potentially affecting the conduct of ACM.

Thinking in Gunderson and Holling's (2002) terms, I begin in Chapter 6 with a broad-scale issue—devolution, which addresses national-level policies. Chapter 7, on forest type and population pressure, describes the environmental parameters; and Chapter 8, the management goals. As large-scale issues, these are likely to change comparatively slowly.

Chapter 9, on human diversity, and Chapter 10, on conflict and social capital, focus on issues that seem most amenable to local action—though it will become clear that even these are typically closely linked with broader scale issues: gender and diversity issues are affected by ethnic and global politics; national-level conflicts affect local conditions; and social capital (creating links between local communities and their broader policy and institutional context) can be critical in bringing about change. Similarly, of course, the broader scale issues have local links and implications.

Chapters 6 through 10 thus describe the 30 communities and their forest management contexts through the lenses of these seven variables—all widely acknowledged as relevant for community involvement in forest management. This examination demonstrates the diversity of the world's forests at the same time that we ask how it relates to ACM impacts. A second way in which I present these data is more conventional and reductionist. I allocate the different sites to different types within each dimension and make overall assessments of ACM success on each site (see Appendix 1-1).

But my primary interests are in conveying the different angle that each dimension provides and the overall global variety in forests and forest peoples (see also the Appendix of Cases). The importance of a particular dimension will vary for different sites. The management dimension, for instance, is very important in sites where powerful timber companies or international conservation projects are operating; it seems less significant where power differences are less pronounced. Appendix 1-2 provides brief synopses of all the sites, for easy referral.

In Chapter 11, the characteristics and approaches of the research teams are presented. This chapter, together with Chapters 5 and 12, presents the most directly practical lessons that have been learned from the ACM experience.

Throughout these chapters, the researchers have contributed short items about particular events in particular places to provide a better feel for the kinds of processes and impacts in their sites. I hope these boxes convey the excitement and enthusiasm that emanate from these communities and our researchers.

Chapter 12 provides the conclusions, including the ostensible effects of each of the seven variables on impacts. This chapter also returns to the theoretical issues raised in this chapter and the next; it brings up a set of recurring concerns or questions pertaining to the implementation of ACM; and it provides a practical summary of what has been learned about the ACM process.

The Appendix of Cases describes the sites in the 11 countries (Brazil, Bolivia, Cameroon, Ghana, Indonesia, Kyrgyzstan, Madagascar, Malawi, Nepal, the Philippines, and Zimbabwe), with information organized according to the dimensions we initially anticipated might be important in determining ACM success. An acronym list and glossary complete the volume.

I hope this book will convey the diversity that characterizes the world's forests and the people who live in them. I also hope to demonstrate the importance of moving beyond the conventional approaches of reductionist science when looking at complex adaptive systems like forests, oceans, wetlands, deserts, and other contexts emphasizing people's interactions with their environments. Finally, I hope to lure the reader into supporting and perhaps engaging in the kinds of community processes described here. In this same connection, I hope to provide useful warnings about what to avoid and suggestions about what seems to work under what conditions.

The theoretical discussion and the examples from our work with communities hold potent lessons for policymakers. The conclusions suggest that if we really want to affect forests and forest communities in a positive way, we need to reorient our approach to forest conservation and rural development. We need to trust the capabilities of rural people; we need to develop more flexible and responsive bureaucracies; we need to acknowledge failure as an indispensable learning tool; we need, in short, to let go of our illusion that we can plan and control our—and especially anyone else's—destiny. The best we can do, I argue in this book, is work with other people, developing common goals, creating monitoring and feedback mechanisms, and promoting flexible institutions that can respond to the dynamism and diversity that our world so abundantly provides. In that way, we can cope with the inevitable surprises.

Appendix 1-1. Site Selection Matrix

Site	Devolution status	National conflict	Forest type	Population pressure	Management type	External diversity	Internal diversity	Local conflict	Social capital	Overall success
Baru Pelepat, Indonesia	L	H	HT	L	LT/ST	H	M	H	L	H
Guarayos, Bolivia	M	M	HT	L	LT/ST	H	M	M	H/L	H
Rantau Layung/Rantau Buta, Indonesia	L	H	HT	L	LT/ST	M	L	H	L	H
Bulungan, Indonesia	L	H	HT	L	LT/ST/LC	M	M	H	L	H
Andheribhajana, Nepal	H	H	SM	H	MP	L	H	H	L	H
Bamdibhirkhoria, Nepal	H	H	SM	H	MP	L	H	M	L	H
Deurali-Baghedanda, Nepal	H	H	SM	H	MP	L	L	M	L	H
Jaratuba, Recreio, Nova Jerusalem in, Muaná, Brazil	L	M	HT	M	MP	L	M	L	L	H
Mafungautsi, Zimbabwe	L	H	DT	H	MP	L	M	M	H	H
Manakamana, Nepal	H	H	SM	H	MP	M	H	M	L	H
Ottotomo, Cameroon	M	M	HT	L	MP	M	L	M	M	H
Porto Dias, Brazil	H	M	HT	L	ST	M	L	M	H	H
San Rafael–Tanabag–Concepción in Palawan, Philippines	H	H	HT	M	ST	H	L	M	M	H
Basac, in Lantapan, Philippines	H	H	HT	M	LC	H	L	H	M	M
Dimako, Cameroon	M	M	HT	L	LT	H	M	H	M	M
Lomié, Cameroon	M	M	HT	L	LT	H	M	H	M	M
Campo Ma'an, Cameroon	M	M	HT	M	LT/LC	H	M	H	L	M
Chimaliro, Malawi	M	L	DT	H	MP	L	L	M	M	M
Canta-Galo, in Gurupá, Brazil	L	M	HT	M	ST	M	L	L	H	M
Ranomafana, Madagascar	M	H	HT	H	LC	H	L	?	M	L
Akok, Cameroon	M	M	HT	M	LT	H	L	H	M	L
São João Batista/Nova Jericó, in Tailândia, Brazil	L	M	HT	M	LT	M	L	M	L	L

Achy, Kyrgyzstan	M	L	T	H	MP	M	M	M	M	L
Adwenaase, Ghana	H	L	HT	H	MP	L	L	M	M	L
Arstanbap-Ata, Kyrgyzstan	M	L	T	H	MP	M	M	M	M	L
Namtee, Ghana	H	L	HT	H	MP	L	L	M	M	L
Ntonya Hill, Malawi	M	L	DT	H	MP	M	L	M	H	L
Ortok, Kyrgyzstan	M	L	T	H	MP	M	L	L	H	L
Uzgen, Kyrgyzstan	M	L	T	H	MP	M	L	M	H	L

Note: L = low, M = medium, H = high, HT = humid tropical, DT = dry tropical, T = temperate, SM = submontane, subtropical, LT = large-scale timber, ST = small-scale timber, LC = large-scale conservation, and MP = multipurpose reserve.

Appendix 1-2. Snapshots of ACM Sites

This appendix provides snapshots of each of the research sites, organized by country, in alphabetical order. Appendix 1-1 presents details on the characteristics of each site in tabular form.

Bolivia

The ACM research in Bolivia took place in four communities: Salvatierra, Cururú, Urubichá, and Santa María. All are part of the Guarayo TCO (or indigenous territory), a humid tropical forest inhabited by indigenous people (Salvatierra, Cururú, Urubichá) and settlers (Santa María). The Guarayo communal indigenous territory claim is for 2,205,370 ha. Our work has piggybacked on the Bolivia–U.S. Sustainable Forestry Project (BOLFOR) project, which is funded by the U.S. Agency for International Development, and was generously funded, also by U.S. AID. Peter Cronkleton, who coordinated this work, joined CIFOR in late 2000, after having worked with ACM in Acre, Brazil. He was assisted by students and faculty from the University of Florida. The participatory action research had two goals: stimulating communities to prepare themselves for involvement in the more formal, timber-oriented forest management made possible by recent national policies on decentralization; and training government officials and BOLFOR staff to deal more equitably with community members. ACM personnel are now using some ACM tools in other areas of Bolivia.

Brazil

The Brazil work was divided into two somewhat separate efforts, one in Acre (western Amazonia) and one in Pará (eastern Amazonia). Both sites suffered from very low levels of funding ($10,000 to $20,000 each per year), and researchers were therefore unable to complete all the studies specified in the ACM umbrella approach.

The Acre work took place in the Porto Dias Agroextractive Reserve, a tropical rainforest area near the border with Bolivia, beginning in early 1999. This site covers 22,345 ha of land and is specially designed for use by rubber tappers, who are expected to conserve the forest while maintaining their forest-based way of life. Our work was initially led by Marianne Schmink of the University of Florida (who has maintained her involvement in our work), subsequently by Magna Cunha dos Santos of the Agroforestry Research and Extension Group of Acre (PESACRE), an NGO. Cunha dos Santos worked closely with another NGO, the Center for Amazonian Workers (CTA), which was trying to implement a small-scale timber-harvesting project with the Porto Dias Rubber Tappers Association. The work there focused on efforts to make timber harvesting profitable in this remote setting and involved learning cycles that coincided with the yearly timber harvesting cycle. One important success of the project was certification by the international Forest Stewardship Council in early 2003. Another was the decision by faculty at the

University of Florida to seek funding for continued monitoring at this site and elsewhere.

In Pará, the work was led by Benno Pokorny and covered villages in three municipalities, all in the humid tropical forest. In Tailândia, there was ACM work in São João Batista and Nova Jericó, villages located primarily on *terra firme* (land not subject to seasonal flooding), which had been logged up to three times. The population in these communities was extremely fluid, making sustained advances in collaborative work particularly difficult. Guilhermina Cayres led a collaborative inventory of forest resources and a study of the feasibility of using participatory methods with local communities. Despite encouraging progress with current inhabitants, the instability of community residence meant that the progress was not sustained. Links between the communities and the timber company, however, improved dramatically.

In Gurupá, where we worked in the community of Canta-Galo, the work was led by Doerte Segebart and Rozilda Drude, who worked in close collaboration with a long-established NGO, the Federation of Organizations for Social Assistance and Education (FASE). The community (all descended from an ex-slave) had an effective leader and had already demonstrated considerable strengths at collective action. ACM efforts focused on evaluation of participatory methods by local people, more systematized data collection at the field level, and substitution of more participatory methods for the extractive ones proposed in the ACM plan for context studies. This process resulted in greater skills at local-level analysis by the community, with some negative implications for FASE; such criticisms were not entirely appreciated by the NGO, which had worked in the community for a long time.

The final Pará site was in Muaná, where we worked with three communities: Jaratuba, Recreio, and Nova Jerusalém (led by Westphalen Nuñes and Guilhermina Cayres). Here, one interesting focus was on heterogeneity analysis within the communities (with communities having somewhat different interpretations than researchers) as a means to make forest management more responsive to different community members' needs and interests. The approach with the most apparent potential was the development of "local researcher groups." These community-selected groups most closely approximated the intent of the participatory action research process, working together to develop, implement, and evaluate their own plans. These communities were also involved in the effort to systematize data collection and storage in the field and evaluate the use of participatory methods by community members.

Cameroon

The Cameroon team, led by Chimere Diaw, began its work by identifying the policy issues considered important by the government in 2000. Sites were then selected to respond to these concerns, on the assumption that if the ACM research addressed an existing policy problem, our results would be more likely to be adopted. This work was well funded—primarily by the European Union, World Resources Institute, and the Alternatives to Slash-and-Burn (ASB) program of the Consultative Group on International Agricultural Research—and all sites were in humid tropical forests.[12]

Ottotomo is a forest reserve of 2,950 ha, managed by the government timber agency (ONADEF), since 1930. Population growth in the surrounding villages had increased pressure on the forest, as had conflicts between the people and the government. The ACM team was invited to help develop a comanagement arrangement between the communities and the government, in cooperation also with the Land and Development Association (ATD), an NGO. Beginning with a multistakeholder workshop that used a discussion of criteria and indicators for sustainable forest management as an entry point, the team (led by Cyprain Jum and Anne Marie Tiani) eventually built effective partnerships among the primary stakeholders. One of their principal accomplishments has been a reduction in conflict and an increase in cooperative behavior among these stakeholders.

The Campo Ma'an management area, located on Cameroon's southern border with Equatorial Guinea, spans 7,710 km², including protected areas (a national park), production forests, oil palm and rubber plantations, and (considerably reduced) areas reserved for local communities. Our work in this area has been led by Anne Marie Tiani, Joachim Nguiébouri, and George Akwah. Intended to contribute to Cameroon's policy on protected areas, the plan was to work closely with Tropenbos, the Netherlands Development Organization (SNV), and the Ministry of Environment and Forests (MINEF), the primary stakeholders officially responsible for managing Campo Ma'an; however, the intended collaboration was marred by strife. A first step in the process was a workshop on criteria and indicators, which brought together local community members, timber company personnel, local officials, and a few formal managers to discuss sustainable and collaborative management of the protected area. This was followed up with discussions on differences in impacts and desires for the future, by segments of selected communities. In 2002, the government transferred responsibility for management of the area to the Worldwide Fund for Nature (WWF) and is rethinking the existing management plan. The ACM team is now working with the Circle for the Promotion of Forests and Local Development Initiatives (CEPFILD), an NGO, to provide community capacity building on Cameroon's forestry law.

Lomié is a municipality in the eastern part of Cameroon, selected because five community forests there were among the first to be legally recognized under Cameroon's 1994 forestry law. These communities were Ngola, Kongo, Bosquet, Koungoulou, and Echiembor/Malen. One of the community forests had also been granted to Baka pygmies, a group recognized as a minority (with important policy ramifications) within Cameroon. The work was coordinated by René Oyono, Samuel Assembe, and Charlotte Kouna and conducted in collaboration with the World Resources Institute, as well as the Dutch SNV. Emphases included ongoing decisions about whether logging in community forests should be industrial or artisanal, and how benefits from these forests are distributed.

Dimako, also in eastern Cameroon, was selected because of its success in defining part of the area as a communal forest (to be managed by the *commune*, or municipality). The ACM team, led by René Oyono, cooperated with a project of MINEF and the Center for International Cooperation on Agronomic Research for Development, *Forêt et Terroirs,* as well as the World Resources Institute. Emphasis has been on rights and rules of use and benefit sharing among the communal council, the mayor, and the surrounding communities (which like Lomié includes Baka pygmies).

Akok and Makak were initially selected to represent the forest-agriculture interface and landscape mosaics; Makak was subsequently dropped because of conflicts. The interest in Akok was in examining trade-offs between agriculture and forest uses that would be socially and environmentally sustainable: could the typical mosaic of agricultural land uses be improved by perennial agroforests interspersed with annual cropping and multistrata systems, complemented by community-based forest management? The existing project leaders—from ASB, the Institute for Agronomic Research for Development (IRAD), and the International Institute of Tropical Agriculture (IITA)—actively sought the involvement of ACM, which was perceived as a bridge that could help facilitate and collaboratively monitor the induced changes. William Mala coordinated this research. Work continues in Cameroon and has expanded to the Democratic Republic of Congo.

Ghana

The Ghana research, which began in 2001 with full funding from the European Union, took place in two community forest reserves: Adwenaase and Namtee, in Ghana's Assin Fosu District, Central Region. The communities of Worakese and Akin Akenkausu manage the Namtee Community Forest Reserve, and Assin Akropong manages the Adwenaase Community Forest. The communities have firm tenure rights over these humid tropical forests. ACM efforts, led by Dominic Blay and Kofi Diaw, have focused on conflicts over distribution of forest benefits and related feelings of injustice on the part of the community volunteers who monitor and protect the forest without any direct benefit. The communities and the ACM team developed a set of criteria and indicators to help them monitor the condition of their forests, though efforts to catalyze an adaptive process among the communities have so far borne little fruit.

Indonesia

Indonesia-based ACM research took place in three humid tropical forest areas: Baru Pelepat in Jambi, Sumatra; Lumut Mountain Protected Area in Pasir, East Kalimantan (in the communities of Rantau Layung and Rantau Buta); and Bulungan Research Forest, also in East Kalimantan. All these sites were fully funded, variously by the International Tropical Timber Organization, the Asian Development Bank, the McArthur Foundation, the U.K. Department for International Development (DFID), the Consultative Group's Participatory Research and Gender Analysis program, and CIFOR. Linda Yuliani has played an important role in coordination for the Indonesia research since 2001.

The research in Bulungan[13] differs somewhat from most of the other sites discussed in this book, having begun in 1995, when CIFOR gained access to this 321,000-ha research site. Eva Wollenberg, who leads the ACM-related research there, was instrumental in conceptualizing the program, based partly on her team's findings. Given this long history, there was no need to conduct the context studies, and action research was already under way. The area represents a transect from quite remote, densely forested hinterlands to areas closer to the sea where oil palm plantations, timber extraction, transmigration, and mining are in progress or being considered. Numerous indigenous groups (including the hunter-gatherer Punan) as well as immigrants, come here to seek their fortunes. During the period covered by this analysis, efforts in Bulungan have centered on participatory mapping of community boundaries, on conflict resolution, and on the implications of the new decentralization policies—all within an empowerment framework.

Our work in Baru Pelepat, Jambi, began in early 2000 under the leadership of Trikurnianti Kusumanto. Located in the buffer zone of the Kerinci Seblat National Park, the village is ringed by development activities (timber, transmigration, oil palm plantations). The population includes indigenous Minangkabau, the hunter-gatherer *Orang Rimba*, as well as settlers from Java and other areas of Jambi. The research team intended to focus on social learning and equity issues, but as in the other Indonesia sites, they were drawn into the decentralization process. Interesting accomplishments included participatory mapping of local boundaries, the development of more transparent and democratic processes for selecting community representatives, and progress on other governance issues.

Rantau Layung and Rantau Buta, in the Lumut Mountain Protected Area, are small, homogeneous, indigenous communities, with territories of 18,913 ha and 16,546 ha, respectively. Inhabitants here, as in the other Indonesia sites, practice swidden cultivation. Comparatively inaccessible, these communities are surrounded by timber companies in the immediate vicinity, and oil palm plantations and transmigration are not too far removed. The initial orientation of Stepi Hakim's team in late 2000 was on environmental protection and livelihood issues, but as in the other Indonesia sites, the team wound up also doing participatory mapping and conflict resolution, focusing on governance issues. Work continues in Balungan and Jambi.

Kyrgyzstan

The four sites in Kyrgyzstan are located in the southern part of the country, where temperate walnut forests abound. Each site represents a forest reserve *(leshoz)*, which was once part of a very top-down, centrally planned Soviet forestry establishment. The government is trying to decentralize, for almost exclusively financial reasons—to devolve financial responsibility for forest management. Kaspar Schmidt coordinated the ACM research in Kyrgyzstan, with full funding from the SDC (Swiss government) since 2001.

Arstanbap-Ata *leshoz* covers about 7,489 ha of mainly broadleaf forests in an upper valley in the mountains of southern Kyrgyzstan. About 40 percent of the forested area is covered with the valuable walnut. Overuse of the forests is due to population pressure. Seventy-seven percent of the population is Uzbek, with most of the remainder Kyrgyz. Arstanbap-Ata *leshoz* started leasing forest plots near the village to private farmers in the late 1980s to ensure forest protection and improve the forest belt around the village. Leases were for one year.

Achy *leshoz* borders on Arstanbap-Ata *leshoz,* along the Kara-Unkur River. Its forested area constitutes 6,996 ha, with only 10 percent of the forest in walnut, and its population is 44 percent Uzbek and 55 percent Kyrgyz. The Achy *leshoz* started to give out leases in the 1980s, partially to comply with the requirements of the state production plan in the face of drastic cuts in state funding for forestry activities. Leases were for 10 to 15 years.

Ortok and Uzgen differ from the first two sites in being part of a government experiment, called Collaborative Forest Management, with longer term leases. Under the new arrangement in Ortok and Uzgen, leases are for 49 years (following a 5-year probationary period). Ortok has 47 percent of its 10,282 ha of forested area in walnut. Of Uzgen's 21,777 ha of forested area, only 25 percent is in walnut.

The team's emphasis has been on strengthening the participatory element of forest management in Kyrgyzstan. Several studies focused on local knowledge and forest use practices, biodiversity, and silviculture, but the planned participatory research process was cancelled.

Madagascar

Ranomafana National Park, which was created in May 1991, is in Fianarantsoa Province of southeastern Madagascar, some 90 km west of the Indian Ocean. The humid tropical forest has globally significant levels of biodiversity. Research began under the leadership of Louise Buck of Cornell University in 1997, and fieldwork was finished by 2000. The site thus served as a well-funded precursor to the full-blown ACM sites described in this work. The emphasis of the action research was on building partnership agreements between local communities and the park management to protect the park from primarily agricultural incursions from the surrounding communities. Population was expanding, and the communities were dependent on shifting cultivation for their livelihoods. The effort to develop partnerships was adversely affected by park managers' unwillingness to consider the proposition seriously and problems (death, dishonesty) within the team.

Malawi

Malawi has two ACM sites where research has been under way since late 2000, funded by the European Union and the U.K. Department for International Development. The work was coordinated by Lusayo Mwabumba and Judith Kamoto.

The Ntonya Hill site is 370 ha of land (once miombo woodland, now covered with *Eucalyptus camadulensis*) adjacent to the city of Blantyre in southern Malawi. Our work involves 11 communities that had been part of the Blantyre Fuelwood Project. Working with local communities, James Milner, of the Center for Social Research in Zomba, has focused on establishing a more sustainable management system for the eucalyptus plantation, which had been planted to supply fuelwood for the city of Blantyre. Historically, the area had seen considerable conflict over land use, and much of the plantation had been degraded when the ACM project began. Efforts to work with the existing committees in charge of management proved futile, and very little was accomplished.

The Chimaliro Forest Reserve covers 160 km² in the Kasungu District, between Lilongwe and Mzuzu. The forest is semideciduous miombo woodland dominated by *Julbernardia paniculata, Brachystegia utilis, B. spiciformis,* and *B. floribunda.* Judith Kamoto of the Bunda College of Agriculture has led the research, in which 10 villages are learning to comanage three blocks of forest with the forestry commission. Each block has a management committee with whom Kamoto has worked closely to implement the management plans and more effectively protect the forest. User groups, for whom honey and mushrooms are income-generating possibilities, were formed and have been progressing well in their efforts. This site has been hard hit by AIDS, drought, and general poverty.

Nepal

In Nepal, the participatory action research has involved two groups, one in Kaski District in the central foothills of Nepal, coordinated by Netra Tumbahangpe, and the other in Sangkhuwasaba District in the East, coordinated by a research NGO, New Era (first under Shibesh Regmi, then Laya Prasad Uprety). Forests are subtropical montane. ACM work began in early 2000, with funding from the Asian Development Bank, coordinated by Cynthia McDougall. All ACM teams in Nepal worked closely with both the community at large and the forest user group committees to expand the involvement of marginalized groups in decisionmaking about forest management. All sites were marked by comparatively high amounts of internal diversity, by caste and ethnic group, and strong gender differentiation. All teams also maintained close ties with government forestry personnel, structured their input to comply with governmental regulations about community forests, and thereby increased the likelihood that their findings could be immediately taken up by other communities. ACM efforts in the Nepal sites have been successful.

Bamdibhirkhoria's 48-ha community forest is in Kaski District, not far from the city of Pokhara. The forest is an even-aged, mature stand of *Schima castanopsis,* with more than 70 percent crown cover. A devastating landslide some years back strengthened the community's commitment to conservation. Our research there has been led by Sushma Dangol, with Chiranjeewee Kadkha.

Deurali-Baghedanda's 181 ha of community forest, also in Kaski District, is composed of natural, multistory, deciduous broadleaf forest and plantation (at the pole stage), with crown cover of 70 to 80 percent. ACM research was initiated by Mani Ram Banjade and subsequently led by Raj Kumar Pandey.

The Manakamana Community Forest covers 132 ha of primarily *sal (Shorea robusta), chilaune (Schima castanopsis),* and *botdhairo (Lagerstomia parviflora)* in the Sangkhuwasaba District. Fire has been a serious threat to the forest. Work was led by Narayan Sitaula of New Era.

Andheribhajana's community forest, also in Sangkhuwasaba District, covers 113 ha, about half of which is the valuable *sal.* This site has suffered from both landslides and fire. The community also accepted migrants between 1958 and 1963, which put additional pressure on their resources. The community was characterized by unusually high levels of conflict. Kalpana Sharma, of New Era, led the ACM work in this site. Work continues in Nepal.

Philippines

Research began in the Philippines in early 2000 in San Rafael–Tanabag–Concepción, Palawan. Conflicts within Mindanao interfered with our original plan to open a second site there soon after. Ultimately, research began in Basac in 2001. Both sites were fully funded by the Asian Development Bank and coordinated by Herlina Hartanto.

The San Rafael–Tanabag–Concepción Multi-Purpose Cooperative has a community-based forest management agreement with the Department of Natural Resources in Palawan to manage 5,006 ha of forestland. When the ACM efforts began (led by María Cristina Lorenzo), the cooperative's efforts were focused primarily on a salvage logging project that was encountering difficulties. The cooperative worked at defining criteria and indicators for sustainable management in efforts to involve a wider segment of the three communities in conflict resolution with the neighboring Batak community, and in negotiations with various levels of Palawan's complex governmental and NGO bureaucracies. They strengthened their abilities to deal effectively with other stakeholders dramatically and made some progress on diversifying sources of income from their forest area.

Basac, Lantapan, Bukidnon, Mindanao began in mid-2001, led by anthropologist Erlinda Burton of Xavier University. Her team has worked with the Talaandig, an indigenous group residing in the buffer zone of Mount Kitanglad Natural Park. Their efforts have been focused on a community-based forest management area covering an area of 2,800 ha of mostly agricultural land. This community forest was granted by the government to the Basac Upland Farmers Association. Initially stymied by a major conflict (see Box 10-5), the team and the community eventually formed user groups interested in specific topics, such as medicinal herb gardens and income from small-scale cut-flower sales. These groups have visited other communities, undergone training, and implemented plans that built on and strengthened indigenous cultural practices. Work continues in Palawan.

Zimbabwe

The work in Zimbabwe focused on the people living in the communal areas surrounding the Mafungautsi Forest Reserve and began in 2000. Leadership of this work shifted from Ravi Prabhu to Frank Matose, allowing Prabhu to assume overall coordinating responsibilities for most of the Africa and Asia work. The Zimbabwe team works with resource management committees, which were created by the forestry commission as part of a previous experiment in comanagement. The communities formed three kinds of user groups (thatch grass, broom grass, and honey) and planned and implemented improvements in the management and income-generating potential of these products. In the course of these activities, they have significantly improved the relationships between community members and the forestry commission, enhanced women's involvement in formal forest management, and strengthened their self-confidence and willingness to act on their own behalf. This work continues and has scaled up significantly.

CHAPTER TWO

ACM's Intellectual Underpinnings
A Personal Journey

*I*N THE 1990s, IN HER *COMPOSING A LIFE*, Catherine Bateson likened professional women's lives to a patchwork quilt. Using case materials from individual professional women, she showed how their lives had been shaped by meshing their personal interests and capacities with the opportunities and constraints they encountered. These women took a bit from here and a bit from there and put these bits together creatively, forming full, satisfying, productive lives. In this responsive and evolving way, they were able to combine meaningful relationships, children, leisure, and work.

In looking back on my own life, many of her observations apply—as I've moved in and out of a marriage and a partnership, borne two children, married again, gained two stepchildren, and moved repeatedly for different jobs—both my own and my partners'. But this analogy of life as a patchwork quilt can also be applied in dealings with theory and method.

In this chapter, I examine pertinent theoretical patches that together form the quilt that is adaptive collaborative management (ACM), focusing on my own experience, knowing that different ACM team members have come to similar perspectives by different routes. Each patch—or object of study—links to the concerns of forestry. This review begins with the study of cultural anthropology, highlighting cultural and gender differences and health issues. The next sections describe bodies of literature from, and experience with, farming systems research and development, social and community forestry, and integrated conservation and development projects. All these—plus the criteria and indicators for sustainable forest management, adaptive management, and systems and complexity theory discussed in the next chapter—have informed the development of ACM, but the quilt is still not finished.

The Importance of Culture in Forestry

In 1955, when I was nine years old, my parents (an anthropologist and a psychologist) took me to live in Ankara, Turkey, for seven years. During that time I learned a great deal about cultural difference and about global inequities. It was a time when the idea flourished that America had only to replicate the post–World War II Marshall Plan (under which American money and expertise had effectively contributed to rebuilding Europe) in the rest of the world, and development goals would be met. Scott's (1998) notion of "high modernism" was in full swing. Part of this idea included assumptions that development simply involved a unilineal process that would culminate in an American lifestyle. Such notions failed to recognize the cultural differences between the United States and most other areas of the world. My observations about the inaccuracy of this unilineal and ethnocentric view—combined with books like *The Ugly American,* by William J. Lederer and Eugene Burdick—led me to study anthropology and the variety in human cultures. I reasoned that such study was needed to turn the development process into something benign. In the following section, I introduce some of the literature supporting the idea that attending to people and their cultural systems is important for good forest management—that is, forest management that takes into account the well-being of the people in and around the forest.

People and Their Forests

Anthropology shows us how people participate in complex cultural systems that interweave kinship with local economic transactions, local-level politics with religious beliefs, forests with symbolism. The same people are also almost inevitably involved to greater or lesser degrees in other, large-scale systems. Their actions link to events and actors in the wider world (cf. Vayda 1983).

A huge body of anthropological literature documents the uses to which people put the forest, the meanings they attach to it, and their impacts on it. Collections by Kunstadter et al. (1978), McCay and Acheson (1987), Fortmann and Bruce (1988), Croll and Parkin (1992), Redford and Padoch (1992), and Padoch and Peluso (1996) are representative of this growing treasure trove. My own long-term ethnographic research, particularly among forest dwellers in the United States (Colfer 1977; Colfer with Colfer 1978) and in Indonesia (Colfer 1991b; Colfer with Dudley 1993; Colfer et al. 1997) convinced me of the intimate ties between forest dwellers and their environment, and the pressing need to integrate anthropological findings with forest management. Indigenous people are already managing forests—they are simply managing them for different goals than are foresters, ecologists, and government officials.

Anthropological studies also document the rationality of much human behavior.[1] What outsiders may interpret as irrational often simply reflects their misunderstanding of how local systems function. An outsider may judge a Dayak in central

Borneo to be profligate as he gives away a large proportion of the money he earns working for a logging company. Yet he lives and dies in a context where access to money is a rare boon, and tomorrow he may well be dependent on the goodwill and help of those receiving his money today. Rather than being irrational, he is in fact thinking ahead and strengthening his links to those whose help he is likely to need tomorrow (Dove 1988; Mauss 2001). A Dayak woman's preference for rice production over cash crops, in apparent irrational rejection of opportunities to make money, reflects her awareness of the rationality of "wasting" money on social ties (not to mention the fact that rice production is her profession). In that context, rice, something everyone grows, represents a much more reliable contribution to family livelihood (Colfer et al. 1997). Excess rice functions as a savings account that need be shared only under emergency conditions, whereas money is expected to be shared as needed by others. The different context creates a different starting point for reasoning and a different set of underlying assumptions.

Studies of cognition also support the rationality of much human behavior. Psychologist George Kelly (1963) wrote of "man [sic], the scientist," making a good case for the view that people are naturally programmed to try to predict and understand the world around us, and that we engage in a continual process of hypothesis testing in our daily lives. Observations in the field suggest that forest peoples share this propensity. Freire's (1970) widely accepted arguments in favor of "education for critical consciousness" build on a similar perception of rural people's capabilities (see also Freire and Freire 1994, and Nyirenda et al. 2001, for a recent example).

Understanding of the potential contribution of indigenous knowledge has been growing in recent years. Beginning modestly with the elicitation of local taxonomies of plants (Conklin 1957), firewood (Metzger and Williams 1966), colors (Berlin and Kay 1969), and other domains, this field of study has grown to address complex issues like representing the underlying logic of indigenous knowledge on computers (Colfer et al. 1989; Joshi 1997; Sinclair and Walker 1998; Walker and Sinclair 1998; Haggith and Colfer 1999). A journal, *The Indigenous Knowledge Monitor,* is devoted entirely to documenting such knowledge. The Intermediate Technology Development Group in the United Kingdom publishes a series called Studies in Indigenous Knowledge and Development, and Iowa State University hosts the Center for Indigenous Knowledge for Agriculture and Rural Development. This kind of information has relevance both for understanding local human systems (the anthropological concern) and for linking local people in a more mutually beneficial way with outsiders (a more pragmatic need related to protecting the environment and human well-being).

Jordan (1991; 1997; Jordan with Davis-Floyd 1993) has documented one source of the difficulties integrating indigenous knowledge: the concept of "authoritative knowledge." Those in power have knowledge that is generally recognized as authoritative; the knowledge of those without power is not recognized in this way (cf. Foucault 1980; Escobar 1995). Although Jordan writes about childbirth and air traffic control, this concept is equally relevant to forest people (Wollenberg et al. 2001a; Banuri and Apffel-Marglin 1993).

Besides indigenous knowledge, indigenous organizations and institutions have been identified as important bases for improving management (Clay 1988; Warren et al. 1996; Sarin 1993; Peluso 1994; Colfer et al. 1997). People in all societies group together to accomplish particular purposes (many of which have natural resource implications). Human beings also live in groups that share norms and rules, many of which can be strengthened to improve management. Many of the studies with particular relevance for forests pertain to people's management of the commons (McCay and Acheson 1987; Ostrom 1990; see also the Journal of the International Association for the Study of Common Property).

External recognition of forest people's unique knowledge of their environments can serve both to enhance management and to strengthen the voice of local people in making policies more appropriate to their and the environment's needs. Many formal forest managers remain unaware of local people's potential contribution to their work in managing tropical forests. We have been ignoring a huge human resource in not recognizing forest people's abilities to participate in development processes (Vayda et al. 1980; Vayda 1983; Clay 1988). The ACM process recognizes this knowledge and builds on it in a collaborative way.

Turning for a moment from the potential of individuals and communities to contribute to better forest management (and improvement of their own conditions), anthropology's long-standing attention to human adaptation and cultural change is worth noting—and is also clearly central to ACM. Cohen's (1968a, 1968b) somewhat dated two-volume collection reflects this broader interest within the discipline. Harris (1968) contrasts two threads within American anthropology—the idealists and the techno-environmental determinists, with Harris representing the latter. The idealists are those who prioritize the way people look at the world (values, ideals, perceptions) in their analyses of culture and culture change; the techno-environmentalists are convinced that technology and the environment play determining roles in the shape and trajectory of human culture. Within ACM we acknowledge both forces to be operative, expanding our interest in the "ideal" to include a focus on conscious learning—something not addressed by Harris.

Women and Marginalized Groups

Throughout the 1970s and 1980s, many researchers were struck by the marginalization, particularly of women, but also of other groups, and by the lack of homogeneity in communities. Any community has women and men and different age groups, and it often has different ethnic groups, classes, castes, and occupational categories, with differing access to resources, differing involvement in local decisionmaking, and differing health status. Much of the large body of literature that considers gender differences applies equally to other marginalized groups, with implications for both equity and management. I was convinced that we ignore these differences at our peril.[2]

Anthropologists have thoroughly documented women's different roles in society. Sanday (1974) used the Human Relations Area Files to examine women's position in societies around the world—a central issue for the well-being of at least half of the population. She identified conditions under which women's status varies, with involvement in productive activity—often closely connected to forest management in the contexts we are discussing here—being one prerequisite for high status. Many authors have explored the different roles of men and women in various parts of the world, documenting the ubiquity of women's (often unpaid) contributions, inequitable access to power, and reduced access to both decisionmaking and the benefits of their labor (Rosaldo and Lamphere 1974; Goodman 1985; Atkinson and Errington 1990; Sigot et al. 1995). Insofar as we see human well-being as central to sustainable forest management, we need to address such issues.

The late 1970s and early 1980s marked a growing awareness of the situation of women in farming communities. Michigan State University developed a working papers series, "Women in Development," and an increasing number of analyses showed the importance of women in family livelihoods around the world (Siskind 1973; Staudt 1975–76, 1978; Colfer 1991b; Wickramasinghe 1994; see also the collection by Poats et al. 1988). Eventually tools were produced to help agricultural scientists better address women's concerns in agricultural projects (Feldstein and Poats 1990; Feldstein and Jiggins 1994). Although the emphasis in most of this literature was on farming, some of those farmers were forest dwellers. Numerous studies have shown women's close links with forests as well (Colfer 1981; Murphy and Murphy 1985; Leach 1994; Townsend et al. 1995).

At the same time, there was an increasing awareness that disaggregating data by gender (or even gender and other markers of diversity), having diverse teams, and interviewing a more diverse range of community members, though all important, were not enough—particularly where greater equity was a goal (Rocheleau et al. 1996; Colfer et al. 1997; McDougall 2001b). Greater attention would need to be paid to more systemic (and more difficult) issues like the distribution of power, interaction patterns that reinforce inequitable power relations, and group introspection and analysis, to identify locally practical solutions (Tsing 1993; Leach 1994; Sarin 1997, 1998; Porro and Porro 1998; McDougall 2001b; Porro 2001; Sithole 2001; see also the abundant materials on this subject available from Future Harvest's gender program, www.genderdiversity.cgiar.org).

In recent years, gender and diversity have received more attention, both as separate issues (Sigot et al. 1995; Townsend et al. 1995; Wilde and Vainio-Mattila 1995) and as integral parts of other issues, like participation (Rocheleau and Slocum 1995; Carter 1996; Borrini-Feyerabend and Buchan 1997; Guijt and Shah 1998a) or land tenure (Fortmann and Bruce 1988). These scholars' findings made it clear that we would have to hear the voices of women and other marginalized groups if we wanted to benefit both forests and human well-being.

As researchers became more sophisticated in their attention to the question of gender, they also noted important parallels between women's experience and the experience of lower status ethnic groups, castes, and other marginalized groups.[3] Like women, these groups tended to have less access to employment and other benefits, muted voices (Ardener 1975), and less power (Colfer 1983b).

ACM focuses on including diverse stakeholders in the management process. We have gone beyond the broad categories of stakeholders to look at intracommunity variation, seeking to amplify the voices of the communities that are marginalized by more powerful outside stakeholders, as well as those who are marginalized within communities.

Human Health

By the late 1970s, I was convinced that knowing about anthropology and culture was insufficient. The holistic, anthropological approach needed application to some particular sphere. The health field struck me as central to human well-being.

Human health is rarely tackled by foresters and others involved in forest management, although it has been identified as important by expert teams working on sustainability issues in forestry (CIFOR 1999b). More recently, a number of international institutions have identified pertinent parallel concerns (e.g., WEHAB 2002).

I see health as closely linked to forest management. The most fundamental way is in the links among women's status, health, population, and forests (Sen 1994). There have been recent attempts to link human health and environmental concerns conceptually and in the field (Vogel and Engelman 1999; Engelman 1998; Gardner-Outlaw and Engelman 1999). One concern of many biophysical scientists and managers of forested areas is the pressure that increasing populations, directly dependent on local natural resources, can put on a landscape. This leads in some cases to local communities' forcible removal—an approach that has in general proved neither humane nor effective in improving environmental sustainability (Brockington 2002). A more systemic, constructive, and humane way to address such concerns would involve attending to local reproductive health issues. Such an approach also holds the promise of catalyzing the efforts of local women in addressing local community and environmental problems.

More specific health issues link directly with forest management. A clear link between AIDS and forests in West Africa (van Haaften 1995) can be traced to the road network that logging companies create, combined in many cases with sex ratio imbalances due to the use of male labor in logging. Kunstadter (2002) explains some of the links between malaria and forests. The relationship between malaria and logging is more tenuous (Singhanetra-Renard 1993), but increases in malaria are arguably linked to forest clearing and ponding (van Haaften 1995). Logging is considered one of the most dangerous occupations, and worker safety is a well-recognized problem for timber companies and governments (Poschen 1993, Strehlke 1993).

Threats to mental health are also serious under conditions of severe environmental degradation. Feelings of depression, hopelessness, lack of initiative, and consequent adverse effects on communities' capacity for collective action and problem solving are often noted in conditions of environmental stress. See van Haaften and van de Vijver's (1996) study in Côte d'Ivoire, in an area where the forests had been largely cleared and there was a continuing influx of refugees (due to war in Liberia, poverty in Ghana, and environmental devastation in Mali); or my own study in East Kalimantan, Indonesia, which was ravaged by the El Niño–induced fires of 1997–98 (Colfer 2001b). Where local people's lives have been intimately connected with forests that are now disappearing, some blame serious mental health problems on the loss of "cultural integrity" (Laksono 1995; Prabhu et al. 1996).

Human health is likely to be ignored in any conventional forestry endeavor, perceived as remote from a forester's concerns. Yet forest management that is truly adaptive and takes genuine account of issues of human well-being cannot ignore human health any more than it can disregard forest health. As we move toward

Box 2-1. Stereotypical Views of People and Forests

View 1: Forestry focuses on trees. Foresters learn about forest management in school and have expertise—about species, silviculture, harvesting methods, and other management strategies—that is not available to the general public. Decisions about how to manage forests should be left in the capable hands of these trained professionals.

View 2: The forest is a complex ecosystem, including plants, animals, insects, and other biota. It is aesthetically beautiful and intellectually fascinating. Ideally, local people are kept away from the forest, since many of their activities disturb the natural processes at work there. They are likely to destroy it.

View 3: The forest is a national resource and must be managed for the good of the country. It can be managed for timber production by large operators that bring in more foreign exchange; or it can be managed for conservation purposes, in which case local people should be kept out. Certainly "primitive" techniques like swidden agriculture or hunting and gathering are an embarrassment to the nation and indicative of the backwardness of forest people.[4]

adaptive and collaborative management of systems, health emerges as an important issue in many contexts.

Box 2-1 provides stereotyped synopses of some common views about the roles of local people in tropical forests, based on my ethnographic experience working in forest contexts globally. These snapshots, like Roe's (1994) policy narratives, are stylized versions of different stakeholder perspectives—recognizable by those with experience in tropical forests—with important implications for local people's roles. In these stereotypes, local people range from an embarrassment to a positive menace. Despite the oversimplification, the prevalence of these views makes such caricatures an important part of the context in which ACM has been evolving.

Those views ignore a fundamental ethical consideration: the question of justice. Often the people who live in the forest are indigenous and derive their livelihoods from it. They are often marginalized within their own countries, perceived by their compatriots as primitive or backward. They often have little wealth, education, power, or prestige to protect the resources that they have inherited—particularly against large-scale timber and plantation companies, conservation projects, and resettlement programs. Ignoring the longstanding claims of such people is quite simply unjust.

Other people who live in or near the forest are immigrants, pushed from their previous homes by economic necessity, war or internal civil strife, environmental degradation, or government-sponsored settlement programs. They are almost always economically and politically disenfranchised and dependent on the forest for some significant component of their livelihoods. Like indigenous peoples, they have little leverage against powerful outsiders and thus suffer blatant injustice.

The next three sections—drawing from the fields of agriculture, forestry, and conservation—examine the most important initiatives in these fields to integrate local people into their respective concerns. Each has provided important insights for ACM and demonstrated gaps that need to be filled if we want people and forests to prosper.

Agriculture, Forestry, and Conservation

Beginning in the 1980s, I turned to more activist approaches. Along with many others, I recognized the static nature of much of our writings on culture. We had tended to focus on the stable, conservative aspects of culture, not giving sufficient attention to its dynamic aspects and the role of human agency in bringing about desired changes. The next three patches in the ACM quilt come from the world of applications: farming systems research and development, social and community forestry, and integrated conservation and development projects.

Farming Systems R&D

How can people's lives be improved through increased production and income?

In the early 1980s, while working at the University of Hawaii as a Women in Development specialist, I encountered farming systems research and development (sometimes called farming systems research and extension). This field followed the popular integrated rural development efforts, whose practitioners had recognized the holistic nature of people's existence and tried to put together all the components of people's lives into single—often unwieldy—projects. They made important progress in clarifying the interactions within human systems but failed to meet the world's unrealistic expectations.[5]

In 1982 Shaner et al. (1982) published a widely read book outlining the critical features of farming systems R&D (see also Norman 1982, Hildebrand 1986a). This approach grew out of, and expanded on, the cropping systems research that had been done in Asia in the 1970s, much of it coming from the International Rice Research Institute (McIntosh 1980; Zandstra et al. 1981; Harwood 2000; see Collinson 2000 for a thorough history).

Shaner et al. (1982) identified six characteristics of a farming systems R&D approach; it would be

- farmer based;
- problem solving;
- comprehensive (or holistic);
- interdisciplinary;
- complementary to other approaches; and
- iterative and dynamic.

Farming systems research would identify significant local problems from an interdisciplinary perspective;[6] develop close relationships between farmers and scientists; and plan, conduct, and adapt collaborative experiments on farmers' fields.

I was attracted by that ideal (see McCorkle 2000 for a discussion of the links between anthropology and farming systems research), but one feature that particularly drew me was the focus on the farmer. This seemed like an ideal framework in which to involve biophysical scientists more meaningfully in interaction with farmers. Given my responsibilities for Women in Development, it also struck me as appropriate for addressing women's involvement in agriculture.

One of the techniques widely used in the early stages of a farming systems R&D project is the conduct of a *sondeo* (Hildebrand 1986b)—a precursor to participatory rural appraisal.[7] Interdisciplinary teams visit farmers' fields and talk with farmers and other relevant actors to obtain a holistic understanding of the local context and identify important problems; the site visits also strengthen the team members' understanding of each other's perspectives and contributions. Such experiences permanently imprinted the benefits of field-based, interdisciplinary interaction in my own mind.

During the 1980s and beyond, the *Farming Systems Research and Extension* journal published a steady stream of articles on the subject, and large numbers of projects around the world purporting to use this approach were funded. Simmonds (1985), however, conducted a rather damning evaluation. Many projects that called themselves farming systems research did not involve farmers throughout the process. Many teams were unable to overcome members' disciplinary biases to work constructively together. The focus was on the farm household and its production to the exclusion of environmental and political contexts.

The farming systems approach also suffered from the problem shared by integrated rural development (a problem that ACM also faces): its popularity was ultimately its downfall, as people came to expect more than it (or any other approach) could possibly deliver.[8] But its most crucial failing was that it was increasingly implemented in ways that did not meaningfully involve the farmers.

The conceptual links between this approach in its ideal form and ACM are clear. The elements that ACM adds—which, let me hasten to say, some farming systems efforts also included[9]—are attention to the environment and to the landscape, more systematic and ongoing monitoring, and links between the local community and landscape and the external political context.

Social and Community Forestry

How can forests be maintained without adversely affecting local people's lives?

My involvement with farming systems had convinced me of the importance of attending to environmental concerns; and I began working with foresters, initially through the Community Forestry Program of the UN Food and Agriculture Organization (FAO).

The theme of the Eighth World Forestry Congress, held in Jakarta in 1978, was Trees for People. That signaled the beginning of a sea change in foresters' attitudes toward local communities. Since that time, a wealth of programs and approaches have been created—social forestry, agroforestry, joint forest management, community forestry, community-based forest management, to name a few—to acknowledge and build on the links between people and their surrounding or neighboring forests.[10] Some of these catch phrases have become the names for formal government programs, as well as for approaches used by researchers and NGOs.

In early efforts, forest departments paid local people to plant trees (Poffenberger et al. 1996), often for fuelwood (Cernea 1990). Letting poor people plant food crops among growing tree crops for a few years before the food crops were shaded out, in return for planting and caring for the young trees—variously called *tumpang sari* or the *taungya* system—became popular in many countries (Peluso

1992). Outgrower schemes, wherein small landholders plant fast-growing species for sale to pulp mills, have been popular with both governments and private industry (Momberg 1993; Arnold 1998; Nawir et al. 2002), with variable results.

Agroforestry efforts, such as alley cropping with leguminous trees, were encouraged for decades by the International Center for Agroforestry (ICRAF, now called the World Agroforesty Center). The intent was to obtain ecological services like erosion control and nitrogen fixation, plus products of use to people like animal fodder, firewood. and sometimes some minimal human food (especially with *Leucaena leucocephala*), alongside normal food production. But none of these approaches won full acceptance by local communities.

As the years went by, forestry departments and communities became more sophisticated, and attempts to link forests and people more effectively became more complicated, with more interesting successes. K.B. Shrestha, Don Gilmour, and Robert Fisher in Nepal; Marilyn Hoskins, Kady Warner, and their collaborators through FAO's Forests, Trees and People Program; Mark Poffenberger and the Asia Forestry Network participants; Diane Russell, Janice Alcorn, and others with the Biodiversity Support Network; and many other researchers and NGOs have developed much more sophisticated approaches that look at forest management from a more systemic perspective (Carter 2000 on collaborative forest management; Colchester et al. 2003 on community forestry networks). The Overseas Development Institute's *Rural Development Forestry Network* series, the work of the International Institute for Environment and Development on forest policy issues, the Biodiversity Support Program's studies, and the studies conducted as part of Indiana University's International Forestry Research and Institutions Program represent other useful sources of work on forests and people around the world.

Excellent analyses of this evolution in Asia—where it has had the longest history—are available from Poffenberger (1990), Messerschmidt and Rai (1992), Fox et al. (1993), and Lynch and Talbott (1995). Complex, national-level management systems now exist in India, Nepal, and the Philippines, wherein local communities or user groups can gain access to forest areas claimed by the national government and manage them in collaboration with forest departments (Hobley 1996b; Gilmour and Fisher 1991; Luna 2001; cases 11 and 12 in the Appendix of Cases). These approaches, though promising, have had mixed success (CIFOR 2001; Malla 2001; Sarin et al. 2003), but similar, smaller scale experiments are also under way in many other countries.

Although forestry professionals themselves readily acknowledge that even the recent, more sophisticated approaches to forest management still have problems,[11] these efforts have provided an institutional base on which the idea of adaptive collaborative management could take hold. Our ACM efforts build on this experience, with a greater emphasis on forward-looking, institutionalized learning and on equity.

Integrated Conservation and Development Projects

How can people maintain their environments while improving their quality of life?

In the early 1990s, my involvement in an integrated conservation and development project—to comanage the Danau Sentarum Wildlife Reserve in Indonesia, with funding from the U.K. Department for International Development—added another patch to our quilt: linking human well-being with environmental well-being.

Although conservation has a long history, great public interest was aroused by the dismal projections put forth 30 years ago by the Club of Rome (Meadows et al. 1974). Concern about the state of the world expanded throughout the 1980s and 1990s. Dismaying, even horrifying statistics were compiled and used to demonstrate exceedingly fast rates of forest degradation, biodiversity loss, habitat destruction, and other indications of a global environmental crisis (AtKisson 1999; Terborgh 1999). One result was the Rio de Janeiro conference, often considered the beginning of global concerns about sustainability (Brundtland Report 1992). Meanwhile, conservationists and others were beginning to realize that fortress conservation ("fines and fences," "command and control") often had adverse effects on local people (Brockington 2002). The Biodiversity Conservation Network, for instance, was built on the hypothesis that communities deriving benefits from the use of their environment would act to conserve that environment (Mahanty and Russell 2002).

Considerable funds were made available for integrated conservation and development projects throughout the late 1980s and 1990s. These projects have built on the idea that conservation requires the cooperation of local communities, particularly where the people have few financial resources and depend on local natural resources, and where control over protected areas has been ineffective. "Poor people"[12] were seen to degrade their environments out of need, so integrating conservation and development efforts would make them better off and thus more willing to protect their environments.

Project designers also recognized that many protected areas had been established where local communities had traditional (and underrecognized) claims to the resources (Geisler 2003). They saw the ethical issue relating to displacing peoples from their traditional homelands, and they realized that people who feel they have been treated unjustly might be uncooperative, or worse. These observations went hand in hand, in some cases, with a recognition that local communities often also represented "living libraries" of ecological knowledge that might be useful to biologists (Howard 2003).

Those truths—that local people should receive some benefits from their surroundings, that existing rights to resources should be recognized, that local people can participate in or thwart conservation, and that local people have stores of knowledge and other useful social capital—have influenced the evolution of ACM.

But such truths in no way diminish the problems that have plagued integrated conservation and development projects (Wells et al. 1999; Sayer and Campbell 2003). Many projects focused on income-generating opportunities for local communities (Wickham 1997), unintentionally increasing the pressure on local resources. Local people want financial security as much as other groups of human beings, and when they see a profitable resource use, they are likely to exploit it. To be effective, a project needs more emphasis on institutional controls to balance the acquisitive inclinations of people; see McShane and Wells (forthcoming) for a longer list of lessons learned.

Integrated conservation and development projects were often staffed by biological scientists, some of whom lacked training in the social sciences. In many cases, attempts to include local people in management were implemented with insufficient attention to power relations within and between communities, to intracommunity variation, to local norms of interaction, and with inadequate understanding of the complexity of local resource use patterns (Colfer et al. 1999c, 2000; Brockington 2002; Russell and Harshbarger 2003). These issues were crucial to effective involvement of local communities. Mahanty and Russell (2002), reporting their experience with four Biodiversity Conservation Network sites, stressed the importance of nuanced micropolitical assessments of local conditions and the need for strong facilitation skills when making conservation-related interventions with local communities—both often missing in conservation efforts.

The level of community rights and responsibilities varied considerably among integrated conservation and development projects. Some, such as the Danau Sentarum National Park in West Kalimantan, Indonesia (Giesen 2000), focused on full-scale comanagement (Colfer et al. 1996; Wickham 1997; Claridge and O'Callaghan 1997)—though the option *not* to be involved has rarely been available to local communities living in officially protected areas. In other cases, local people were viewed more as recipients of benefits, in a form of bribery ("you leave our special species alone, and we'll provide you with jobs as guides"). In still others, local communities decided not to cooperate with external actors who made claims on local resources (e.g., Rice 2001). The recent collections by Buck et al. (2001), Hulme and Murphree (2001b), and Brechin et al. (2003) document some of these variations. Lessons learned include the following:

- project workers must deal carefully with the power structures in which local people find themselves (both internal and external);
- the locally acceptable rules, monitoring, and sanctions that Ostrom (1990) identified as critical to good management must be heeded;
- the linkages between different organizations, including checks and balances, are important (Hulme and Murphree 2001b); and
- community conservation efforts take time.

Criteria and Indicators: A Framework for ACM

In 1994, I was invited to join a team looking at criteria and indicators (C&I) for sustainable forest management. As I studied this topic, I recognized its potential. If we could clarify the important elements of human well-being in forests and include significant social criteria and indicators into C&I sets that were widely adopted, we could have a huge impact on the conventional forestry profession.[13]

C&I is a shorthand for a conceptual and hierarchical system involving principles, criteria, indicators, and verifiers (Lammerts van Bueren and Blom 1997), pertaining in our case to forest management. Principles are the most abstract, representing the goals of forest management (e.g., "The health of the forest actors, cultures, and the forest is acceptable to all stakeholders").[14] Criteria are somewhat more specific conditions desired within the forest and forest communities (e.g.,

"The relationship between forest management and human health is recognized"). Indicators are ideally (though not necessarily) quantifiable evidence that the conditions specified by the criteria are in fact in place; and verifiers are an even more specific level of evidence (e.g., "Nutritional status is adequate among local populations"). Prabhu et al. (1996) have noted an interesting parallel between principles, criteria, indicators, and verifiers on the one hand, and wisdom, knowledge, information, and data on the other.

From the beginning, we believed that the well-being of both forests and people had to be maintained if sustainability was to be achieved. By 1998, we had developed a generic C&I set, intended to provide flexible guidelines for those interested in assessing the sustainability of timber management in particular forests (Prabhu et al. 1999; see also CIFOR 1999b). We were close to similar sets for forests managed by communities (Burford de Oliveira et al. 2000a, 2000b; Ritchie et al. 2000) and for plantation forestry (Muhtaman et al. 2000; Sankar et al. 2000). All of these sets included explicit attention to human beings in forests, as well as policy, ecological, and production considerations.

Looking at the C&I set prepared for forests managed for timber, for instance, we had identified three sustainability principles pertaining to social issues, through a systematic process of interdisciplinary evaluation and field testing by experts from various countries (Prabhu et al. 1999):

- forest management maintains or enhances fair intergenerational access to resources and economic benefits;
- concerned stakeholders have acknowledged rights and means to manage forests cooperatively and equitably; and
- the health of forest actors, cultures, and the forest is acceptable to all stakeholders.

Assessment methods—for use by formal forest managers—for the first two principles had been tested internationally in the field (Colfer et al. 1999a, 1999b, 1999c; Colfer and Byron 2001; Salim et al. 1999). The third issue, health, has not yet received adequate attention (Margoluis et al. 2001; Melnyk 2001; Vogel and Engelman 1999), though plans to address this gap are currently in the works at CIFOR.

The C&I tests provided convincing evidence of the importance of human factors in sustainable forest management and the increasing sensitivity of formal forest managers to local communities' needs. The various C&I sets that have been produced—by CIFOR, the International Tropical Timber Organization, Lembaga Ekolabel Indonesia, and many others—signal a major step forward in persuading forestry professionals to attend more systematically to human issues in forestry. The C&I concept has been widely accepted within the forestry field. The sets serve to remind users, many of whom have been trained and further steeped in reductionist scientific methods, of the interconnections among the parts of the systems being examined. The C&I emphasis on the importance of local people's access to resources and necessary involvement in forest management also serves as a global reinforcement for local communities' efforts on the ground. Finally, we reasoned, the C&I themselves, if simplified, might serve as monitoring devices to help communities and other stakeholders ensure more responsive management in their own areas.

Summary

The various intellectual "patches" that form the ACM "quilt" have included anthropological evidence about the knowledge, creativity, and institutional capacities of rural people. Particularly significant have been the potential contributions of marginalized groups (such as women, untouchables, and pygmies), whose voices in forest management are typically muted or not heard at all, and health as a part of human well-being. Other patches—including work in agriculture, forestry, conservation, and on criteria and indicators for sustainable forest management—have provided a historical and partial conceptual basis for ACM.

CHAPTER THREE

Complexity, Change, and Uncertainty

IN 1994, AT THE SUGGESTION OF Jack Ruitenbeek, I read *Complexity: The Emerging Science at the Edge of Order and Chaos* (Waldrop 1992). Although the scientists involved in this field tended to be computer whizzes and number crunchers, I still found their conclusions appealing. They recognized the interconnections among parts, feedback, and the importance of seeing things holistically—issues that any anthropologist finds consistent with ethnographic experience. Complexity theory eventually came to influence our work.

Kai Lee's *Compass and Gyroscope: Integrating Science and Politics for the Environment* (Lee 1993), which Ravi Prabhu recommended, also played a seminal role in the development of our thinking about adaptive collaborative management (ACM). We began looking at adaptive management research conducted by others (e.g., Holling 1978; Hilborn and Walters 1992; Schindler et al. 1996; Stankey and Shindler 1997; Stankey and Clark 1998). The emphasis on experimentation and institutionalized learning was an important precursor for ACM and has become a central component of our approach (Wollenberg et al. 2001a; McDougall et al. 2002a; Guijt and Proost 2002; Prabhu 2003; Mutimukuru et al. 2003; Guijt forthcoming; Mutimukuru et al. 2005).

Our concepts evolved. We now think of collaborative management rather than comanagement; the latter implies to some people an equality of rights among stakeholders, and to others, necessary partnerships between governments and civil society, neither of which may be warranted. The concept of adaptive management has also evolved in our usage to more intimately involve local-level stakeholders and pay more attention to differences within stakeholder groups (such as communities) than does most U.S.-based adaptive management. Our version of adaptive management depends less on having a benign and democratic government or vast quantities of data than the version practiced in the United States.

Systems Theory and Complexity

Although the anthropological literature is replete with references to the interconnections among parts (holistic analysis), feedback, and lags between causal factors and their impacts (see Bateson 1972 or Richardson 1991), many anthropologists (and others) take issue with systems theory because of its determinism. With roots in engineering (Richardson 1991), some elements of systems theory are overtly deterministic—something I see as inconsistent with the evidence about human beings. This theme is taken up in *Complexity and Management: Fad or Radical Challenge to Systems Thinking* (Stacey et al. 2000). These authors err, in my view, by seeing *all* systems orientations as deterministic, but they do make compelling arguments against deterministic approaches.

Gunderson and Holling (2002) have recently brought together systems and complexity theory in their collection, *Panarchy: Understanding Transformations in Human and Natural Systems* (see Box 3-1 for some pertinent extracts). We have also found congenial lines of inquiry in the work of AtKisson (1999), Axelrod and Cohen (1999), Stacey et al. (2000), Anderson (2001), and Ruitenbeek and Cartier (2001), all of whom build on complexity theory.

Axelrod and Cohen (1999) focus specifically on ways to "harness" complexity. This has also been part of our thinking within ACM, though our ambivalence has been evident in the work of Ruitenbeek and Cartier (2001), who question whether the kind of adaptive collaborative management we seek can be created or is emergent.

Box 3-1. Some Features of Ecosystems and Human Systems

"Episodic behavior is caused by interactions between fast and slow variables.

"… [S]caling up from small to large cannot be a process of simple aggregation; nonlinear processes organize the shift from one range of scales to another.

"On the one hand, destabilizing forces are important in maintaining diversity, resilience, and opportunity. On the other hand, stabilizing forces are important in maintaining productivity and biogeochemical cycles.

"Ecosystems are moving targets, with multiple futures that are uncertain and unpredictable. Therefore, management has to be flexible, adaptive, and experimental at scales compatible with the scales of critical ecosystem functions" (Holling and Gunderson 2002, 26–27).

"These three properties shape a dynamic of change. *Potential* sets limits to what is possible—it determines the number of alternative options for the future. *Connectedness* determines the degree to which a system can control its own destiny, as distinct from being caught by the whims of external variability. *Resilience* determines how vulnerable the system is to unexpected disturbances and surprises that can exceed or break that control" (Holling and Gunderson 2002, 51) [italics added].

"Sustainability is the capacity to create, test, and maintain adaptive capability. Development is the process of creating, testing, and maintaining opportunity. The phrase that combines the two, sustainable development, is therefore not an oxymoron but represents a logical partnership" (Holling et al. 2002b, 76).

In our work, we have taken a view of systems thinking that builds on nondeterministic assumptions about systems. Systems thinking has provided a framework for looking at the interactions among the networks or patterns of behavior at various scales. Systems thinking serves both as a communication device for people (scientists and villagers) looking at phenomena from their different perspectives and as a mechanism for specifying interactions and feedbacks among actors and phenomena. Stacey et al. (2000, 14) acknowledge the importance of purpose in human behavior. Systems thinking is helpful for considering possible future directions in a coherent way: it allows people to consider what might happen and develop scenarios to cope with changes they cannot in fact predict (see Gallopín 2002 for an example) or plan strategies to bring about certain effects on their surroundings (Axelrod and Cohen 1999).

Although the ideas in Gunderson and Holling's collection on *Panarchy* (2002) are theoretical, they do provide a starting point that is consistent with many of the findings reported in this book. We can see many social and ecological processes moving through the four phases described in Figure 1-5. Figure 3-1 presents a panarchy, and Figure 3-2 provides a visual representation of the interplay between scales.

Holling et al. (2002c) consider two features to distinguish panarchies from traditional hierarchies. First, represented in Figure 3-1, a panarchy is a cross-scale, nested set of adaptive cycles. The second, Figure 3-2, shows the connections between levels. The two kinds of connections that Holling et al. identify as the most important can be labeled *revolt* and *remember.* Specifically, they say,

When a level in the panarchy enters its Ω phase of creative destruction and experiences a collapse, that collapse can cascade up to the next larger and

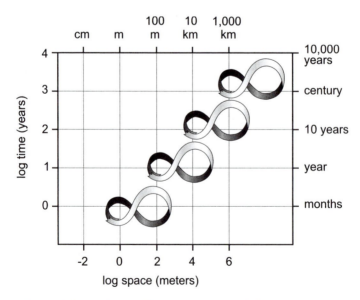

Figure 3-1. *A Stylized Panarchy (from Holling et al. 2002c, 75)*

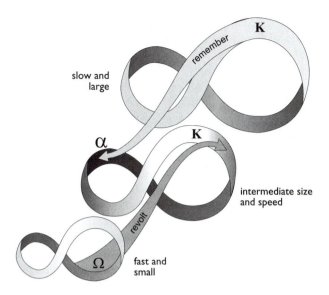

Figure 3-2. *Diagram Showing Panarchical Connections, Showing Interplay Between Scales (Holling et al. 2002c, 75).*

slower level by triggering a crisis, particularly if that level is at the K phase, where resilience is low. The "revolt" arrow suggests this effect—where fast and small events overwhelm slow and large ones. And that effect could cascade to still higher slower levels if those levels had accumulated vulnerabilities and rigidities ... A societal example occurs when local activist groups succeed in efforts to transform regional organizations and institutions because they had become broadly vulnerable ...

The downward arrow labeled "remember" ... indicates the second type of cross-scale interaction that is important at times of change and renewal. Once a catastrophe is triggered at a level, the opportunities and constraints for the renewal of the cycle are strongly organized by the K-phase of the next slower and larger level ... It is as if this connection draws upon the accumulated wisdom and experiences of maturity—hence the choice of the word *remember* ... In a healthy society, each level is allowed to operate at its own pace, protected from above by slower, larger levels but invigorated from below by faster, smaller cycles of innovation.

That summarizes succinctly the heart of what we define as sustainability (Holling et al. 2002c, 76)

Those ideas fit nicely with the processes I observe in the 30 ACM sites.

Although Stacey et al. (2000) argue against systems theory and disagree with that aspect of our approach, they have useful observations. They provide a thorough historical account of the theories of causation (teleologies) that have underlain scientific endeavors. Seeing systems approaches as necessarily deterministic, they argue that scientists have recently built their thinking on two important but different theories of causation: "formative teleology" for the systems of which people are a part, and "rationalist teleology" for the actors who plan and implement systems. These authors see formative teleology as built on the idea of an inherent nature in systems and other phenomena, waiting to evolve or be discovered. They argue that scientists, recognizing human volition and agency, have used rationalist teleology to account for human decisionmaking.

> Rationalist Teleology applies to the choosing manager (theorist, researcher, decisionmaker), from whom the organization itself is split off as a 'thing' to be understood. The organization, that which is to be explained and operated on, is then regarded as an objective phenomenon outside the choosing manager (theorist, researcher, decisionmaker), equivalent to a natural phenomenon, to which … Formative Teleology can be applied. There are two major problems with this move, problems that many think have bedeviled management thinking for decades. First, managers and researchers are humans participating in the very phenomenon their approach splits them off from: they cannot be objective observers in the manner of the natural scientist, but they proceed as if they can. Second, and closely related to the first, the split locates human freedom entirely in the manager (theorist, researcher, decisionmaker) and reduces other members of the organization to inhuman parts without freedom, just as Kant warned (Stacey et al. 2000, 57–58).

Although the authors are discussing management specifically, it is not difficult to see these views of causation in much development and conservation discourse. Stacey et al. argue that *trans*formative teleology is more appropriate for understanding human behavior. In an argument relevant to ACM, they make a strong case against our (human) ability to predict in human contexts. They paraphrase Prigogine (1997), for instance, who

> … sees the future for every level of the universe as under perpetual construction and … suggests that the process of perpetual construction, at all levels, can be understood in nonlinear, nonequilibrium terms, where instabilities, or fluctuations, break symmetries, particularly the symmetry of time … [N]ature is about the creation of unpredictable novelty where the possible is richer than the real … [Prigogine] sees life as an unstable system with an unknowable future in which the irreversibility of time plays a constitutive role. He sees evolution as developing bifurcation points and taking paths at these points that depend on the micro details of interaction at those points. Prigogine sees evolution at all levels in terms of instabilities, with humans and their creativity as part of it… These features, unknowable futures emerging in here-and-now interactions, are essentially what we have defined as the causal framework of Transformative Teleology (Stacey et al. 2000, 97).

Similar thoughts are echoed, for instance, in Uphoff (1996) and Gunderson and Holling (2002). Even Axelrod and Cohen (1999), who are trying to *harness* complexity, acknowledge the difficulties of prediction. Prediction is about the future, as are the questions of emergence that we have raised. As Holling et al. (2002b, 10) have pointed out, "Neither ecology, nor economics, nor institutional theory [all disciplines strongly dominated by reductionist approaches] now deals well with these fundamental questions of innovation, emergence, and opportunity."

We cannot, for instance, really predict whether an effort in Rantau Layung (East Kalimantan) to intensify fruit production in forest fallows will or will not have an adverse effect on forest sustainability. It could strengthen community members' commitment to protecting their forest by making it more valuable to them; or it could make fruit production so profitable that all remaining natural forest would be converted to such agroforests, with negative implications for forest biodiversity. Similarly, we cannot know, a priori, whether a formal devolution policy will have the hoped-for effect of strengthening local people's rights to their resources and their voice in decisionmaking; or whether, as seems to be happening in some areas, it will bring greater intrusion of governments into the day-to-day lives of rural people and the devolution of larger scale corruption from the center to the periphery. These uncertainties are not shortcomings of our particular team or approach; they are characteristics of the inherent uncertainties in the systems we are studying. And they justify institutionalizing monitoring and social learning at all scales.

Again drawing from Prigogine, Stacey et al. (2000, 105) emphasize the importance of diversity:

> Managers seek to remove or suppress the conflicts that arise when people differ, seeing such conflict as disruptions to orderly processes of change. It is all part of a framework of thinking, drawn from Newtonian logic and systems theories which equate equilibrium and harmony with success. The work of Prigogine and his team, in focusing on a notion of transformative causality incorporating difference, challenges this perspective, suggesting that the very difference managers seek so strenuously to remove is the source of spontaneous, potentially creative change. Living beings, including humans in organizations, need to evolve in novel ways in order to survive. … [M]any of the current ways of making sense of life in organizations are completely antithetical to this need. Managers may be struggling to change their organizations in ways which ensure that they stay the same.

If our analogy between the management and development-conservation contexts is appropriate, that suggestion explains the lack of success that has characterized the latter (see, e.g., the discussion of "high modernism" with its emphasis on standardization, head counts, simplification, and the devastating effects of such development efforts, in Scott 1998). Our efforts to devise standardized solutions are in fact likely interfering with the efforts of people in forested areas to address their problems creatively.

To deal with such philosophical quandaries, Stacey et al. (2000, 52–53) propose their transformative teleology, which focuses on relationships rather than entities: It recognizes a "perpetual iteration of identity and difference, continuity and transformation, the known and the unknown at the same time … processes of micro interactions in the living present forming and being formed by themselves." They argue, building on the ideas of G.H. Mead, that human beings cannot be separated from their social environment, and that we need one theory of causation that will apply to both the actors who plan systems and the systems of which they are a part.

Their concern with relationships is reflected also in Uphoff's (1996) fascinating account of an irrigation project in Sri Lanka. Concerning the ways that positive change occurs in communities, he writes, "Chaos concepts are … more suited than classical ones for understanding and acting within an expanding and continually surprising universe, one that nevertheless exhibits some repeating principles of pattern and purpose" (Uphoff 1996, 387).

He emphasizes the roles of ideas, ideals, and friendships, showing how each has affected the remarkable, long-lasting success of the irrigation project he describes. To separate the project from political conflicts, for example, the field-based people observed that water is colorless—unlike the political parties, each of which was symbolized by a color. Successful project groups emphasized and built on traditional values of cooperation and caring for others in their efforts to implement the project. And extra efforts were made by individual researchers and government officials to help others with whom they had developed strong personal ties of friendship—efforts that clearly helped the project succeed.

Uphoff also makes a convincing and philosophically grounded argument for expanding our scientific efforts beyond reductionist scientific approaches, without suggesting we abandon them altogether. All of this echoes Gunderson and Holling's emphasis on connectedness as an abstract feature affecting systems.

The questions—how to recognize abundant human agency along with the persistence of cultural patterns, how to deal with stability and change, whether change can be planned or emerges spontaneously—reflect ongoing uncertainty within the scientific community. We are also unsure.

But we have proceeded with the following thoughts in mind. Complexity, change, and uncertainty are omnipresent wherever human beings live. Human beings—whether community members or timber company personnel or government officials or scientists—have views and experience, and they affect their environments, both human and environmental. Human behavior has systemic features (interactions, feedback, lag times) while also demonstrating the unpredictable human agency and creativity that result in cooperation, conflict, and change. Although we have made attempts to harness complexity, our fundamental goal has been to work with local people to create the conditions for the emergence of more adaptive and collaborative management of forests.

We are taking a very different approach than that taken, for instance, by the International Forestry Resources and Institutions Program at Indiana University (Poteete and Ostrom 2003), which emphasizes the acquisition and manipulation of a standardized set of data from many environments (also called for by Collinson and Lightfoot 2001, for farming contexts). A long and honorable history supports the utility of such databases for scientific analysis, and that approach may ultimately be the best strategy. But my own conviction, at this stage, is that extracting social and environmental data from their context is counterproductive because it ignores the changing, interacting effects of local, regional, national, and global systems on each other. Moreover, it ignores the power of human agency to bring about change and render one's distant, disaggregated, reductionist analyses irrelevant. The deteriorating situation of forests and forest peoples may require a more radical, field-based approach.

CHAPTER FOUR

Seven Analytical Dimensions

*I*N 1998, THE ADAPTIVE COLLABORATIVE management (ACM) team selected seven dimensions on which to focus in the cross-site analyses: devolution, forest type, population pressure, management goals, human diversity, level of conflict, and social capital. In this chapter, I discuss the issues that led us to these choices and introduce the ways in which we use them.

Devolution

Devolution refers to the degree to which rights and responsibilities for forest management have been handed over to institutions—such as communities—closer to the forest. We differentiated between formal devolution status, which is a matter of laws and formal recognition, and informal devolution status, which refers to the de facto rights local communities may have when they may have no or few de jure rights. Informal devolution has been important in Indonesia, where many communities maintained day-to-day management of their forests simply because officials rarely visited and did not implement the formal laws of the land or interfere in any meaningful way with longstanding community management.[1]

Although informal devolution still exists, the intrusion of the state into forested areas has proceeded with great vigor over the past decade, spurred by various global trends. Decentralization,[2] the loss of natural forests suitable for profitable logging, population increase, and various financial crises have prompted laws designed to devolve rights and responsibilities for forest management to communities.

The effects of devolution and the related decentralization legislation have been widely discussed in the literature of late. Ribot's survey of African decentralization processes (2001) generally found the form but not the content. He concluded that decentralization has not yet been implemented. Shackleton et al. (2001), focusing on southern African countries, found a range from little meaningful devolution (e.g., parts of Zimbabwe, Malawi, Tanzania) to substantial devolution (e.g., Botswana, Namibia, Lesotho). Edmunds et al. (CIFOR 2001) conducted a more focused, in-depth analysis of devolution in three countries in Asia—the Philippines, China, and India—and found that devolution served to extend state control over local forest management. From local people's perspectives, devolution has led to some benefits and control but fallen short of their expectations. Often, benefits were captured by local elites. These authors recount both positive and negative impacts of devolution processes but conclude that the effects have not been as positive as hoped, partly because of unwillingness or inability of governments to actually devolve power and resources. Sarin et al. (2003) describe three cases in India where joint forest management appears to have functioned to retain state control while ignoring local institutions that manage forests. Reports from Cameroon (Mvondo and Sangkwa 2002), Guatemala (Wittman 2002), India (Hobley 1996b), and Indonesia (Peluso 1994; Resosudarmo and Dermawan 2002) echo these conclusions. In some cases, devolution is intended to empower local people and communities. In others, devolution processes are more explicitly oriented toward reducing costs at the center (Pacheco 2002; Nhira et al. 1998; Schmidt 2002b). The Interlaken Workshop on Decentralization held in April 2004 considered these experiences in different countries (Colfer and Capistrano, forthcoming).

Still, Don Gilmour (2002, 2) reports:

> A recent analysis of experiences in Thailand, Vietnam, Lao PDR, and southern China (Dupar and Badenoch 2002) concluded that even the most cautious forms of decentralisation led to more appropriate natural resource management decisions. They further concluded that the best chances of improved livelihood security and environmentally sustainable development come from a range of innovations including:

> - granting sufficient control of productive resources to local people; and
> - establishing multiple ways to hold local authorities accountable to their constituents and to national environmental standards.[3]

Based on research in the Philippines, Magno (2001) notes the importance of social capital (see below), both within communities and between civil society and government, in efforts both to decentralize and to manage forests more effectively. Looking at African examples, Dubois and Lowere (2000, 33) caution that "Decentralisation is a facilitating factor, not a precondition for successful [collaborative natural resource management]."

Forest Type

Many have written from the point of view of environmental (or technoenvironmental) determinism (see Harris 1968 for an older review of this literature), a view that is more popular among biophysical than social scientists [cf. Jared Diamond's (1999) controversial *Guns, Germs and Steel: The Fates of Human Societies*, which provides a more recent version of environmental determinism]. The perspective, simply put, is that environmental conditions cause certain cultural forms. Not an

Box 4-1. Devolution in Africa and Asia

- Most "devolved" natural resources management reflects rhetoric more than substance, and is characterized by some continuation of substantive central government control and management over natural resources rather than a genuine shift in authority to local people.
- The ways in which local people realise the benefits of devolution differ widely, and negative trade-offs, mostly felt by the poor, are common.
- States, communities and other stakeholders have different visions of devolution and its mode of implementation. A shared framework, more accountable to local livelihood needs and people's rights to self-determination, is required. Careful reassessment of the state's claim to be protecting the wider "public interest" forms part of this process.
- Organisational models that devolve authority directly to disadvantaged resource users are more embracing of local interests and priorities than those that allocate control to higher levels of social organisation.
- More powerful actors in communities tend to manipulate devolution outcomes to suit themselves. Checks and balances need to be in place to ensure that benefits and decision-making do not become controlled by élites.
- Strong local organisational capacity and political capital enhance outcomes for local people by enabling them to mobilise resources and negotiate better benefits. NGOs, donors, federations and other external actors have a key role in moving devolution policy and practice towards local interests.

—Shackleton et al. (2002).

environmental determinist, I was shocked on my first visit to South America to see the similarities in material culture that linked the Peruvian Amazon and Borneo. Following this line of reasoning, one might expect to see cultural similarities, and thus similarities in the way ACM might play out, in areas with similar environments. Conversely, different forest types might herald differences in the utility of ACM for people who live in these divergent environments.

Another interest of ours is the role of deforestation or environmental degradation. Some have argued that as the forest disappears, local people's benefits from it, and thus interest in it, will also disappear (e.g., Dudley 2002). Others argue that as people perceive the loss of something that has benefited them, they will be motivated to slow or prevent further loss. We wondered how these kinds of possibilities would play out when communities of people began to make more systematic choices about how to manage their resources together.

Population Pressure

How would population density affect people's interest in participating in a process like ACM? Would the utility of the process vary? Would people who were intimately bound up in forested contexts—typically sparsely populated—be more interested in working together to maintain the forest; or would people who had to compete among themselves for scarce forest products and landscapes try harder to work together to share the resources more fairly, more transparently, or with less conflict?

Few researchers have paid much attention to population and forests beyond the sorrowful complaints of ecologists, some of whom see humans as trespassers. Demographers and other population specialists, recognizing that forests typically have small populations, have been less interested in forest settings than in areas of high population growth, which can play a much bigger role in global population stabilization. But I share the concerns of ecologists about the impacts of growing populations on forests—in the West as well as in developing countries—and see increasing population as potentially adversely affecting the well-being of forest peoples. This concern in no way implies any lesser concern with consumption in the West. High population growth also has implications for individual women (Colfer et al. 2002). Recent literature on population and the environment (Engelman 1998; Vogel and Engelman 1999; Gardner-Outlaw and Engelman 1999; Cincotta and Engelman 2000) is more sophisticated than what was written in previous decades and is consistent with the approaches taken in this book.

Management Goals

Our previous research on criteria and indicators—as well as long-term ethnographic research—had convinced us that timber management, conservation management, local community management, and other types of management were characterized by different goals, traditions, and requirements. We anticipated that the problems and opportunities offered in each of these contexts differ as well.

Given the widely divergent power and wealth of local people vis-à-vis large-scale timber operations, we wondered whether we would be able to interest timber companies in working with communities. Or perhaps communities' attempts to influence timber management would be consistently rejected by the more powerful actors. It seemed, on the other hand, that the opportunities for compromise with timber companies might be greater than in community interaction with conservation groups, which might seek to restrict local people's access to wildlife and other nontimber forest products, with impacts on local diets and incomes. Although the power and wealth of conservation groups would likely be less than those of timber companies, the potential for direct conflicts of interest seemed greater, especially where conservation areas had core preserves that were off-limits for local communities, with restricted access in buffer zones.

In areas without strong timber or conservation concerns, we expected that the communities would interact with other communities and with government. We anticipated conflicts between communities over boundaries and access to forest areas and products, on a comparatively level playing field. The role we imagined for the government was one of rulemaker, arbiter, and large-scale planner with varying legitimacy in the eyes of the communities.

Human Diversity

Women and other marginalized groups make extensive use of forests in many areas yet are often sidelined in the management process. We have argued consistently for gaining access to the views and preferences of marginalized people in our research sites, on both pragmatic and ethical grounds. From the pragmatic perspective, effective and beneficial forest management requires the cooperation of the local population. Such cooperation is more likely if those people's needs and concerns are incorporated into any planning and implementation process. Ethically, we argue that those who have traditionally used a resource should have some voice in decisions about how that resource will be used in the future. The ACM process that we have sought therefore requires the involvement of many segments of the forest communities.

But the way to gain access to such input in an effective manner remains a question. There are at least two contradictory perspectives about the desirability of diversity. On the one hand are those who argue in favor of planning and action in contexts of diversity (including CIFOR and our own internal organization of the ACM team). Such sentiments are explained and rationalized in books like *Gifts Differing* (Myers and Myers 1990) or *Please Understand Me* (Keirsey and Bates 1984). From this perspective, diversity is good. A diverse group of people brings to any task synergies deriving from different perspectives and experiences, complementary skills, wider networks, and more creative problem solving.

On the other hand, some argue that the power differences that accompany diversity—men over women, the North over the South, Bantus over pygmies—inhibit the less powerful from speaking (Ardener 1975; Colfer 1983b). Different languages and ways of life form barriers to effective communication and cooperation. What might be gained in diverse and valuable input may be counterbalanced by a loss of social capital.

Merrill-Sands et al. (2000, vii, 9), discussing diversity in organizational settings, summarize the important issues:

> Diversity is a complex concept. While diversity efforts have the potential to strengthen organizational effectiveness and efficiency and to advance social justice, experience has shown that realizing the full benefits of diversity is neither a simple nor a straightforward process. It is one thing to create diversity by recruiting people of different nationality, cultural background, race, gender, sexual orientation, religion, discipline or work style. It is quite another to develop a supportive work environment that enables people of diverse backgrounds to perform at their highest levels, contribute fully to the organization and feel professionally satisfied. It is an even greater challenge to integrate fully the varied knowledge, experiences, perspectives and values that people of diverse backgrounds bring into an organization's strategy, goals, work, products, systems and structures ....
>
> While diversity broadens the resource pool of ideas, perspectives, knowledge and work styles, it can also reduce team cohesion, complicate communication and heighten conflict.

In our work with communities, we have encountered many of the same issues. Where diversity within a village offers much potential for the kinds of sharing and synergy sought in organizations, the interplay of power levels often interferes with such a vision. Women's unwillingness or inability to speak up in public settings has been often noted. Dangol (2005), Bolaños and Schmink (2005), and Sithole (2005), writing about ACM in Nepal, Bolivia, and Zimbabwe, respectively, find this pattern and describe the risks and dangers to women of stepping out of their culturally prescribed roles. Diaw and Kusumanto (2005) describe the effective silencing of a lone pygmy voice in a multistakeholder workshop in Cameroon (see also Oyono 2003b). Anau et al. (2005) and Hakim (2005) discuss similar issues in intervillage interactions in Indonesia. For a more theoretical perspective, see Edmunds and Wollenberg (2001), who describe the real risks of such interactions to the less powerful. Carter (2000, 30) talks about both the size of the group and its homogeneity as being potentially important:

> As far as group size is concerned, small groups are considered more conducive to [collaborative forest management] agreements, since it is easier for a limited number of individuals to meet regularly and reach agreement. However, larger groups have other potential advantages (such as commitment through force of numbers, and a sharing of the workload), although practical problems of communication may arise if such groups are spatially highly scattered (this can be particularly pertinent in areas of difficult terrain ...). With regard to group homogeneity, what seems to matter most is homogeneity of interest, rather than other social criteria.

Krishna (2002a), in a quantitative study, looked at the heterogeneity of 69 villages in northern India (close to the Nepali sites described in case 11 in the Appendix of Cases) and found that diversity made little difference for village "development." The number of castes was a variable of minor importance in explaining high levels of conflict until he included social capital in his equations. Then the significance of number of castes was lost. He also reported the findings of other researchers on the importance of heterogeneity—basically mixed results.

In short, there is truth in both views. Diversity can be a valuable asset in collective action; and it can be an effective silencer of nondominant views. We have approached the work reported here from a pluralist perspective (Anderson 2001; Wollenberg et al. 2001a), convinced of the value of incorporating diversity yet aware of the difficulties of doing so.

Communities are often discussed in opposition to other, more powerful stakeholders such as timber companies or conservation groups—as I sometimes do in this book. Though justified at larger scales, stressing such an opposition can imply that communities are homogeneous entities, whose members share similar perspectives and have common interests. The inaccuracy of this assumption has been pointed out in recent years (Shiva 1989; Braidotti et al. 1994; Rocheleau and Slocum 1995; Venkateswaran 1995; Guijt and Shah 1998a; Antona and Babin 2001; Baviskar 2001; Engel et al. 2001; Leach and Fairhead 2001; McDougall et al. 2002b). We began the ACM research with a resolve to address such internal community differences in our work. This was particularly important, given our interest in empowerment and equity issues.

Level of Conflict

Conflict has been important at both macro and micro scales. Lee's (1999, 3) comments on conflict in protected areas are pertinent for other contexts as well:

Conflict is an essential element of governance. Differences are inevitable, and an orderly approach to resolving those differences is essential; over the time periods needed to establish conservation practices, conflict and turbulence must be expected and should be welcomed. But conflict needs to be bounded—disputes should be conducted within the boundaries of a social process that the disputing parties perceive as legitimate. Unbounded conflict can tear apart the social fabric, thwarting learning. The difficulty is that conflict is a situation in which control of the rules of engagement are themselves contested …. In practice, even parties with little power can delay the resolution of conflicts enough to frustrate experimentation and learning.

Many of the conflicts we have seen in ACM sites are comparable to what Brockington (2002, 187) describes for pastoralists in Tanzania when he talks of the resistance of Meru villagers to the Arusha National Park:

> [R]ural groups resort to "everyday forms of resistance" that effectively nullify the regulations. These [are] viewed in the context of a moral economy that views threats to livelihoods, posed by exclusion or raiding by wildlife, to be a greater wrong than poaching or illegal grazing. Peasants do not encroach on parks; rather, "the park is seen to be encroaching on the Meru villages."

He quotes Neumann (1998), who says, "the 'crime' of poaching is not a crime at all, but a defense of subsistence, and the 'real crime' is that park animals are allowed to raid crops with impunity" (Brockington 2002, 187).

Such comments on protected areas show that differences over the uses and management of natural resources are probably inevitable. The task is indeed to develop modes of governance that can deal with those conflicts in a bounded manner. Ramírez (2001) points out the frequency with which collaboration and conflict represent phases in a process. Schmink and Wood's (1992, 14) perspective on the difference between conflicts and resistance is also worth bearing in mind. Conflicts are contests between people of equivalent power and position; in resistance, the subordinate challenge attempts by those in power to appropriate resources or impose their will. Much of what has been called conflict in this book can be seen as resistance.

One ACM facilitator suggested that we also attend to the conflicts that emerge between the present generation and those generations to come:

> Given that the generations in question [past, present, and future] have never "lived together", in the same historical period of time and within the same material space, they entertain symbolic conflicts, somehow a "from a distance confrontation." Thereon, these conflicts are moral in priority, as it is a question of security and equity in the successive access of generations to forest and its resources, considered as a common heritage or … as a common property. Such "from a distance conflicts" are marked by an "historical sedimentation" as they clearly raise the moral—and why not the ethical—problem liberated by access of human societies to natural resources, either in a degrading and egoistic/competing way or in a sustainable and equitable way (Oyono et al. n.d., 7).

Social Capital

This term conveniently encapsulates some positive community traits we believed might be helpful in attempts to catalyze collective action and institutionalize social learning. Important criticisms of the concept have included the idea that social capital can be used for negative as well as positive purposes. Sciarrone (2002), for instance, analyzes the Mafia from the perspective of social capital and concludes that social capital is not necessarily a good thing. This idea was also widely discussed at the Collective Action and Property Rights Workshop on Collective Action held in Nyeri, Kenya, in February 2002 and at Cornell University's Workshop on Social Capital and Civic Involvement in September 2002.

Uphoff (2000) suggests defining social capital to exclude that possibility by focusing on two elements that contribute to the development of mutually beneficial collective action: (1) structural social capital, or the roles, rules, procedures, and precedents as well as social networks that establish ongoing patterns of social interaction; and (2) cognitive social capital, or the norms, values, attitudes, and beliefs—like trust, reciprocity, truthfulness, solidarity, belief in fairness—that predispose people to cooperate. Both elements contribute to cooperative behavior and the mutually beneficial collective action that he considers fundamental to the concept of social capital (see also Uphoff and Wijayaratna 2000). Although this is a major step forward in making the concept useful analytically, there remains the problem of defining "mutually beneficial collective action." Are the demonstrations villagers organize to protect their claims to lands also claimed by the state mutually beneficial? The answer depends on your point of view.

Several authors have objected, some strongly, to the use of an economic metaphor to refer to social relations and networks. Greenwood (2002), for instance, sees the popularity of the concept as deriving from the hegemony of neoliberal economics. He, like Durrenberger (2002) and Schafft and Brown (2002), complains that a social capital interpretation ignores social class and issues of power and domination. Greenwood and Durrenberger further argue that social capital is a rerun of the "culture of poverty" concept, which they see as essentially blaming the victim.

In my view, much of the criticism of social capital derives from overblown expectations (and perhaps claims) for the concept. Schafft and Brown (2002, 3), for instance, talk about "the use of 'social capital' as a primary causal factor in understanding development and underdevelopment." They believe this risks "obscuring and bypassing more critical analyses of how resources, power and privilege are embedded within dynamic, historically developed power structures." In our work, we have not seen social capital as a primary causal factor in understanding development and underdevelopment but rather as (1) a potential resource on which to build and (2) a possible lack, which might be strengthened through participatory action research and other methods. The concept of social capital also seemed a shorthand way of recognizing the strengths of traditional or indigenous grouping mechanisms and networks, perhaps encouraging outsiders to build on what already exists rather than creating new (and often ineffective) organizations.

We recognize and support the need to address inequity, privilege, and power dynamics generally. We believe that communities and others have the potential to make effective and beneficial use of social capital, not that communities without it are doomed. Inkeles (2000, quoted in Krishna 2002a) echoes our own interest in "added value" and sheds light on the two basic definitions of social capital (one focusing on community, as we have done, and one on the individual):

> … what makes the study of social capital compelling is the assumption of added value … it permits communities to do what they could not do … without the addition of social capital. By contrast, the studies focused on the individual generally say nothing about *added value*, and concentrate rather on *competitive advantage* in the gaining of shares from a fixed pie. They almost invariably deal with a win-lose situation. It is a much more attractive proposition for human-kind if we can establish that adding social capital at the communal level is a win–win situation for both the individual and for the community at large.

The most interesting and pertinent examination of social capital in a Third World context has come from Krishna (2002a; Krishna and Uphoff 2001). In an unusually systematic and careful examination of 69 Indian villages (with more intensive case studies in 16 communities), he sought to evaluate the role of social capital (as a community characteristic) in communities' success in three spheres: economic development, community harmony, and democratic participation. He hypothesized three possible scenarios: Social capital alone is responsible for success; institutions are responsible for success, with spinoff social capital gains; and social capital is most effective in combination with effective agency. After measuring a wide variety of parameters, he concluded that social capital was important in all the three spheres but was significantly more important when combined with effective agency. Specifically, in issues pertaining to economic development and political participation, the involvement of young, more educated, new elites serving as brokers was central; and in maintaining community harmony and resolving conflict, an effective traditional council was the key. Whether homogeneous or heterogeneous groups are more likely to form and maintain social capital, however, has not been resolved.

The differentiation between social capital formation in formal and informal groups in relevant contexts has not, to the best of my knowledge, been systematically studied. But my own conclusion, based on both long-term fieldwork and comparisons of the ACM sites, is that at least in villages in developing countries, informal groups are much more likely to have high levels of social capital. Krishna (2002) also notes the comparative irrelevance for collective action of many formal groups in India. In Bushler Bay, an American village I studied in the 1970s, on the other hand, formal and informal groups were equally likely to be characterized by high levels of social capital—but the "public employee" segment of the community opted for formal groups; the "local" segment, for informal ones (Colfer 1977; Colfer with Colfer 1978).

Bonding and bridging social capital are yet another distinction, both of which Scheffer et al. (2002) consider important for social transformation. Bonding social capital involves strong bonds among members of the same group; bridging social capital refers to bonds between groups. These concepts are useful for studying connections both among villagers (bonding) and between villagers and "higher" levels of relevant bureaucracies (bridging).

These dimensions will surface again in Chapters 6–10 and in the Appendix of Cases.

CHAPTER FIVE

Creativity, Learning, and Equity
The Impacts

*A*DAPTIVE COLLABORATIVE MANAGEMENT (ACM) is built on a belief in the abilities of people—in forest villages, in government offices, in research institutions—to act in the interests of themselves, their communities, and future generations. It assumes the necessity for people to maintain their environment for their own survival. It is also built on the conviction that precise prediction of change in panarchies (multiple interacting complex systems) is impossible, which in turn implies the centrality of learning as an element of effective management—indeed, of living. ACM also builds on the existence of, or the latent potential for, social capital, as a critical prerequisite for collective action.

From that follows the desirability (and inevitability) of diversity of impacts. And that is what we have. Although the impacts—large and small—that those using ACM have generated are myriad, this chapter provides the flavor of what has been accomplished. The most important impacts to date are not how much cash was generated, or how many hectares of forest were protected—figures that are unlikely to be impressive in the short run anyway—but rather the human and institutional capabilities that were strengthened. These capabilities are a potentially enduring legacy that can help these communities deal proactively with future surprises and trends. We anticipate, though we can't yet prove, that these new capabilities will raise incomes and improve forest management over the long haul.

Our research design initially included context studies that could serve as baselines against which to measure change: pre-ACM studies would be compared with periodic updates to tell us what we had accomplished. Some colleagues found this approach appealing; some found it unlikely to provide believable results. In practice, although the context studies yielded useful data, the staggered onsets of ACM activities in the different sites meant that some updates were available, some not, when this analysis was done. The ACM facilitators' ever-growing understanding of what was happening on each site also conspired against comparability: what was actual change and what was a different level of understanding on the part of the reporter was difficult to disaggregate.

So my approach to impact assessment is an inductive and largely qualitative one. I have used anthropological methods of data analysis, immersing myself in the ACM teams' written outputs, visiting sites, interviewing and e-mailing field-based team members, and serving as a communication node among sites. In this way, I have been able to pull together and categorize the positive changes that our activities appear to have brought about. I have made extensive use of Bob Fisher's "plausible, causal links" in my search for the effects of our work, and the result is a preliminary assessment.

This chapter looks at ACM impacts from two perspectives. The first examines impacts according to the three-way differentiation explained in Chapter 1:

- horizontal collaboration among stakeholders;
- vertical collaboration between communities and policymakers; and
- adaptiveness, focusing on our efforts to catalyze social learning.

See Box 5-1 for more detail on how this played out in Nepal.

Impacts, from this perspective, are closely linked both to the way in which local stakeholders applied an ACM approach and to the strategies that the researchers used to facilitate or catalyze it. The second section involves a more conventional take on the same sites, looking at livelihoods, environment, and empowerment.

Box 5-1. Elements of an ACM Approach Used in Nepal

The two main components of ACM can be understood as learning (shared and conscious/intentional) and stakeholder relations (including communication and power and negotiation). We have broken these down into smaller elements:

1. Stakeholders effectively communicate and transfer knowledge and skills (in multiple directions).
2. All relevant stakeholders are involved in decisionmaking and negotiation, and have the "space" and capacity to make themselves heard.
3. Stakeholders effectively manage conflict.
4. Stakeholders implement actions together, as appropriate.
5. There is intentional learning and experimentation in the forest management process (and forest managers/decisionmakers are constantly increasing their understanding, knowledge and skills).[1,2]
6. The learning is not only individual but is shared—that is, there is the cocreation of understanding and knowledge (social/transformative learning).
7. Institutions consciously apply learning as the basis for refinements in their community forest management activities and processes.
8. Institutions consider relationships within and between human and natural systems.
9. Planning clearly reflects links to the "desired future" and takes into account actual past and present trends in their planning.
10. Institutions identify and deal effectively with uncertainties[3] (including risks) in their planning processes.

—Cynthia McDougall et al. (2002b)

Our two-stream approach (described in Chapter 1) is important to remember here. On the one hand, we were trying to catalyze adaptive and collaborative management in specific communities; and on the other, we were trying to assess the effectiveness of that effort in maintaining or improving human and forest health. The latter goal is not yet fully possible, since significant changes in human well-being and environmental health do not typically happen overnight. We anticipated at least 5, perhaps 10 years would be needed to thoroughly test this approach. In this chapter, instead, I provide some sense of the opportunities and pitfalls that the ACM approach offers.

Appendix 5-1 provides an overview of the impacts in tabular form.

Approaches and Impacts: Adaptiveness and Collaboration

In this section, I describe the approaches and impacts that have resulted from our interest in adaptiveness or social learning, in collaboration among stakeholders, and in links between communities and policymakers, typically acting at wider scales. These are the diachronic, horizontal, and vertical elements, respectively.

Adaptiveness: The Diachronic Element

We have seen adaptiveness as a naturally occurring process whereby people and systems evolve, changing their behavior over time, to adjust to changing circumstances. But adaptiveness can also be a more conscious attempt to structure feedback, by means of ongoing monitoring, into a given system so that changes can be made purposefully in an institutionalized manner. We have recognized that the natural process is under way all the time, but we have been uncertain of the degree to which we and the communities with whom we work could institutionalize such a process.

We have tried several methods to bring about the social learning that would be necessary for such adaptiveness to exist (Wollenberg et al. 2001a; Guijt forthcoming). The ACM teams found various approaches useful, and participants were also enthusiastic (indeed, if they were not, the effort was abandoned). Teams were generally free to select the methods that fit best with their communities and their own skills. Mechanisms and skills for strengthening adaptive processes were instituted clearly in Bolivia, the Philippines, Nepal, Indonesia, Brazil, Cameroon, and Zimbabwe. Success is less clear (perhaps partly in some cases because of lack of documentation) in Ghana, Malawi (Ntonya Hill), and Kyrgyzstan. These sites were also among the last to begin the ACM process.

The methods we employed were the use of future scenarios, training of community members by capacity development and cross visits, criteria and indicators, qualitative approaches to social learning, and modeling.

Future Scenarios. Throughout our work—both with communities and in our research framework—we have valued a diachronic component, beginning with historical context studies, and keeping in mind likely future trends. In the community context, trends should include the visions that local people and other stake-holders have for their forests and communities. Clarity about such a vision is also important for the process of monitoring; one needs a star by which to navigate.[4] Ultimately, successful collaboration also depends on a common vision or, barring that, complementary visions.

In Zimbabwe, Cameroon, Malawi, Madagascar, Bolivia, and Brazil, teams drew on the work of Wollenberg et al. (2000) to develop a systematic process for leading communities and mixed groups in discussions of their visions of the future. This process of clarifying the stakeholders' visions performed two impor-tant functions. It served as a mechanism whereby stakeholders (or community subgroups) who did not normally discuss such things together could learn about each other's views; and it served to build their support for shared (or comple-mentary) goals for the future. This was a useful step in the adaptive process, all teams agreed. The Bolivia team found the process effective in integrating wom-en's views into the forest management process—something that had been found difficult previously (Cronkleton 2005).

In Malawi, the community expressed a largely agricultural vision of the future, with little attention to their forest resources. The future scenarios process revealed hopes for access to a nearby block of land that had once belonged to the commu-nity but had been taken over by Malawi's previous president; people hoped it might be returned to them. Other visions (more funds to pay for funerals and care for orphans) reflected the incidence of HIV/AIDS, the serious hardships at this site, and the minimal economic benefits anticipated from the forest. Progress included work on developing indicators for assessing progress toward their goals and on conflict resolution (Kamoto 2003c).

In heavily forested Nkoelon, in Campo Ma'an, Cameroon, an egg—with the yolk representing society, the albumin representing nature—served as a starting point for a discussion of the future. The mixed group (old and young, men and women, pygmy and Bantu) identified indicators for their vision of human and natural well-being that reflected their ongoing dependence on and value attribut-ed to the forest. Their past efforts to protect their environment emerged, as did the features they considered desirable for their future (Nasi et al. 2001). The process was a first step in improving communication and reducing conflict in the area.

The team in Chivi, Zimbabwe, used the approach to circumvent the slow and ineffective legal mechanism for hearing the views of local communities—normal-ly from village development committee to ward development committee, and then via the chairperson of that committee to the various subcommittees of the council.[5] Revolutionary visions emerged from these groups of villagers and gov-ernment officials.

Campbell et al. (2000) stress the importance of the long-term involvement of the facilitators in Chivi, the use of the local language, and the preliminary, smaller scale meetings and orchestration that preceded the large workshop. They also had some warnings for others: they had allocated too little time to the process (though partic-ipants' time constraints made a longer meeting impractical); and the more powerful participants made it difficult to give everyone an opportunity to contribute. Such equity concerns also arose in Malawi (Kamoto 2003c) and Madagascar; facilitation and process skills help considerably, but the problem remains a "wicked" one.

The use of system dynamics to encourage a similar visioning process in Pasir, East Kalimantan (Purnomo et al. 2003a) can be seen under "Modeling," below. The ACM team's modeling group also produced a software package, Co-View (Collaborative Vision Exploration Workbench). This software leads facilitators through a process to articulate and explore a shared vision of the future and to develop strategies to achieve it (http://www.cifor.cgiar.org/acm/index.htm).

Training of Community Members by Capacity Development. Institutionalizing an adaptive process requires that the local people be willing and able to act in their own and their environment's interests. In Mafungautsi, Zimbabwe, the ACM team found a worrying amount of inertia and passivity within the communities. To increase motivation, they used Training for Transformation (Freire and Freire 1994; Freire 1970). This approach is built on the assumption that education can either domesticate or liberate, and trainers work toward inculcating an interest and ability in critically analyzing one's own condition. Following the training, community members began contributing their ideas about improving forest management much more readily. Women's involvement in decisionmaking fora increased dramatically (Mutimukuru et al. 2005). Nyirenda et al. (2001) report that collective action was "substantially stimulated" and give examples:

- The participants in one of the Training for Transformation workshops were offered per diems, as recompense for their time. After one group had received half the training, they decided on their own initiative to reject the per diem (despite assurances from the ACM facilitators that they could keep it), saying they were already benefiting from the workshop.
- One group attended the ACM meeting (without incentives) rather than attend a grain marketing board meeting where they would have received handouts. They also persuaded the grain marketing board official to change the next meeting venue so that it was more convenient for them, rather than only for the official.
- One group declined a small amount of food aid offered by the government because they decided that the potential inequity in distributing it, as requested, was not worth the small amount of food offered.
- One community concluded that some of its problems derived from not having a councillor to represent them in district government. They organized to make an official request, and as an interim measure they selected a development coordinator from within the community to perform some of the needed functions.

In Nepal, one ACM team conducted facilitation training for district-level personnel and forest user group members involved in the quarterly range post meetings (see Uprety et al. 2002b). And in the Pará, Brazil, sites, community members received training in participatory methods as part of an assessment of community members' abilities to take on the facilitation and other roles of the ACM researchers. The research team concluded that community members could, with training, profitably use many of the participatory methods; of course, the community members increased their skills in the process. Box 5-3 refers to training in Chimaliro, Malawi, on participatory forest resource assessment, somewhat similar to the collaborative inventory of forest resources carried out in Tailândia, Pará, Brazil. All of these training opportunities were determined on site, in response to local conditions, needs, and interests.

Cross Visits. Cross visits, from very local to international, were arranged for those who expressed a shared need for outside information and insight. A group of indigenous women from Basac (Mindanao, Philippines) went to a neighboring village to meet with a women's group. The Basac women's purpose was to learn from the other women, to reflect on the reasons for their own past failures as a group, and to clarify their future plans (see Arda-Minas 2001). A group from Palawan made a longer trip to Nueva Viscaya to visit the innovative community work in Kalahan, where the indigenous community has argued against collaboration in favor of exclusive rights to their traditional territory (see Rice 2001). The group from Palawan went home energized. What they learned led them, in fact, to strengthen their collaboration with government agencies and start several small income-generating activities (see Lorenzo 2001b). Cross visits were also used in Nepal (Uprety et al. 2002a) and all the Indonesian sites. In Pasir (Indonesia), a community group went to Central Kalimantan, famous for rattan production and artisanal processing; in Jambi, ACM team members visited a community forestry project in neighboring Lampung Province. The ACM team in Acre, Brazil, participated in a multistakeholder visit to Mexico, organized through our partners, the University of Florida, to compare forest management in the *ejidos* with local Acre attempts. This resulted in a series of high-level meetings to discuss improved timber marketing arrangements and other serious policy issues—contributing to "vertical collaboration" as well (Cunha dos Santos et al. 2002). In each case, the participants learned valuable information, directly pertinent to their ongoing efforts at collective action.

Criteria and Indicators. In our preliminary thinking, the role of criteria and indicators (C&I) was central[6] and built on our previous research. In Indonesia, the Philippines, Nepal, Zimbabwe, and Malawi, we used scientific C&I in our baseline context studies (Hakim 2001b; Kusumanto 2001c; Hartanto et al. 2000; New Era Team 2001a, 2001b; Vermeulen 2000; Kamoto 2002d, 2002e).

But more importantly, we anticipated that locally developed, simplified C&I would be useful to communities for monitoring their own actions (Tiani et al. 2001). We also reasoned that institutionalizing such a process might be useful in strengthening the bargaining position of communities vis-à-vis other, more powerful stakeholders (governments, conservation agencies, and timber and plantation companies). And we hoped that the communities would help us simplify C&I; the complexity of our own sets had been criticized, particularly in the ecological sphere.

Community reactions to C&I were ultimately important in decisions about whether to use them, both in the context studies and in the participatory action research process. In our Acre sites, for instance, C&I were perceived as alien ideas (Schmink 2000), emanating from those advocating "sustainable forest management" (an idea often linked in South America to oppressive government policies). In Pará, on the other hand, the ACM team used the concept to monitor the process of empowerment (a use not originally envisioned, see Table 5-1). Formal C&I sets were produced in Nepal, Indonesia (Jambi, Pasir), the Philippines (Palawan), Zimbabwe, Malawi, Ghana, and Cameroon (Ottotomo, Campo Ma'an, and Lomié). In

Table 5-1. *Potential Empowering Effects of Participatory Methods*

Categories	Empowering effects
Organization	Planning and distributing tasks, learning how to make decisions
Confidence building	Gaining self-confidence, respecting the group, respecting local knowledge, gaining technical training, guaranteeing ownership
Learning processes	Enhancing creativity, getting access to new ideas, learning to learn, finding stimulation for own initiatives
Sensitization	Critical perception of one's own environment, self-positioning, reflection
Communication	Information exchange, articulation of one's own opinions, constructive discussion

Source: Pokorny 2005.

Nepal and the Philippines, governmental agencies were also interested in developing C&I,[7] and our long-term, community-level efforts were much appreciated as contributing to a broader national effort—consistent with our initial thinking. The links between the ACM and C&I processes in Indonesia were less clear. This was one of the few instances where I pressured the field staff to undertake specific studies—justified in my mind by our commitments to donors. Consistent with our overall philosophy, such pressure was not a good idea. Indonesian policymakers and managers in other contexts have, however, made extensive use of the C&I software we produced, called CIMAT (Criteria and Indicators Modification and Adaptation tool).

In Nepal and the Philippines, locally generated sets of C&I served as the engine of the participatory planning process (see Box 5-2). Pandey (2002), for instance, describes how the C&I have spawned healthy competition among community residents of Deurali-Baghedanda. One young woman was prompted to learn the community forestry rules and regulations—and proved it—when her own ignorance came up in her hamlet's reflection meetings; another hamlet noted in January its own poor hygiene compared with that of other hamlets in the village, and by the next monitoring session (in March) had exceeded its own target by building

---

Box 5-2. Local Indicators Used in Bamdibhir, Nepal

In Bamdibhir, the community developed a system of self-monitoring, using the five phases of the moon for measurement, in which they periodically assessed their progress on several indicators they had selected themselves (initial, during the participatory action research process, and at the end of this phase). These indicators were

- relationships with other forest user groups and institutions;
- forest management and soil conservation activities;
- distribution of benefits;
- community forestry development work;
- participation in decisionmaking and institutional development;
- fund mobilization and income generation activities (IGA).

On all these indicators, there was movement from the new moon towards fuller moon phases, indicating progress in the areas they were working on.

—Cynthia McDougall et al. (2002b)

26 toilets; some households did so without the planned support from their collective fund.

In Malawi, several income-generating activity (IGA) groups were formed and began using C&I to monitor their own activities (see Box 5-3). Monitoring indicators were also developed for the mushroom IGAs (discussed below, see Box 5-8): These include the number of individuals involved in the activity, the quantity of mushrooms collected, the variety of species collected, the variety of species sold

---

Box 5-3. C&I and the Monitoring Process in Chimaliro, Malawi

The Block Natural Resource Management Committees have been involved in beekeeping activities since comanagement in the forest reserve began. The idea was to provide regular benefits to the communities in recompense for their involvement in forest management. People stole the honey (in all three blocks), so committee members removed the beehives and hung them at their homes. This defeated the purpose of comanagement, and the members were asked to return the hives to the forest. Most committee members complied.

To solve the theft problem, the three committees merged into one, which assigned patrol duties to community members on a regular basis: first every Thursday, later changed to fortnightly. They also agreed that the main committee and the Chimaliro Forest Compound staff should organize reflection and feedback meetings every other Tuesday. They agreed to monitor colonization of the beehives, any insect attacks on the hives, and any thefts of hanging wires, tar, or honey. Monitoring indicators were developed, with the main objective being to obtain honey and wax to generate income for household food security. The specific indicators were availability of beehives and tar at all times, frequency of patrols, colonization of hives by bees, time for harvesting, volume of honey harvested, quantity of honey and wax produced, storage facilities and place, market availability, retail and wholesale price, sales per harvest, income from marketing, banking procedures, and income to be used for community assistance and development.

The unsupervised sharing of benefits from beekeeping among block committee members was another problem. To address this, the group developed the following rules: harvesting should include a community representative in addition to the committee members; storage of honey should be at the field collaborator's house along the tarmac road; and contributions towards funeral services from sales of honey should be announced at the funeral.

Monitoring and reflection, after some time, revealed that the hives in the forest were never colonized and thefts were still a problem. The committee attributed this sad event to the regularity of their Thursday patrols, and the fact that the wider community knew their movements. They agreed to have unannounced patrols in order to apprehend the culprits.

Another challenge involved the placement of hives. The communities were putting hives wherever a tree had nice branches on which to hang the hive. The ACM facilitator decided to train the group in participatory forest resource assessment, which would help in implementation and monitoring of forest resources. After the training, the participants realized that they had put hives in areas where animals grazed and where there were primarily regenerating stems, with few trees to attract bees. Realizing that they could find more suitable places, they selected sites close to water to increase the chances of colonization, rather than close to the main road, as had been done before. They realized that the location was the main reason for the colonization failure. They also agreed to move some of the beehives back to homestead woodlots and village forest areas. This was done to increase the volume of honey produced.

—Judith Kamoto (2003a)

quickly, the quantity sold, the price variations, market availability, and the changes in numbers of user group members involved (Kamoto 2003c).

Qualitative Approaches to Social Learning. This approach was taken where the ACM facilitators were knowledgeable about social learning as a discipline and participant observation as a method. In Jambi, Sumatra, for instance, the main facilitator (Trikurnianti Kusumanto) was working toward her dissertation with a major focus in the field of social learning. All community members—but particularly women—were most comfortable speaking out in ordinary settings where interactions were informal. Participatory action research activities became part of people's daily lives, "as if nothing new had been introduced" (Yuliani et al. 2002). The team described their activities to local people as *mengalir* (to flow with the stream) in people's lives—using participant observation skills (another actor, Yayan Indriatmoko, was trained in anthropology). By means of focus groups, facilitation, informal meetings, and discussions, the idea of social learning was discussed, shared, and encouraged to good effect. Kusumanto refers to social learning as snowballing, with later communication and relationship building more significant than early attempts. She specifically notes the political implications of such learning, with improvements in leadership, consensus building, conflict management, and negotiation. One previously passive group, for instance, began taking on an active role in interactions with a neighboring community, improving two-way information flow (Yuliani et al. 2002; Kusumanto et al. 2005).

In Bolivia, in close interaction with community members, Cronkleton discovered that the plan for distributing monetary benefits from the forest management project was unclear. He found unanalyzed data regarding the distribution and used them to prompt discussion, enhance transparency, learn from past failures, and plan for the future with community members. In this way, he was able to catalyze an iterative, learning process within the community about a topic that held considerable potential for disruption if not handled carefully (Cronkleton 2005).

The researchers in Bulungan (Indonesia), Brazil, and Cameroon also used such qualitative approaches to supplement their other, more formal, methods.

Modeling. We experimented with several kinds of formal models (system dynamics, multiagent, and participatory systems analysis) to generate understanding among experts, create shared or complementary visions for the future, and stimulate social learning (see Box 5-4).

System dynamics modeling allows for the development and testing of various scenarios; the user can specify feedbacks and lag times and thus anticipate possible opportunities and problems related to a particular line of action. Teams in Indonesia, Zimbabwe, and Cameroon have used FLORES (Forest Land Oriented Resource Envisioning System) to improve understanding of how the relevant systems work.[8]

Purnomo and his colleagues (2003a) employed system dynamics (using the software platform SIMILE) to help community members, government and timber company officials, and NGO representatives develop a common understanding of forest management around the Lumut Mountain Forest Reserve (Pasir, East Kalimantan). One purpose was to integrate stakeholders' different perceptions of forest management. Scenarios built on this understanding were examined by participants in the modeling workshop, resulting in shared learning, increased mutual understanding, and some concrete planning, particularly for institutional and legal needs.

Box 5-4. The Role of Modeling in ACM

Modeling is a difficult process to manage well, and the more people involved, the more difficult it gets. However, it has the potential to facilitate rapid insights into complex issues and be a tool for mutual or social learning. Etienne Wenger (1999), in his book *Communities of Practice*, proposes that a well-designed learning process needs to have three phases—participation, imagination, and alignment—and that simulation models are a good tool for the imagination phase. This suggests that learners need to be engaged in a process first (participation), before modeling is introduced, and that after modeling, people need to have an opportunity to understand how they can use this new knowledge (alignment). There are at least five potential roles for modeling in participatory action research.

1. Models can be representations of what is known. This role uses modeling as a process for pooling baseline knowledge in a way that is useful for comparison in the future. The FLORES (Forest Land Oriented Resource Envisioning System) modeling process has been primarily used in this way. The modelers are the research team, and they are usually facilitated by outsiders.

2. Models can be used for planning actions. This is the role of conceptual models in soft systems methodology, which is an action research methodology for addressing complex human problems. The systems model is a conceptual tool for considering possible interventions or actions to bring about a change in the interactions between people, which have been the cause of problems or concerns. Modeling is used as an intermediate step between defining a problem, or issue, and intervening. It is a team process for deciding on an appropriate intervention. It is a good tool to use after visioning, as a way of exploring potential strategies for achieving the vision. The ACM Zimbabwe team and part of the ACM Indonesia team have been using modeling this way, and we have developed a toolkit, called Co-View, to support the process of visioning and modeling together. The research team includes both modelers and facilitators.

3. Models can be used as interventions. This is what is usually thought of as participatory modeling. The ACM Zimbabwe team has used modeling in this way. The team worked with a group of women from a village next to Mafungautsi forest, who built their own model of how they manage broom grass in order to explore the impacts of different harvesting methods on their income and on the grass. Insights from this process have led these women to devise a new kind of decorated broom which they can sell for much higher prices. The modeling process is designed and facilitated to enable local stakeholders to share knowledge and generate insights into a problem or issue that they have identified. The modelers are the stakeholders, facilitated by the research team.

4. Models can be used for monitoring. Modeling can be a process for deciding which data need to be collected, what indicators need to be measured, or whose knowledge needs to be acquired. A process of developing a set of criteria and indicators is a modeling process of this type. A model that includes links between indicators can be used in a process of integrating and synthesizing results from monitoring in order to develop a big picture of the system. The modelers are the monitors, facilitated by the research team.

5. Models can be used for reflection. Modeling is used as a means of posing a question in a rigorous or formal way and exploring various potential theories that may provide answers to the question. It can involve exploring alternative hypotheses to explain or understand the results of an intervention. Alternatively, it can be used to explore the implications of a theory before embarking on interventions to test it. I have used modeling in this way to explore issues, including communication, social capital, social networks, and how knowledge and "infectious ideas" spread across these networks. The modelers are the research team, facilitated by the ACM team or outsiders (see also Haggith 2001).

—Mary Haggith (2001)

Purnomo also used CORMAS (Common Pool Resources and Multi-Agent System) to examine the hypothesis that comanagement of forests by all relevant stakeholders would improve outcomes in Bulungan Research Forest. The purpose of this model was to represent important processes under way and thus provide a basis for discussion. Its utility arises from its ability to enable exploration of people's assumptions (Purnomo et al. 2003b).

Team members from Cameroon and Zimbabwe were trained in participatory systems analysis, a more qualitative method, in South Africa (Grundy et al. 2002; Cwebe Working Group 2001) and used their insights to strengthen ACM data gathering and analysis in their respective countries.

Horizontal Collaboration

We considered two types of collaboration: horizontal and vertical.[9] Horizontal collaboration refers to cooperation and complementarity among people who are or could be in regular contact with each other, in a particular community or forest area. Efforts to strengthen horizontal collaboration were motivated by concerns to "harmonize" and rationalize the diverse management systems typically present in any given forest.

The kinds of horizontal collaboration deemed important in the different sites varies enormously. Collaborative problems that struck me most forcibly when we were planning our research were those between local communities on the one hand and timber companies and conservation projects on the other. At a finer scale, intracommunity problems of representation in decisionmaking also have been worth our attention. Three main approaches were taken to strengthening collaboration among stakeholders: workshops, user groups, and networks.

Workshops. Efforts to strengthen horizontal collaboration often started with workshops. We used workshops a lot within the ACM program itself, both to enhance our own collaboration among the various sites (e.g., a planning workshop in February 2000 and a writing workshop in November 2001, in Bogor, Indonesia; International Steering Committee meetings in Zimbabwe and Nepal in 2000, and in the Philippines in 2001) and among collaborators on a particular site (e.g., training in participatory action research by Bob Fisher and Mohammad Emadi in Bogor in 1999; ACM training in Bolivia in 2001 (Cronkleton 2001a); and an ACM writing workshop in Cameroon in 2002, among many others).

Workshops provide a venue where representatives of multiple villages, various government departments, different industries, and research organizations can come together. These were particularly important in Indonesia, Cameroon, and Bolivia, where the problems (and corresponding power and wealth differentials) between local communities and other forest managers were obvious.[10]

The teams were cognizant of the potential pitfalls of bringing together groups with such radically different power, and they gave considerable thought to structuring interaction in ways that would minimize its relevance. Specifically, we wanted to reduce the likelihood that local people would hesitate to express their ideas in the presence of individuals with the power to harm them. In all cases, workshop participants agreed on a set of principles that would maximize freedom of expression; in all cases, the workshops alternated between small groups and plenary sessions. Where possible, participants were coached beforehand by team members on both the importance of a free exchange of views and the basics of communication (greater listening skills for some, more assertive speaking for others). All workshops had themes.

In Cameroon, the teams used the C&I framework to structure discussions about forest management and human well-being—what it meant to participants and how they might be able to measure it. They used simple pebble games to bring out volatile issues pertaining to access to resources, voice in forest management, and distribution of forest benefits among these various stakeholders (Diaw and Kusumanto 2005; Tiani et al. 2002a; Oyono and Efoua 2001). The results were increased respect and understanding on the part of government and industry officials, improved confidence on the part of community participants, and increased understanding of the complexity of local conditions by all parties. There was a corresponding increase in the legitimacy granted to local concerns within the region.

In Pasir, Indonesia, the workshops were structured around priority setting and division of labor at the district (kabupaten) level, at a time when local governance was undergoing dramatic change because of Indonesia's decentralization process. There was a conscious decision not to begin with the difficult problems dividing communities and timber companies—though both participated in the workshops. These conflicts were deemed too volatile in early 2001, but by mid-2002, it was possible to raise them in one of the workshops and some plans were made to begin addressing them. Meanwhile, significant progress was made on the plans from the 2001 workshop—particularly intensification of community use of swidden fallows. Hakim identifies the following qualitative improvements on this site as the research progressed: gradual increase in local stakeholders' efforts to address local problems, greater willingness of other stakeholders (timber companies, government officials) to listen to local people, greater ability of local people to express their views, and increased awareness among the stakeholders that they could plan and learn from their efforts in a deliberative way (Hakim 2005).

One important workshop in Jambi, Indonesia, focused on the principles of people's representation under decentralization. It brought together local community members, district and subdistrict government personnel, and district-level legislators. One failure was the team's inability to include the hunter-gatherers, the Orang Rimba, whose mobile lifestyle made communication with them sporadic and unpredictable. Nevertheless, the group established a shared set of principles, which differed from the requirements of representation in current policy but were more appropriate for local conditions (Kusumanto et al. 2001). Subsequent field activities built on these principles through a continuing process of shared learning. A truly democratic election resulted in a much more representative group of leaders than had characterized village politics in the past (though the facilitators note with dismay that the one woman elected has been given the stereotyped role of secretary (Kusumanto and Indriatmoko 2003).

In Bulungan, Indonesia, the team organized workshops on community mapping and decentralization. Early on, they noticed the striking differences in information obtained when they broke large meetings into smaller ones composed of men, women, and youth, respectively. Among the young and the women, they found much greater antagonism to the destructive logging that was under way than they had uncovered in plenary, where older men dominated the discussion (Wollenberg 2001). Through such meetings the team was able to identify the need for more legislative literacy among the population and obtained funding to address that issue (Moeliono and Djogo 2001).

In Madagascar, the team used a workshop on medicinal plants as an alternative strategy when their efforts to finalize the partnership arrangements between communities and national park managers were blocked by the latter's persistent failure to respond. In February 2000, the team brought together a wide range of stakeholders (communities, traditional practitioners, pharmaceutical industry representatives, academics, NGOs, government and project officials) to discuss the potential for a medicinal plants industry in Sahavoemba. Notable results included enhanced communication and joint plans to create an association between traditional and "modern" practitioners, called FIMARA (Fitsaboana Malagasy eto Ranomafana) (Rakotoson et al. 2000; Buck 2003).

The teams were able to skirt many of the problems that can arise when representatives of government and industry meet with community representatives who have less power, wealth, education, and prestige; they found that if the workshops were structured and facilitated with care, the gains outweighed the risks. However, teams in Cameroon and Indonesia noted the near absence of women in many of their workshops and, in some cases, problems with representation of marginalized ethnic groups (particularly hunter-gatherer groups).

The gender issue was addressed directly in a workshop in Bolivia, where BOLFOR and others had already noted the lack of involvement of women in forest management (e.g., Bolaños and Schmink 2005). Important results included the adoption of a comprehensive gender policy by BOLFOR, the development of a set of "best practices," and apparent buy-in on the importance of this issue from previously skeptical (and powerful) project personnel (Cronkleton 2001d). Some success has been realized in including women by scheduling meetings at times convenient for women, waiting for women before beginning, looking at and expressly soliciting input from women, and using the local language rather than Spanish (see Cronkleton 2005).

User Groups. Another approach to enhancing collaboration involved user groups (Nepal; Zimbabwe; Acre, Brazil; Pasir, Indonesia). Forest user groups were institutionalized in Nepal by community forestry legislation, and the Nepal ACM teams blended their efforts into these existing institutions. This allowed them to follow up on areas that were recognized locally as problematic (e.g., equity, transparency, authoritarian leadership), and it strengthened the possibilities for wider application of research findings. One solution that emerged was the decision to devolve many of the rights and responsibilities of the forest user group committee to a lower, neighborhood level (toles, or hamlets). The relative caste and ethnic homogeneity of the hamlets in the four ACM sites reduced the workload and authoritarianism of the user group chair and central committee, while increasing transparency and representation within the overall group. The teams show impressive increases in involvement by women and marginalized castes and ethnic groups in various activities, fora, and decisions (McDougall et al. 2002a; Dangol 2005; Sharma 2002; Sitaula 2001). Uprety et al. (2001b), for instance, describe workshops held at Andheribhajana and Manakamana in 2001 in which user group members, many of whom were illiterate, reflected on previous plans and solved problems—in one case, successfully reining in a dishonest leader. Some district-level officials attended and noted with surprise the active involvement of women.

Box 5-5. Changes in Access to Markets and Services in Porto Dias

… [T]he local actors involved in the forestry project view the dirt road as an asset of the project, and most of the benefits existing in the Agroextractive Settlement Project today result from the road. The road has improved access to the market and basic public services and, consequently, has improved the standard of living of the families living in Porto Dias. Everybody recognizes that all residents of Porto Dias have benefited, to different extents, from greater access to such public amenities, including the state Citizen Project, which works with health and education.

Despite residents' perception of the road as a common good, they initially felt highly dependent on assistance from the state in maintaining it during the rainy season. Families participating in the forestry project and the Rubber Tappers Association, in order to maintain control over the road, took on the responsibility, despite the extra work this imposes on them, for road maintenance. The planning and organization of work to clean up and maintain the road has become a source of frequent conflict. Within the group of rubber tappers, those selected to participate in the forestry project have the road passing near their homes, while others not involved in the project have to travel great distances to reach the road. Other local groups also perceive their potential participation in the forestry project as a possible means to get better access to the road. A process of reflection between local groups is needed both to manage these conflicts and to discuss the benefits and risks of the forestry project …

We analyzed … the ACM project's contribution in promoting collective learning in Porto Dias. We worked with the group of rubber tappers, particularly those participating in the forestry project, applying future scenarios to assess uncertainties and construct new scenarios with the group. One example of a future scenario that we used was the eventual departure of the Center for Amazonian Workers and its consequences for local groups. The rubber tappers are well aware of possible impacts, including the loss of development projects and benefits and, given the group's small number, a weakening in its capacity to act within the community without external assistance. They began to realize that the absence of a social institution within the agroextractive project would diminish the local community's level of organization and empowerment. At the same time, they also recognized the need to monitor the impacts of changes resulting from new community development projects. We used this scenario to encourage the rubber tappers to collectively act and learn with different local community groups and to use this as a strategy for strengthening the community and for learning to handle uncertainties and the complexities of the community's future. This was particularly important for the very practical concern about road maintenance, since the state government's maintenance of the road is not guaranteed.

— Magna Cunha dos Santos (translated by Samantha Stone)

Box 5-6. The Snowball Effect of Collaborative Action in Jambi

As in other ACM research sites, the participatory action research in Jambi has resulted in improvements in social capital. Through continuous learning and action, diverse groups and actors at the site have built abilities to collaborate, organize institutional and community activities, and advocate local issues in wider forums. Interestingly, social capital appears to be cumulative: it starts with a minimal amount and grows as people build their experience in creating "new capital." For example, the women's groups at our site started with relatively simple institutional improvements aimed at better collaboration among local and settler women in running their savings and loan activities. This in turn enhanced the women's individual and institutional abilities to deal with wider and more complex problems that were beyond the boundaries of the women's groups. Other examples exist, showing how community groups accumulated social capital just by initiating relatively simple activities. At this stage, community groups, facilitated by the ACM field team, hope to embark on building vertical relationships and structures with formal institutions—like the government (especially at the district and subdistrict level)—in the years to come. [this has, indeed, happened]

—Trikurnianti Kusumanto (2003)

Most sites did not initially have formal user groups. In Zimbabwe, after the training for transformation described above, team members had little trouble working with the communities to form what they called action groups, based on particular natural resources of importance in the site (broom grass, thatch grass, and honey). These groups were able to improve cooperation significantly with the Forest Commission and served as fertile ground for collaborative, social learning within the group.

In Porto Dias (Acre, Brazil), a small group of rubber tappers had already formed a user group when ACM began, but their activity was dominated by personnel from an environmentally oriented NGO, the Center for Amazonian Workers (CTA), which was coordinating the small-scale timber project there. The ACM emphasis on joint learning from experience involved the rubber tappers more directly in planning and decisionmaking while prompting more open attitudes from CTA personnel. In Box 5-5, Magna Cunha dos Santos reports the successful collective action that the rubber tappers organized to maintain a road important to them; but it also brings up some of the conflicts and learning that have accompanied this process in the Porto Dias Agro-Extractive Reserve.

The women of Basac, in the Philippines, also formed themselves into several groups. One concerned itself with the production of cut flowers for income generation (Arda-Minas 2002a); another with the development of a medicinal herb garden for community use (Valmores 2003). The latter revived and built on an indigenous form of collective action (*pahina*). Overall, there were four action groups, each having about 15 members (with little overlap between group memberships). See Box 5-6 for a discussion of similar groups in Jambi, Indonesia.

In Pasir, East Kalimantan, Indonesia, the team and the community formalized three user groups in Rantau Layung and two in Rantau Buta, based on clusters of swidden fields. These groups followed up on the multistakeholder workshop recommendations for improving forest fallows after rice cultivation (specifically to incorporate rattan, durian, coffee, and rambutan). The followup process was an

Box 5-7. Establishing Networks in Palawan

While struggling to increase collaboration and gain trust from their members and other community members in the three barangays, the Board of Directors of the San Rafael–Tanabag–Concepción Multipurpose Cooperative in Palawan effectively applied different mechanisms to increase their voice and power with other external stakeholders.

This was apparent in the way they dealt with the complexities of procedural requirements to get their management plans approved by various government institutions, to get forest charges/fees for extracted fallen timber reduced, and when they applied for a permit to transport lumber samples to Manila.

The Board of Directors sent resolutions (letters) to appropriate government institutions and diligently followed up. When dealing with the Department of Environment and Natural Resources, in most cases they had to address their letters to the highest authorities (beyond the provincial level), such as regional directors, the undersecretary, and even the secretary himself. They also lobbied the provincial governor of Palawan, congressional representatives, and members of the city council. These required not only good knowledge about the functions and powers of different institutions and levels of authorities, but also strong lobbying and advocacy at the local and provincial levels to get support. Their good documentation system certainly helped as, without that, papers could fall through the cracks along the way.

They also established networks and strategic alliances with various organizations. The most important one is the Palawan Community-Based Forest Management Federation, a federation of all formal community forest holders in the province. The former chair of the cooperative played a crucial role in the formation of this federation and remains very active. In the case of forest charges, based on the lobbying activity of the cooperative, the federation wrote a resolution supporting this request and putting forward rationales for such reduction.

The capability to establish networks and linkages was very much related to the relatively high education and influence of the cooperative's board members. Four (out of nine) either held political positions in the past or have a college education. At least one of them stayed in Puerto Princesa City, the locus of provincial power, while their villages are very accessible by a good road system (see Devanadera et al. 2002 for a fuller description). This is in contrast with the situation in Lantapan [the other ACM site in the Philippines]. Despite having better collaboration and trust within their own community, members of the Basac Upland Farmers Association still need to improve their networks with external stakeholders. Being an indigenous community, they need to overcome their low self-confidence, the language barrier, and low communication skills.

—Herlina Hartanto

excellent example of the need for a dynamic approach to problem solving. When the government did not deliver on its promise to provide seedlings, the user groups decided to seek other sources, even trying to persuade CIFOR to provide funding. Eventually, the group worked with the ACM facilitators and obtained the seedlings by writing a successful proposal (Hakim 2005). More important than getting the seedlings was working together to deal with the Indonesian bureaucracy. The groups also improved their knowledge of rattan cultivation, processing, and marketing, in cooperation with other villages and outside stakeholders.

Networks. The establishment or strengthening of networks often has both horizontal and vertical implications. Links with formal networks were important in Nepal (FECOFUN) and in Acre (the National Rubber Tappers' Council) as a potential way to strengthen local people's voice in policymaking through pressure groups. But increased networking with local officials in larger scale institutions was important in most sites, including the Philippines (see Box 5-7).

In Madagascar, in hopes of following up on the activities proposed in Anjamba's draft forest management plan, the Manirisoa Association was created, with help from NGOs. It was intended to be a "mother association," including all the villages of Anjamba and the interest groups the people formed. A similar organization was formed in Sahavoemba, called FAFSA Association (Fampandrosoana ny Faritra Sahavoemba), for the economic and sociocultural development of the community—focusing specifically on medicinal plants, environmental protection, and crop and livestock improvement (Rakotoson et al. 1999). But the failure of the hoped-for partnership between individual communities and park management led to a more regional approach focusing on partnerships between local medicinal practitioners and the formal medical establishment. Although we do not have recent information on this process, things looked good as the funded research drew to a close in 2000 (Buck 2003).

Vertical Collaboration

Vertical collaboration involves interaction and two-way communication between individuals with less power and those with more, between people operating in community-level structures and those operating in larger scale, "higher" level structures. Efforts to strengthen vertical collaboration were motivated by the desire to empower people whose lives were being adversely affected by actors operating at larger scales. Our approach included explicit attention to the links between forest communities and the outside actors and events that affect their lives. Empowerment, to us, implied a greater ability to influence events at the larger scales, and we began by searching for mechanisms that might make this easier.

One of the first mechanisms we tried was the creation of national-level steering committees in Africa and Asia (except Kyrgyzstan). These steering committees typically had five to seven members, drawing from government, academia, NGOs, and projects. As we had hoped, steering committee members have guided us on national priorities and opportunities to affect national-level policy, and at the same time they serve as outlets for the results.

The steering committees have certainly been a positive addition to the research in Asia. In the Philippines and Nepal, members were instrumental in encouraging adoption of various ACM methods within the formal governmental context (community forestry in Nepal and community-based forest management in the Philippines). The Indonesian steering committee has contributed to our analyses and products (Salim 2002; Sardjono 2004; Soemarwoto 2003), but perhaps because of the recently chaotic Indonesian political scene and the size of the country, we have made little headway in integrating our approach into any national-level programs. Most of our Indonesian successes have occurred closer to the field.

Another approach involved more direct involvement in policymaking settings. In Zimbabwe, for instance, the leader of our ACM team was seconded from the Forestry Commission, and what he learns will be available to the national agency when the project ends.[11] In Bolivia, the team was integrated into a large, U.S. AID-funded forest management project (BOLFOR), with strong ties to the Bolivian government's forestry establishment and to the timber industry. The Kyrgyz team has a similar situation: a bilateral Kyrgyz-Swiss program (KIRFOR) was closely tied to the Kyrgyz forestry establishment. This approach gives field researchers access to a wide range of researchers, practitioners, and resources. In Bolivia, for instance, BOLFOR has been under way for nearly a decade and has built up a fund of experience, project reports, analyses, and community ties that have been available to our team. The ACM team also has access to the policy network of this long-standing project. The disadvantage is that the network is fairly clearly established, and attempts to expand it to involve more NGOs and people with social science expertise have been difficult.

The Cameroon team began the research with a thorough investigation of the policy issues in the country—forest minorities, forest margins, community and communal forests, conservation, and timber management. Members then selected sites that would provide useful input into the debates, reasoning that this would ensure interest in their results. Although the team is engaged with policymakers in thinking through the issues, concrete impacts remain difficult to pin down. Akwah (2003a) lists the Campo Ma'an policy issues on which ACM team members—and many other actors—have expressed views. Although the following examples cannot be called impacts per se, they do provide some sense of the kinds of issues ACM is addressing:

- Proposed modifications (of unknown dimensions and location) on the limits of the national park now await government approval.
- The area of the park's agroforestry zone, available for community use, has been increased, pending official confirmation.
- HFC (a logging company) and the Ministry of Environment and Forestry have signed a partnership document (*cahier des charges*) outlining, among other things, their roles and responsibilities in combating poaching; 10 guards, recruited, equipped, and paid by HFC, are posted at the two national park gates.
- HFC and another logging company, Wijma, are seeking certification.
- At an April 2003 meeting, most Campo Ma'an stakeholders (the Ministry of Environment and Forestry, local communities, NGOs) rejected the management plan proposed by the outgoing Campo Ma'an project, calling for more consultation on its contents.
- SNV, the Netherlands Development Organization, changed its strategy of employing local extension workers and is now encouraging the emergence of local NGOs. Former employees are enthusiastically forming NGOs, which will probably increase the number of village organizations.

In Asia and Africa, ACM findings have been presented to policymakers, forestry officials, NGOs, and academics at national-level workshops; in Asia, regional-level workshops have been held as well (Prabhu et al. 2002b).

Some teams developed strong links with NGOs, on the theory that NGOs would have better rapport with local communities and be freer to pressure governments and other stakeholders as needed. NGOs had very prominent roles as research partners in the work in Acre, Brazil; in Nepal; and in Pasir and Jambi,

Indonesia—where important field personnel were seconded or contracted from NGOs to work as facilitators. NGOs have also been important as collaborators in Cameroon and in Pará, Brazil. Their involvement has broadened the applicability of ACM findings. NGO personnel—often young, passionate, and motivated—impart vibrancy and relevance to our work. The downside is that some NGO personnel have been suspicious of CIFOR as an international organization (like the World Bank or other consultative groups), and we spend valuable time on reassurance, communication, and conflict resolution.

The above mechanisms for communicating community-level views to policymakers are all indirect, however; we wanted to identify or develop mechanisms that would directly strengthen local people's involvement in policymaking. One route has been to develop their skills. The Training for Transformation curriculum, for example, includes role playing, analytical skills, and self-confidence building and has helped the people of Mafungautsi, Zimbabwe, level the playing field. Matose et al. (2002) found antagonistic relations between local communities and the Forestry Commission, but by December 2002, the two sides had collaborated in drawing up a template for how monitoring committees should work. The template was then adopted by Forestry Commission officials, who in fact adopted the ACM approach more generally, taking over many of our facilitation tasks. The local forestry official recently allocated funds to Gababe for firefighting that would previously have paid for Forestry Commission tractors. He is also exploring the communities' idea of instituting a resource management committee board to facilitate mutual support and sharing of successful approaches.[12] Two-way communication has thus been strengthened.[13]

Other ways to help communities shape policy include C&I workshops in Cameroon; legal literacy training in Madagascar; priority-setting workshops in Pasir, Indonesia; mapping, decentralization, and legal literacy sessions in Bulungan, Indonesia; decentralization workshops in Jambi, Indonesia; and workshops on gender, future scenarios, and forest benefit distribution in Bolivia. The facilitation process, including developing shared rules and breakout groups for more equitable participation, allowed reticent rural people to speak up. In Pasir, the facilitators informally coached community members before meetings and mentored them in subsequent dealings with the bureaucracy and with timber companies, gradually building up the communities' skills and self-confidence.[14] In Jambi, women were coached before community meetings for similar reasons. The Madagascar team also coached community members before workshops with outside stakeholders concerned about Ranomafana National Park.

The Pará, Brazil, team, not wanting to dominate the process, experimented with training community members in facilitation and interactional and analytical methods. Community members also self-selected to undertake research on important issues, such as marketing of forest products. This exposed them to the wider world and enhanced their ability to deal with other stakeholders, besides gaining the valuable information the community needed.

Adequately facilitated, with attention to the mix of participants and agreement on rules of the game, workshops provided effective fora for interchange among stakeholders. When community members were well prepared, they had opportunity to directly affect the policymaking process and work with others to solve policy dilemmas. This occurred dramatically in the Indonesian sites, where the wider society was also struggling with the implementation of decentralization. The workshops provided a forum wherein diverse views could be expressed, in a problem-solving and comparatively safe atmosphere.

In the Nepal sites, the communities initially focused on improving their community forest management within the structure established by the community forest legislation. The improvements gained the attention of other stakeholders, who became more attentive to the communities' needs and more interested in working with them.[15] The ACM team then identified the Forest Range Post's quarterly meetings with forest user groups as a possible place to strengthen both horizontal and vertical collaboration. Using facilitation skills, the ACM researcher turned one of these meetings into an opportunity for real, substantive exchange among the participants. The eventual result was more effective transfer of governmental information, better information exchange among forest user groups, and more creative use of time and resources (Uprety et al. 2001b, 2002c). At the end of 2002, Nepal's study communities were strengthening their ties to FECOFUN (a national network of forest user groups) and participating in new fora bringing together the village development committees with the forestry personnel (two separate bureaucratic segments with overlapping jurisdictions and without previous links).

Although the efforts to strengthen the links between communities and policymakers have been worthwhile, it is important to acknowledge and plan for the extra time this takes. Steering committee members are busy people, and even arranging the meetings can be time consuming. Forest communities are in inaccessible places, and meetings to strengthen their ties with policymakers often mean arduous travel and loss of time that may be needed for subsistence. Community members who have least access to policymakers are often those whose livelihoods are the most precarious and who need their time to supply their daily food (Dangol 2005; McDougall et al. 2002a; Sitaula 2001).

Livelihoods, the Environment, and Empowerment

Our interest in stimulating more adaptive and collaborative management was based on our hypothesis that this could empower people to develop a better livelihood and improve their environment. We cannot yet tell whether these changes will occur—we are dealing with complex systems, changing at varying rates, and interactions among variables may have serious lag times—but in the following sections, I extract some hopeful indications.

Livelihoods

Livelihood is the topic that community members themselves are expected to feel strongest about and be most willing to commit time to address. Although livelihood issues were definitely important on our sites, people did not seem disproportionately interested in working on them. The actual implications of ACM for people's

Box 5-8. Generating Income with Mushrooms in Chimaliro, Malawi

Disappointed with the small amounts of income emerging from the beekeeping activity, a group of 11 women in Block III formed a user group and decided to try seasonal products like mushrooms (*Parinari curatefolia*) and *Uapaca kirkiana* fruits. Initially conceived as a group effort, they quickly switched to an individual approach. Each person collects, processes, stores, and sells her mushrooms.

Six Block III women collected three species, which were processed differently. Some species are sun dried soon after harvesting; others are boiled and dried using fire or sun. The sun-dried variety is soft, easy to cook, nonpoisonous, and can be eaten fresh. The boiled varieties are a combination of species, including poisonous ones. Once boiled and dried, the mushrooms are safe to eat.

The dried mushrooms from the 2002 harvest were snapped up at MK5* per cup on site or MK10 per cup in town; a total of 60 cups was harvested (mostly dried, though processing made no difference in price) by the six women, who began late in the rainy season of 2002. Their gross income was MK300, all of which they were allowed to keep, as this was considered a startup activity by the district forestry officer. The District Forestry Office expressed its willingness to help with marketing of the mushrooms in Kasungu town on the women's behalf, in 2003, and the women expect to get an earlier start this year.

—Judith Kamoto (2003c)

*90 Malawian kwacha (MK) = U.S.$1 (early 2003)

livelihoods to date have been quite modest, though I remain optimistic that these are tip-of-the-iceberg phenomena and that as the communities develop capabilities for joint planning and reflection, collective action, and networking, their lives will improve.

The Mafungautsi broom grass user group's improved broom design (see Box 5-4) enabled the women to charge Z$50 a broom at the provincial agricultural show, compared with Z$30 for a traditional broom. The broom sellers have also become more confident and now go in groups to distant towns to sell their brooms. Group travel has also removed a significant marketing constraint: their husbands' fears about their wives' traveling alone (Matose et al. 2002).

In Chimaliro, Malawi, the members of one income-generating activity group (see Box 5-8) collected 55 liters of honey from the three forest blocks, in two harvests, from hives in both village forest areas and the comanagement forest blocks during the first year. The honey was sold at MK750 per 5-liter container. This achievement spurred growth in the number of privately owned beehives—from one to five in Block II and from three to seven in Block III (Kamoto 2003c).

In Pasir, in East Kalimantan, Indonesia, the communities wanted to make their rattan production more profitable. They had heard about rattan processing in South Kalimantan, and as part of their ACM activities, a small group made a "look and learn" visit and then arranged for training under a cost-sharing agreement with CIFOR and another village. One family each in Rantau Layung and Rantau Buta sold rattan baskets and furniture; other families expanded their rattan cultivation to,

Box 5-9. Managing a Boundary Conflict in Palawan, Philippines

In Palawan, Philippines, the ACM approach and processes were applied to assist the San Rafael–Tanabag–Concepción Multipurpose Cooperative to manage a conflict over forest boundaries. The cooperative received rights from the Department of Forestry and Natural Resources to manage a piece of forestland (expanded from 1,000 ha to 5,005 ha in 1997) under a community-based forest management agreement. The new area was neither surveyed nor marked on the ground. The neighboring indigenous community, the Bataks, also formally manage a much smaller piece of land nearby under another community-based forest management agreement. The conflict emerged in 2000 when the Bataks assumed that the cooperative expansion included their village and an area that had traditionally been their harvesting area for almaciga (*Agathis damarra*) resins. At the same time, the Bataks also faced another problem, as they were not allowed to extract resins from their tenured forests unless they could meet the Department of Natural Resources' technical requirements.

To solve the boundary conflict, the ACM team facilitated the cooperative to reflect on the causes of the problem and strategize how to handle it. Their first proposed solution was to show the Bataks a map, demonstrating that the two forest areas did not overlap. They decided to ask the Department of Natural Resources to check the existing map to see whether or not the boundaries of the two community forest areas overlapped. However, there was no such map in the agency.

The cooperative then decided to seek assistance from the Palawan Tropical Forest Protection Project to obtain GPS coordinates of their forest boundaries. This was done, and the survey produced a clear delineation of the forest area, which was marked on the ground (using natural features, such as rivers, caves, etc.) and on the map. Unfortunately, neither Bataks nor village leaders were able to participate in the survey.

Once the map of the boundaries was available, several representatives from the group conducted a dialogue with the Bataks and presented the results of the survey to them. The dialogue was facilitated by a Department of Natural Resources representative and also attended by a local NGO. From the ensuing dialogue, the cooperative realized that Bataks had a different concept of boundaries. It appeared that the map would not be useful or effective in explaining to the Bataks where the forest boundaries were located. On reflection, the cooperative members realized that they could solve this conflict best by helping the Bataks to solve their livelihood problems.

The cooperative proposed an arrangement whereby the Bataks could extract almaciga resins in its area as long as they sold the resins to the cooperative, and the latter would purchase the resins for a price at least one-half to one Philippine peso per kilogram higher than the price offered by local traders. The Bataks considered this arrangement to be better, as the existing practice had been to barter the resins for rice, coffee, or tobacco, which resulted in very low "prices" for their resins. Furthermore, both parties agreed that they would be partners in forest protection activities and would collaborate in other livelihood efforts.

This conflict was resolved by increased collective action within the group and across stakeholders to solve the boundary issue. An important contributor to successful resolution was the level of trust between the cooperative members and the Bataks, despite the conflict, that allowed them to discuss the problem openly. Furthermore, they were able to critically analyze the situations and creatively find a way to solve the boundary conflict and the underlying livelihood problem.

—Herlina Hartanto

on average, 0.25 ha each (Hakim 2005). Actual benefits have, however, so far been small. These communities have also sought to intensify fruit and other tree crop production in their fallow areas.

In Palawan, the Philippines, a cooperative was able to persuade the government to reduce the fees on the timber harvested. And the same group negotiated with the neighboring Batak group to obtain higher prices for, and retain secure access to, the almaciga resin, which had formed an important part of their supplementary income, in a disputed area of the cooperative forest management unit (see Box 5-9)—thus also contributing to their ability to retain the community forest agreement with the government.

The Environment

CIFOR, as a forestry research institution, has a mandate to focus on environmental issues while also emphasizing human well-being. Our enduring interest in the environment is reflected in the large number of biophysical scientists involved in the research (see Table 11-1). However, we have been wary of pushing this interest too strongly with community members. To test our hypothesis that more adaptiveness and collaboration would benefit forest management in an enduring fashion, we had to work on issues the communities and other stakeholders considered worth their effort. Our results genuinely reflect the interests of the local communities.

Hartanto describes, in Figure 5-1, how illegal almaciga extraction was reduced in the Palawan site. The process suggests communities' increasing awareness of the importance of monitoring and management in maintaining sustainable supplies of almaciga.

In Mafungautsi, Zimbabwe, the willingness of the Forestry Commission to support local-level firefighting suggests at least the potential for improved fire management. At a workshop organized by the Forestry Commission prior to the grass-cutting season, an official encouraged the resource management committees to measure the areas where grass grew, determine the amount of grass they had, and monitor the resource. The people of Gababe measured two grass areas, Wadze and Lutope. They also divided each *vlei* into equal parts of 20 ha each, and drew a map of the plots, with the intention to mark them with pegs. Their plan is to collect records of the quantities, height, and thickness of grass from each plot so that they can take corrective action if needed (Matose et al. 2002, 68). The degree to which this plan is implemented will, of course, be the proof of the pudding.

To harvest broom grass in Mafungautsi, one can either uproot or cut the plant. The latter method has been clearly shown, in local experiments, to result in much better regeneration, but broom buyers preferred the brooms made from uprooted grass because they last longer. After considerable discussion (Mutimukuru et al. 2005) and modeling (Standa-Gunda et al. 2003b), the broom grass user group experimented with ways to strengthen and beautify the brooms made from cut grasses and held a workshop to share the best method. The new products proved popular at the annual provincial agricultural show, and the approach seems to hold promise for strengthening the sustainability of broom grass harvests (Matose et al. 2002).

Nepal had already witnessed the environmental benefits from community forestry when ACM began, and there was general agreement that the communities were doing a better job environmentally than the government. The forest problems on the ACM sites were, rather, equity issues. However, the progress the teams and communities have made in equity seems likely to spill over to the forests. The greater involvement of more community members in the forest management process (see Figure 5-2) should mean more people obeying forest-friendly rules that are more equitable, more people monitoring to ensure compliance with regulations, and more people supporting sanctions when rules are ignored. And there is some evidence that this is happening (see McDougall et al. 2002b). Some of the specific activities the study communities have undertaken—a bamboo nursery project, a forest protection system (involving passing of a stick among rotating guards), a community forestry nursery, planting of broom grass, fine-tuning of regulations on forest product harvesting—suggest continuing and beneficial attention to environmental issues.

In Madagascar, where laws devolving rights to local communities have been passed but barely implemented, communities in Anjamba worked on management plans for three products (anguille, pandanus, and ecrevisse). These plans included inventories, boundary refinement, analysis, evaluation of pressures, and alternative income-generating mechanisms. Followup community meetings in 1998 created community action plans, divided tasks among stakeholders, and set a schedule for completion (e.g., Rakotoson et al. 1999). The improved management that would likely have emerged from this analysis of environmental conditions and income-generating opportunities was thwarted, however. Despite the communities' persistent efforts to develop the partnerships allowed by the law, park managers were ultimately unwilling to let go of any of their authority (Buck 2003).

In Porto Dias, Acre, Brazil, the conflicts between the rubber tappers, for whom the extractive reserve was initially created, and illegal settlers with a more agricultural bent is described in Chapter 10. Some settlers illegally subdivided their *colocações* (rubber tapping areas) and sold them to be cleared for agricultural purposes—with obvious negative effects on the forest. The rubber tappers were able to expel one such individual from the area with the help of the Brazilian Institute for the Environment and Renewable Resources. The agency's presence has increased since the forestry project began because it is required to audit the timber management plans of the Rubber Tapper Association. Additional illegal settlers were recently expelled and resettled (Cunha dos Santos 2002; see also case 2 in the Appendix of Cases). This is a good example of the kinds of dilemmas one inevitably encounters in the field: The illegal settlers have lost livelihood opportunities; the environment has won time. The Forest Stewardship Council certified the timber project in Porto Dias Agro-Extractive Reserve as a sustainably managed forest in January 2003.

No dramatic changes in environmental conditions have occurred in Bolivia, but the fact that the voices of women are being heard may hold some promise (Cronkleton 2005). The women showed much more interest in maintaining wildlife and other nontimber forest products than did the men, whose interests, like those of outsiders, have centered on timber production (Bolaños and Schmink 2005). The women's apparently more holistic view of the forest may prove instrumental in maintaining habitats if they continue to have opportunities to speak.

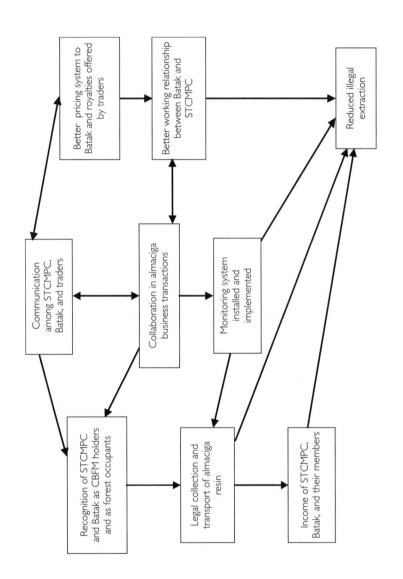

Figure 5-1. *ACM Processes Reduced Illegal Almaciga Extraction*

In Chimaliro, Malawi, Judith Kamoto conducted training in participatory forest resource assessment for community members. This training included information on how to select areas for thatch grass collection and how to monitor the availability of trees of different sizes (to reduce illegal harvesting, particularly for construction poles and firewood, which are in high demand, Kamoto 2003c).

Indigenous views of biodiversity were the focus in Bulungan, Indonesia, where a complementary project has been under way (cf. CIFOR 2002b). An ACM village, Setulang, was recently one of the 150 finalists (of 850 entrants) for the World Water Prize (Campbell 2003, 8), and the village won the prestigious Indonesian Kalpataru Award on World Environment Day, 5 July 2003.

Empowerment

Early on, we discussed the central role of empowerment in our own thinking. We had seen many forest communities who had little say in their own futures or in the management of their environment. We saw this as both inequitable and wasteful of valuable human and cultural resources. But we feared that an emphasis on empowerment might put off some of the stakeholders we hoped to enlist in our efforts; we also found the concept a bit fuzzy. Much of what has been discussed under collaboration, particularly vertical collaboration, pertains to empowerment: on most sites, local people have developed better facilitation and negotiation skills for conflict management, lobbying, and coalition building. Pokorny et al. (2005b) specified what they considered the empowering effects of the participatory methods we have used (see Table 5-1).

In Jambi, the Baru Pelepat election of a village council, *badan perwakilan desa*, after a collaborative effort to make the process more inclusive of women, settlers, and nonelites, was pronounced "the most democratic" in a local Jambi newspaper. Limberg et al. (2002), however, working at a meso scale, recount the substantial barriers to empowerment of communities in the newly decentralizing district of Malinau in East Kalimantan (Bulungan).

Formal training in legal literacy—in Bulungan, Indonesia; in Campo Ma'an, Cameroon; and in Madagascar—is intended to make citizens aware of their new, legal rights and strengthen their abilities to negotiate fairer deals with government and other stakeholders. Actual effects, though, are not yet known.

The increased transparency of benefit distribution in the Guarayo TCO, Bolivia, and the explicit involvement of women and youth in community discussions suggest increased empowerment for some. But the work of Bolaños (2003), who looked at gender issues in Santa María, suggests that real empowerment for women in forestry issues will remain difficult because of cultural norms about women's appropriate roles.

The newfound voices of women and lower castes in the Nepali ACM sites seem impressive. The results in Cameroon, where efforts have been focused at the mesolevel, are less encouraging. Tiani et al. (2005) document the adverse effects of conservation efforts on women in Campo Ma'an, and increasing marginalization of pygmies has been noted by several Cameroon researchers (e.g., Oyono 2003b).

Box 5-10. Outcomes That Strengthen Forest Management in the Philippines

1. Increased self-confidence and self-reliance of the People's Organization in their own skills and resources, and to start initiatives within their own capacity rather than waiting for external assistance.
2. More democratic decisionmaking and planning processes that engaged more organization members, different community groups, and other key stakeholders.
3. Increased joint action by members (in establishing nursery, herbal gardens, newsletter production, proposal making), and across different stakeholders (controlling illegal activities, resolving a boundary dispute, developing a local monitoring system).
4. Increased communication and feedback provided by the People's Organization to policymakers, identifying policies that hinder effective implementation of community-based forest management and recommending alternatives.
5. Increased level of trust between the organization and key government institutions, leading to increased transparency and increased resource sharing.
6. Increased participation and support from various government institutions to the organization in implementing community-based forest management.
7. More active People's Organization and community members, and more functional committees participating in forest resource management.
8. Management of conflicts through various conflict resolution mechanisms, rather than avoidance. A boundary dispute between the organization and the neighboring Batak community, for example, was resolved and led to even better collaboration in the commercialization of almaciga resins.
9. More structured and conscious reflections among members on their actions and experience for learning and improvement.
10. Improved skills of several organization members in proper documentation, expressing and communicating their views and opinions, effectively using different mechanisms for information sharing (billboards, bulletin boards, newsletters, different forums), networking, proposal writing, and managing small enterprises.
11. Increased income generated by the organization from a wider range of forest resources (from lumber to nontimber forest products); increased income of several women who engaged in handicraft making.
12. Monitoring of members' livelihood activities and the impact on the environment.

—Herlina Hartanto et al. (2002b, 60–61)

Hartanto, in the final report for the first phase of activities in the Philippines, listed the important outcomes (Box 5-10) for forest management. I have reordered them to show how many of them (1–10) in fact also relate to issues of empowerment. Although these outcomes are difficult or impossible to measure, they have been observed and reported or implied in most ACM sites.

In sum, from the more conventional perspective—looking at livelihoods, the environment, and empowerment—our successes are modest. This is consistent with our expectations, given the amount of time that the teams have been in the field.

The groundwork has, however, been laid for more impressive results, if the process is allowed to continue.

Conclusions

In Appendix 5-1, I have arranged the important impacts of our work in tabular form. Four broad categories emerge.

- Livelihoods has two components: knowledge about livelihoods, which is relatively easy and fast to achieve, and actual improvements in livelihoods, which are more difficult.
- Environmental impacts, similarly, comprise both knowledge about the environment and actual improvements.
- Empowerment has five components: strengthened social capital (1) within and (2) between groups and (3) improved negotiation skills, all of which can be seen as means to the ends of (4) improved equity, and (5) governance. These five components, as elements of capacity building, also have the potential for significant effects on livelihoods and environmental improvement.
- Social learning is, to some extent, a means to all the previous ends.

The final column in Appendix 5-1 reflects my combined, qualitative assessment of the overall success or impacts (high, medium, or low) obtained on each site. To make this assessment, I consulted the written materials from each site; I interviewed those working on the site in person, by e-mail, and/or by phone; and in many cases I visited the sites. Though qualitative, the assessment is based on careful study of all available data and ongoing involvement with site-based activities for two to four years, in my four-year role as program leader and now as a "copoint person" in a reorganized institution. I have been part of the ACM team since its inception in 1998.

To explain the impressive results obtained in an irrigation project in Gal Oya, Sri Lanka, Uphoff (1996, 347) concluded,

> Changes in behaviour were accomplished by a combination of normative and structural changes, not just by roles and sanctions or by values and norms. Indeed, it was difficult to determine how much influence could be assigned to one or the other, because the effectiveness of each was enhanced by the other in a positive-sum manner, as with the combination of administrative and participatory means used to improve [irrigation] channel maintenance … We had to resist the impulse to explain things in a reductionist way, attributing all improvement to just one set of factors or dividing up influence in a zero-sum way—50/50, 33/67, 90/10, 25/75.

If we look back on the impacts of ACM in 30 different contexts, it becomes clear that our experience was marked by the same positive-sum—not zero-sum—outcomes. An approach initially designed to enhance collaboration among local stakeholders has the additional effect of reducing conflict or strengthening community members' negotiating skills. A field trip designed to increase a user group's

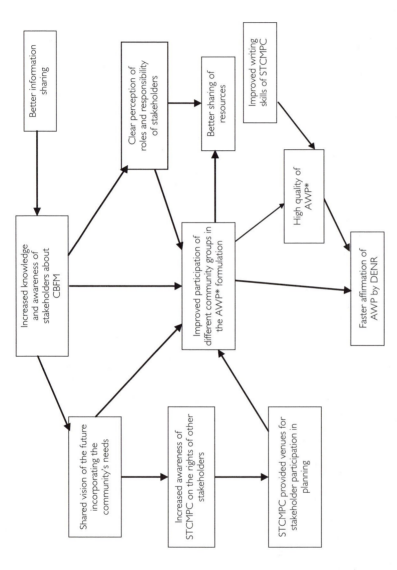

Figure 5-2. *Diagram Showing increased Participation of Stakeholders in Formulating an Annual Work Plan (Martanto et al. 2003, 94)*

*AWP means annual work plan.

knowledge winds up increasing social capital within the group. Increased confidence in one sphere easily expands to other spheres. And so on.

The impacts have been diverse, as have the plans and activities selected by the communities. And herein lies one of the strengths of this approach. Rather than planning for an average outcome from a particular set of conditions (even if such could be found), the ACM approach has sought individually tailored excellence. Following a thorough investigation of the context, each ACM facilitator worked with the local communities to identify the one or two actions that had the most potential, within that context at that time, to yield results. Identifying a locally recognized problem, weighing the likelihood of success, and assessing local people's willingness to follow through on planning, monitoring, and evaluation of their effort, the facilitators targeted their efforts and, together with the communities, quickly achieved locally appropriate solutions to small problems. Starting small allowed communities or user groups to have early success and thereby build self-confidence to take on bigger tasks.

Our successes are not revolutionary: similar goals have been accomplished in different places, at different times, many times over. As our definition of ACM stresses, we are proposing a value-adding approach that builds on the work of others. ACM emphasizes equity, social learning and monitoring, collective action, communication between villagers and policymakers, and human-environmental interdependence. And it is built on assumptions that are uncommon in many settings: that surprise is inevitable, that plans almost always need revision, that failures are an opportunity to learn, that forest people have the ability to act and affect their destinies.

I have argued in the past against getting marginalized groups to "join the mainstream," suggesting instead that each community or group should identify its own path and follow that for full flowering (Colfer 1979). The approach we have taken is based on the assumption that each group of people has its own set of skills and desires that must be taken into account in its own "development" direction. This model is also consistent with the idea of a panarchy, wherein the context of systems operating at different scales both inhibits and provides opportunities for unique paths. ACM has in most cases helped create conditions that are beneficial in two ways: the creativity, energy, and knowledge of local people are tapped for societal uses; and local people use their creativity, energy, and knowledge to approach the conditions they desire. In evaluating impacts, though, it is important to remember the lag times in systems. Small successes can, with time, lead to large impacts.

In the next chapters, I examine the seven dimensions identified at the beginning of our research as variables that might influence ACM success, showing, in the process, the inconclusiveness and potentially misleading nature of reductionist science in these complex and evolving systems.

Appendix 5-1. *Impacts*

	Livelihoods		Environment		Empowerment					Learning	
	K/L	L	A/E	E	SCW	SCB	SN	Equity	Gov	SL	Overall success
Startup: early 2000 or before											
Bulungan, Indonesia (1995)						×	×	×	×	×	H
Ranomafana, Madagascar (1997)		×									L
Baru Pelepat, Indonesia			×		×	×	×	×	×	×	H
Manakamana, Nepal	×		×		×	×	×	×	×	×	H
Andheribhajana, Nepal	×		×		×	×	×	×	×	×	H
Bamdibhirkhoria, Nepal	×		×		×	×	×	×	×	×	H
Deurali-Baghedanda, Nepal	×		×		×	×	×	×	×	×	H
San Rafael–Tanabag–Concepción, Palawan, Philippines	×			×	×	×	×	×	×	×	H
Startup: mid-2000											
Canta-Galo, in Gurupá, Brazil			×			×	×	×		×	M
Tailândia, Brazil						×	×		×		L
Muaná, Brazil	×		×		×	×	×	×	×	×	H
Porto Dias, Brazil	×		×		×	×	×	×	×	×	H
Ottotomo, Cameroon	×		×		×	×	×	×			H
Campo Ma'an, Cameroon			×			×	×	×		×	M
Dimako, Cameroon			×			×	×		×		M
Lomié, Cameroon			×								M
Akok, Cameroon	×		×			×					L
Mafungautsi, Zimbabwe	×		×			×	×	×	×	×	H

Startup: late 2000

Community	K/L	A/E	L	E	SCW	SN	SCB	Gov	Equity	SL
Rantau Layung/Rantau Buta, Indonesia	X		X	X	X	X	X		X	H
Basac, Lantapan, Philippines	X	X	X	X	X	X				M
Adwenaase, Ghana			X	X			X		X	L
Namtee, Ghana			X	X			X		X	L
Ntonya Hill, Malawi										L
Chimaliro, Malawi	X		X	X	X		X		X	M
Guarayos, Bolivia	X	X	X	X	X		X		X	H

Startup: 2001

Community	K/L	A/E	L	E	SCW	SN	SCB	Gov	Equity	SL
Arstanbap-Ata, Kyrgyzstan	X									L
Achy, Kyrgyzstan	X									L
Ortok, Kyrgyzstan	X		X	X	X					L
Uzgen, Kyrgyzstan	X		X	X	X					L

Notes: K/L = increased knowledge relating to livelihoods; L = improved livelihoods; A/E = increased awareness or concerns about the environment; E = improved environment; SCW = strengthened social capital or trust or cooperation within community; SCB = strengthened social capital or trust or cooperation between groups; SN = increased skills at negotiation or expression or representation; Gov = improved internal governance (transparency, checks and balances); Equity = greater equity; SL = improved adaptive ability; H = high, M = medium, L = low.

CHAPTER SIX

Devolution

THIS CHAPTER AND THOSE THAT FOLLOW provide the results of a careful and systematic study—reductionist in nature—of all 30 sites, to determine the conditions under which the adaptive collaborative management (ACM) approach is likely to succeed or fail. I show how misleading it can be simply to link condition x with impact y in complex adaptive systems, but my purpose is constructive, to encourage greater creativity in developing methods that are more appropriate to these complex adaptive systems.

Having identified seven dimensions that might be important—devolution, forest type, population pressure, management goals, human diversity, level of conflict, and social capital—the ACM team selected sites to represent a continuum along each of those dimensions. We then collected extensive case material from each site to be clear about the particular manifestation of each dimension there. For the analyses in Chapters 6 through 10, I have made an assessment of the impacts (success) on each site (presented in Chapter 5), and systematically compared the values for each dimension (usually high, medium, or low) and the level of impact (also high, medium, or low).

In this chapter, I deal with devolution.

Devolution status is part of a macrolevel, policy-related system of which our sites and our own activities are also parts. The policy context on each site varies and changes over time, at its own rate and in its own direction—evolving from the interactions among people and natural systems in each context, and affected by global interactions and systems.

Many authors have noted the relevance of policy in rural areas—often, in my view, attributing excessive importance and effectiveness to policy decrees and national-level regulations. However, there are some impacts (Thomas-Slayter et al. 1996); people living in forests and the forests themselves are certainly affected by decisions made at "higher" levels.

Box 6-1. Definitions of *Devolution* and *Decentralization*

By *devolution*, we mean a purposeful process designed to transfer varying amounts of formal authority from "higher," "centralized" levels of government to "lower," more "peripheral," local-level governmental and nongovernmental institutions. We are particularly interested in the bundle of policies that together serve this function.

Decentralization is a more specific term, referring to the ceding by a central government of formal powers to actors and institutions at lower levels in a political, administrative, and territorial hierarchy—typically lower levels of government (see CIFOR 2001; Resosudarmo and Dermawan 2002; Ribot 2002).

ACM sites were selected partially based on their devolution status (see Box 6-1). We selected sites that ranged from Nepal, with its longstanding and widely hailed devolution of authority over forest management to local user groups, to Zimbabwe, where local communities are granted no formal rights in forest management.

This chapter describes ACM sites where formal devolution is well established and fairly longstanding; sites where formal devolution is a fairly new reality, incompletely implemented; and sites where no formal devolution of forest management exists.[1] For the ACM categories of devolution, see Box 6-2. As with any attempt to extract commonalities and differences from diverse contexts, there are variations in degree, in the availability of information, and in analytical certainty. Additional information on all ACM sites is available in the Appendix of Cases.

Strong, Formal Devolution

Some have argued that real devolution of control would not be possible without support from the formal, legal system, and many have felt that such support was likely to strengthen local people's hand in their efforts to control local resources. Here, I focus primarily on national-level issues, as reflected in local realities. I begin by abstracting commonalities and divergences among the four countries (nine sites) with strong, formal devolution status: Nepal (four sites), Ghana (two sites), the Philippines (two sites), and Brazil (one site).

Box 6-2. Categorizing Sites by Devolution Status

Low: Local communities may have informal control of local resources but no formal rights to lands they have occupied and used (e.g., Indonesia, traditionally).

Medium: There are some efforts to clarify/establish some rights for local communities, but these remain fairly unclear and without much enforcement or "punch" (e.g., Cameroon) (conceivably, communities with traditional informal rights might be considered to fall into this category, if there is virtually no involvement of the central government in day-to-day life).

High: Local communities have both formal and informal control over local resources, and those rights are legally protected (e.g., Nepal).

—Carol J. Pierce Colfer (2000)

Commonalities among Sites with Strong, Formal Devolution

Comparative Clarity about Boundaries. In all of these sites, a fair amount of attention has been paid to the delineation of boundaries. In Nepal, forest user groups and district forestry officers must work together to clarify the boundaries before the designated forest is handed over to the community (New Era Team 2001a). A similar process took place in the Ghanaian sites. By law, the Philippines requires that boundaries be clarified when a community-based forest management agreement is reached. In both Palawan and Lantapan, boundaries have supposedly been delineated, though significant disagreements remain (Hartanto et al. 2000; Hartanto 2001; Lorenzo 2001c), partly related to governmental confusion among devolution programs like certificates of ancestral domain claims, community-based forest management agreements, and certificates of stewardship contracts (all of which grant communities rights to manage land, Lorenzo 2001c). Although in Acre, Brazil, a survey was done to clarify land rights in 1985, as in the Philippines, different preferences and interpretations of government agencies (IBAMA, *Instituto Brasileiro de Meio Ambiente e dos Recursos Naturais Renováveis,* the Brazilian Institute for the Environment and Renewable Resources; and INCRA, *Instituto Nacional de Colonização e Reforma Agrária,* the Federal Institute for Colonization and Agrarian Reform), rubber tappers, and colonists have complicated matters somewhat (Cunha dos Santos 2002b).

Ability to Obtain Support in Problem Solving and Dispute Resolution. In Nepal, there seem to be comparatively many avenues for seeking help. Using Bamdi-bhirkhoria as an example, Dangol et al. (2001b, 24) report the following:

> The District Soil Conservation Office took over the programs initiated by the Phewa Watershed Project in the Bamdibhirkhoria forest user group area and has been supporting and maintaining the fenced area, trail construction, nursery establishment, and riverbank conservation. At the beginning the forest user group had no office building or rooms. The conservation district provided an office building constructed by the Phewa Watershed Project ... The Women's Environment and Development Association encouraged the village women to organize in different women's groups. The women's groups constructed a trail and a temple, and launched the drinking water tap program with the help of the association. The conservation district provided a part of the nursery to the forest user group to run its own nursery. The Bamdibhirkhoria forest user group and women's association jointly donated for school building construction. The forest user group has carried out activities of thinning, pruning, and plantation in collaboration with the district forest office staff.

In the Philippines, people can resolve problems by a variety of external avenues, though the communities still feel the need for improvement. The government bureaucracy is complex but reasonably functional, and the many NGOs, donors, and special intersectoral associations have been helpful (e.g., the Palawan Council on Sustainable Development; Development Alternatives, Inc.; the Ulugan Bay Foundation; and others in Palawan, Lorenzo 2001c). And the populace is comparatively sophisticated about collective action. The ability of the people of Palawan to obtain significant rights to manage the island, at the provincial level, is unique in the Philippines and indicative of their ability to garner support.

In Ghana, community members expect more help from their forestry officials than they get, but they did, for instance, obtain help from the forestry officer in trying to remove an "encroacher" from their community forest (Diaw 2001). The fact that they approached the forestry officer at all indicates a greater likelihood of help than exists in many countries.

In Acre, Brazil, Porto Dias residents, working with the Center for Amazonian Workers (Centro dos Trabalhadores da Amazônia), an NGO, have been supported by IBAMA in their attempts to stop illegal logging by outsiders (ACM Acre Team 2002). They were able to obtain INCRA's help in setting up their agroextractive reserve. The NGO has helped them by serving a liaison function with outsiders and providing technical assistance on logging and community organizing.

Collective Action for Obtaining and Maintaining the Right to Manage Local Forests. In Nepal, many forest user groups are members of the Federation of Community Forest Users, a large network that engages in political action (e.g., Dangol et al. 2001b).[2] In the Philippines, members of the Talaandig Task Force and the Council of Elders (Lantapan) and the San Rafael–Tanabag–Concepción Multipurpose Cooperative from Palawan act together to accomplish community objectives. In Ghana, Adwenaase and Namtee communities, with help from an NGO, acted together to obtain permission to manage their forest collectively. In Acre, Brazil, the political actions of the National Council of Rubber Tappers (Conselho Nacional dos Seringueiros) brought about a new approach to land reform (extractive reserves); and the smaller groups that now function within Porto Dias, Brazil, are affected by this history. The rubber tappers themselves have a management association; and the neighboring colonists have associations of different kinds (e.g., Aspomacre is the community association in the nearby Nueva California settlement, Stone 2002)—all politically active.

Absence of Large Timber Companies Competing to Control Local Resources. Large-scale logging in Nepal is concentrated in the terai, in the south of Nepal; we were encouraged by the government, which was one of our collaborating partners, to undertake research in the middle hills area, where there were fewer conflicts (at that time) and no large commercial logging. In the Philippines, a nationwide logging ban was instituted in 1986 (though with loopholes). The Palawan site was logged extensively in the 1970s and 1980s, but logging was stopped by the Department of Environment and Natural Resources in the early 1990s, when the community-based forest management agreement was being negotiated with our partner organization (Lorenzo 2001c).[3] In Lantapan, no large-scale, long-term logging has taken place (Valmores 2002). In Ghana, there is a considerable amount of logging, but the companies are small (compared with Cameroon or Indonesia, for instance). Acre has not been logged, and in fact, the central purpose of the Center for Amazonian Workers project is to create a viable, small-scale logging industry (ACM Acre Team 2002).

Community Access to "Scientific" Forest Management Information. Such access is a mixed blessing, since it typically brings with it paternalistic attitudes. But in Nepal, by law and custom, the district forest officer is a resource for forest user groups. In the Philippines, Department of Environment and Natural Resources officials are consistent sources of information. Ghanaian community foresters have access to the Collaborative Forest Management Unit personnel. And the Center for Amazonian Workers in Acre has been consistently interacting with the community to provide this kind of information. In all these cases, the information is available from personnel with a "top-down" approach.

Comparatively Clearly Defined and Understood Rights and Responsibilities to Manage. The community forestry legislation in Nepal sets out the rights and responsibilities of various parties in forest management flexibly and fairly clearly. Although the system does not work perfectly (see New Era Team 2001b for a critique), there is scope for interaction between the local community and the district forest officer to work out appropriate local regulations.

The Philippines, similarly, has fairly clear regulations in certificates of ancestral domain claim and community-based forest management agreements, and these regulations are at least partially adhered to. One major problem arises from the diversity of legal management options, which creates uncertainty and disagreement at the local level about which instrument is most desirable or applicable. The ancestral domain claims are ideal for indigenous groups and give permanent rights; the community-based forest management agreements are applicable to anyone but provide limited and conditional tenure. Different government agencies prefer different instruments (including local-level differences of opinion, as between the Palawan Council for Sustainable Development and the local governmental units).

In Adwenaase and Namtee community forests, a management plan was prepared in 1995 with the assistance of the Ghanaian Forest Services Division as the basis for managing the forest. That forest is now operated in concert with the Collaborative Forest Management Unit (Diaw 2001). The collaboration between these two agencies required special attention (to overcome distrust) throughout much of the ACM work.

In Porto Dias, the rights and responsibilities of rubber tappers have been set out in IBAMA's regulations pertaining to agroextractive reserves. In this case, the complicating factors are the lessening interest of rubber tappers in tapping rubber, due to falling prices, and the influx of (and informal invasion by) colonists, encouraged by INCRA's establishment of neighboring settlement areas (Cunha dos Santos 2001a). As in the Philippines, each agency prefers its own instrument.

Pervasive Intrusion of the State into Local Decisionmaking. To some extent, this is the flip side of the second commonality above (the ability to garner support from the state and others). In Nepal, burdensome paperwork (e.g., constitution, operational plans) accompanies a formal community forest. This paperwork is central to the rules that determine forest management and use, yet district forest officers have in most cases been a primary force behind developing them (they also approve them) (Dangol et al. 2001a). See Malla (2001) for a more general discussion of the close links between state representatives and local elites in this process.

In the Philippines, the continuing and excessive intrusion of the state (specifically the Department of Environment and Natural Resources) in decisionmaking was one of the final conclusions about community-based forest management in a recent national policy workshop (Hartanto et al. 2002a).

The reluctance of Ghanaian Forestry Commission personnel to relinquish control has been pointed out by numerous researchers (e.g., Blay et al. 2002; Ribot 2001). Blay points out the top-down approach of the Forest Services Division in its dealings with communities, as well as the National Forest Directorate's unwillingness to devolve powers to the district-level local administration.

In remote Porto Dias, the establishment of an agroextractive reserve (along with the nearby settlement projects) has increased state presence, through IBAMA, INCRA, and indirectly, the Center for Amazonian Workers—complicated by an inherent conflict between the responsibilities of IBAMA and INCRA. As Cunha dos Santos (2001a, 7) notes, "IBAMA, as a federal agency with the power to inspect and control the environment, could not administer land resettlement projects. On the other side, INCRA, a federal agency with the responsibility for agrarian reform, could not form reserves." The personnel involved in the Center for Amazonian Workers timber management project, undertaken with the blessings of the state, have also had a strong voice in local decisionmaking about natural resources.

Differences among Sites with Strong, Formal Devolution

Forest Quality. Nepal was famous in the 1980s for its problems with deforestation (see Hobley 1996b) and is frequently cited, with India, as an example of the devolution of forest management to local communities only after serious degradation has occurred. Although the estimates of widespread deforestation in the area have been widely challenged, the fact remains that compared with the other sites in this category, Nepal's forests have been under considerable pressure. The Philippine and Ghanaian sites are characterized by forests of reasonable quality (Blay 2002b), though the countries at large have experienced serious deforestation.

The excellent quality of Acre's forests, combined with the significant level of devolution of authority to local communities, represents a dramatic exception to the more common situation in which forests are devolved to local people only after the forests have been degraded (see Gomes and Sassagawa 2002 for a recent study of worrying levels of deforestation in the neighboring Chico Mendes Extractive Reserve). The great distance from Porto Dias to markets may function like degradation in making the forest unsuitable for quick profits. Political action by the rubber tappers is also undoubtedly a significant factor.

Conservation-Related Powers. The island of Palawan is unusual in the Philippines in its formal, longstanding, and widespread attention to conservation issues. Guerrero and Pinto (2001, 424) characterize the island as "the country's largest national protected area"—a reference to 1992 national legislation that devolves unusual authority over provincial natural resources to the Palawan Sustainable Development Council. This legislation stipulates a comprehensive environmental protection and sustainable development framework for the island. Basac, in Lantapan, is a formal part of the Mount Kitanglad Natural Park.

Extent of Illegal Logging. In the Nepal and Palawan sites, illegal logging is reported to be minimal (Lorenzo 2001c), though in Nepal generally, during the research period, anecdotal evidence pointed to a significant increase in illegal logging related to the social unrest generated by the clashes between the Maoist rebels and the government and the resulting state of emergency. Although illegal logging is not considered a serious problem, Valmores (2001) names apprehension of timber thieves as one important task of the Talaandig Tribal Task Force, in the Lantapan site (Philippines). Illegal logging is minimal in Porto Dias, Brazil, where IBAMA's presence has restricted the small amount even further.

In Ghana, in contrast, chainsaw operators and timber contractors are major stakeholders (Diaw et al. 2002). Blay et al. (2002, 7–8) say, "a World Bank report indicates that a high percentage of lumber found on the domestic market and for export in the subregion is supplied from chainsawn lumber and that it accounts for about 50 percent of the additional 2.7 million m^3 excess timber outside the permitted annual cut of 1.0 [million] m^3."

Nature of Indigenous Forest Management Knowledge. In Nepal, with its subtropical, montane forests and comparatively high population density, local communities place greater emphasis on agriculture and trade than on managing their forests. In the humid, tropical rainforest settings of the Philippines, indigenous groups (the Batak and Tagbanua in Palawan, and the Talaandig in Lantapan) have longstanding, intimate knowledge of and uses for the forest. In-migrating groups, which are dominant in the Palawan site, have much less forest-dependent ways of life than their indigenous neighbors. Similarly, in Acre, the rubber tappers, though not technically indigenous, have developed a forest-dependent way of life that differs significantly from that of the more agriculturally oriented, recent immigrants in neighboring settlements.

Summary for Sites with Strong, Formal Devolution

We see no identifiable pattern of ACM success or failure. Nepal and the Philippines have had vibrant ACM programs, fully funded and well accepted by national-level decisionmakers. The program in Acre, with minimal amounts of funding and enormous geographical distance from Brazil's capital, has also had significant success locally, though probably very little nationally. ACM successes in Ghana do not appear to have been very impressive: Carter (2000, 28) has remarked of Cameroon, Ghana, India, and Nepal that "good policy does not necessarily equate with good practice, and vice versa."

Looking to the future, Gilmour (2002) has referred to second-generation issues—emerging subjects for debate—in the Nepali context:

- trade-offs between poverty alleviation and conservation interests;
- the impact of commercialization of community forest products on reducing poverty and conserving biodiversity;
- the transaction costs of participation by poorer households;
- equity issues related to membership of a forest user group; and
- financial sustainability of the community forestry program in the absence of donor support.

Similar concerns were voiced by K.B. Shreshta, of Nepal's Ministry of Forestry, in an ACM meeting in October 2001. These would seem to be pertinent in all sites with strong, formal devolution.

Weak, Formal Devolution

In this intermediate category are 13 sites where some formal devolution efforts are under way, often with legal instruments requiring devolution of authority to lower levels of government or other groups: Cameroon (5 sites), Malawi (2 sites), Madagascar (1 site), Bolivia (1 site), and Kyrgyzstan (4 sites). The short time since these legal instruments took effect, combined with bureaucratic inertia and strong vested interests, has typically meant that real devolution has hardly begun.

Commonalities among Sites with Weak, Formal Devolution

Uncertainty about Community Rights in Local Forests. Although in all these countries, legal instruments have been created to strengthen local communities' involvement in formal forest management (see Box 6-3 for an example), ongoing disputes about area, authority, uses, and benefit distribution are more serious than in the countries with strong, formal devolution status.

Confusion about How to Navigate the Bureaucracy. The process of gaining the formal access allowed by the new laws is typically labyrinthine. In Cameroon, for instance, the bureaucratic requirements for formalizing the necessary agreements are widely characterized as complex (Vabi et al. 2000; Fomété and Vermaat 2001; Klein et al. 2001). In Madagascar, one goal of several internationally funded projects is to communicate the new laws to local communities and facilitate legal implementation (Carter 2000; Buck 2002). Similar training is currently under way in the Campo Ma'an site in Cameroon. In Bolivia, Contreras-Hermosilla and Ríos (2002) report, the mandated forest management plan requires the assistance of a

---

Box 6-3. Definition of a Community Forest in Cameroon

A *community forest* is defined as "a forest forming part of the non-permanent forest estate, which is covered by a management agreement between a village community and the Forests Administration. The management of such a forest is the responsibility of the village community concerned, with the help or technical assistance of the Forests Administration" (translation of Article 3(11) of the decree). The communities enjoy the following benefits: forest products of all kinds resulting from the management of community forests shall belong solely to the village communities concerned (Section 37(3) of the law). Hence all products, timber and nontimber resources, animal and plant species, fisheries products, and special products, with the exception of those banned by the law, are considered to be the property of the community concerned.[4]

—Patrice Etoungou (2002, 11)

formally trained forester—expertise most tropical forest villagers do not have. Foresters in Kyrgyzstan are routinely suspected of favoritism in the allocation of forest leases, and the process is anything but transparent (Fisher 1999).

Reluctance of Officials to Relinquish Rights and Sources of Income. Real devolution inevitably involves a realignment of relationships among power holders. Those at the center are expected to hand authority to those on the periphery, along with access to benefits. In some well-forested areas (Bolivia, Cameroon), bureaucrats are reluctant to relinquish access to profitable relationships with the timber industry. Even in less well endowed countries (Kyrgyzstan, Madagascar, Malawi), perks have accompanied their authority. Some bureaucrats genuinely fear what local people might do to the resource or believe that "scientific" forestry is better.

Efforts by Local Elites to Gain Unfair Advantages. In Cameroon, urban elites have returned to their ancestral villages and taken advantage of less sophisticated neighbors. In Bolivia, some leaders have entered into clandestine agreements with commercial loggers (Contreras-Hermosilla and Ríos 2002). In Chimaliro, Malawi, some management committee members share the benefits (from fines and confiscation of illegally harvested forest products) with village heads and forestry personnel, shorting the intended recipients—the local communities (Kamoto 2003a). In Ntonya Hill, Malawi, and Guarayo, Bolivia, elites became intermediaries with outsiders who by law had to be involved.

Differences among Sites with Weak, Formal Devolution

Importance of Indigenous Forest Management Knowledge. In Cameroon and Bolivia, indigenous peoples have a high level of knowledge about traditional forest management. Local people in Malawi, with its high population density, place more emphasis on agriculture, as do the Beteseleo in Madagascar. But the Tanala, also of Madagascar, do have significant forest-based knowledge. Kyrgyzstan has a strong nomadic tradition, despite long years of settlement, and local people have comparatively little cultural connection with forests (Carter et al. 2002).

Forest Type. There is more variability in forest type within this devolution category than the previous one. There are humid tropical forests (Bolivia, Cameroon, Madagascar), dry tropical forests (Malawi), and temperate-zone forests (Kyrgyzstan), with markedly different value. Bolivia and Cameroon have high-value commercial species in sufficient quantity to be profitable. Malawi and Madagascar have few forest resources of commercial value. Madagascar is famous for its biodiversity value. Kyrgyzstan's forests are also touted for biodiversity value, besides the walnut and fruit harvests and valuable walnut burls they provide (Blaser et al. 1998).

Presence of Powerful Outsiders. Cameroon and Bolivia attract timber companies and conservation projects that compete for local resources. Madagascar has a vibrant conservation community, with similarly adverse implications for local communities (Ferraro 2002). No such dominant, nongovernmental stakeholders appear as obvious in Kyrgyzstan or Malawi.

Extent of Illegal Logging. Cameroon and Bolivia suffer significant illegal logging. In Malawi and Madagascar, illegal harvest from forests is smaller in scale. In Malawi, it involves less valuable timber (poles); in Madagascar, the scale is small because of access problems. In Kyrgyzstan, the problem is illegal harvesting of walnuts and walnut burls, rather than logging in the conventional sense.

Summary for Sites with Weak, Formal Devolution

Among the 13 sites with weak, formal devolution, we find great variation in success at implementing ACM. The ACM process in Bolivia appears to have proceeded smoothly and fairly completely. In Cameroon, there has been great variation from site to site. In Ottotomo, for instance, the process appears to have worked well. In Campo Ma'an, the process was stimulating, lively, and forward looking, if slow and marked by a high degree of conflict among stakeholders (including CIFOR). Efforts in Lomié and Dimako have focused more on policy-level issues than originally envisioned, with much clearer understanding of the constraints but uncertain practical outcomes to date. Results in Akok remain unclear. A sixth site, Makak, had to be abandoned altogether. In Malawi, the Chimaliro site has made progress; the Ntonya Hill site, apparently, has made little. In Kyrgyzstan, though steps to understand the context were undertaken, the participatory action research process was never initiated (because of decisions by a doctoral committee member and other factors, described in Chapter 11).

Gilmour (2002) has identified in Asian countries eight common features in the early stages of collaborative forest management that appear to be relevant elsewhere as well:

1. working out the procedural systems for implementation;
2. determining the authority and responsibility for the key stakeholders (generally the forest-dependent community and the forest department);
3. considering benefit-sharing arrangements;
4. designing and carrying out training for community groups and forest department staff;
5. moving from small pilot projects to a large-scale national program;
6. determining the policy and legislative changes needed to support community forest management;
7. creating incentives for collective action; and
8. compatibility between self-governing forest user groups and the interests of and regulations set by local and national government agencies.

No Formal Devolution

In Indonesia, Zimbabwe, and Pará, Brazil, national governments recognize virtually no rights to forest management on the part of their citizens.[5] In Indonesia, we have three sites; in Zimbabwe, one; and in Pará, three.

Commonalities and Differences in Sites with No Formal Devolution

Community Claims on Local Resources. In Indonesia and Pará, people are anxious about their continuing access to the resources on which they depend. The Indonesian populace has been subjected to capricious, top-down policies that have restricted their access to resources since colonial times. Although some believe that having policymakers closer to home may signal an improvement, early indications suggest that leaders at the district level are eager to continue the corrupt practices of the Soeharto regime on a smaller scale. Forest dwellers are reacting to this uncertainty about the law and about their future by harvesting timber to secure resources in the short term—activity that will seriously degrade local forests.

In Pará, multiple layers of ownership (Schmink and Wood 1992; Pokorny et al. 1999) and a tradition of violence have rendered local people very uncertain about their futures in any given place (Hall 1997; Porro 2002). To harvest timber, for example, they must have a forest management plan approved by IBAMA. And IBAMA's inability to regulate timber companies results in frequent abuses, both of the law and of people's rights to their own resources.

Clarity about Rights and Responsibilities. In Indonesia, rights and responsibilities have been unclear for decades. The national government claims most of the nation's land as part of its forest estate, while local communities have continued to manage these same lands (which they have traditionally called their own) on a day-to-day basis. The allocation by the state of specific areas of this land to timber and plantation companies, transmigration projects, and conservation functions has created a conceptual morass of land tenure, use, and management rights. Add to this history the current uncertainty that characterizes Indonesian decentralization policy, and one gets a very confusing concoction indeed.

Pará is representative of much of the Amazon basin in its comparable lack of clarity about rights and responsibilities. As in Indonesia, there are many stakeholders—Indians, long-term communities of people living along the rivers *(ribeirinhos)*, settlers *(colonos)*, timber companies, ranchers, fishers, and various government agencies (Tchikangwa et al. 2001; Porro et al. 2001)—with differing interests and preferences. As a "frontier" area, with poor transportation, the arm of the government is differentially felt. The government's inability to oversee its own regulations results in local-level problem solving, described by Porro (2001, 319): "… colonists were left at the margin of the law, struggling with rights obtained or taken within the context of power relations between them and dominant sectors of the society, such as merchants, loggers, and landlords."

Rights and responsibilities in Zimbabwe are legally clear: forest reserves have been simply off limits, legally. However, the situation in Mafungautsi is changing, as both local communities and Forestry Commission staff attempt to work out the meaning of comanagement.

Access to "Scientific" Information about Forest Management. Partly because the sites are remote, community members in Indonesia and Pará rarely have ready access to reliable outside information. Extension services are minimal, interaction with government officials is infrequent and not necessarily desired, and higher education is rare and expensive. In Zimbabwe, access is easier, but the communication barriers between the communities and the government appear to function like geographic isolation.

Nature of Indigenous Forest Management Knowledge. All of the Indonesian communities where we conducted our research had strong traditions of forest management, focusing on swidden agriculture and its complementary economic activities of hunting, fishing, gathering, and periodic wage labor. Immigrants were a minority on all three sites but manifested less indigenous forest management knowledge and were more oriented to agriculture.

The same split characterizes the Brazil sites. The Tailândia site, inhabited by a constantly shifting population of immigrants, exhibited very little in the way of indigenous forest knowledge. The other two Pará sites, Muaná and Gurupá, are occupied by traditionally forest-dependent residents and exhibit marked similarities to the Indonesian sites in management expertise on swidden agriculture, hunting, fishing, and gathering.

In Zimbabwe, the original inhabitants (the Shangwe) were much more knowledgeable about forest use and management than the immigrants (Shona and Ndebele). Indigenous forest knowledge was not valued here but seen rather as an indicator of primitiveness.

Commercial Value of Forests. Forests in Indonesia and Pará, though being harvested and degraded at worrying rates, remain commercially valuable. Zimbabwe forests are not.

Summary for Sites with No Formal Devolution

We see considerable success in seven sites with no devolution of authority. Activities on all three Indonesian sites and in Zimbabwe were consistently successful and enthusiastically undertaken, with considerable emphasis on improving relations between the communities and the governments in their areas, as well as other, more site-specific emphases. The Pará sites received minimal financial support, but the team was able to leverage collaboration by students and NGOs and implemented significant components of the ACM process, with good results in collective action and links with external markets.

Conclusions

Although there is little doubt that the clearer management roles provided by strong devolution policies *can* provide a stronger basis for agreements and trust, and thus contribute to improved collaboration, the findings from this data set suggest that devolution is but one of many factors influencing the success of adaptive collaborative management.

My reductionist examination of devolution status and ACM impacts suggests that the overall ACM success rate is lower for the sites marked by weak, formal devolution than for those marked by either strong, formal devolution or by no formal devolution (see Table 6-1). The weak, formal devolution category had only

Table 6-1. *Devolution Status and ACM Impacts*

Devolution status	Overall success		
	High	Medium	Low
Strong, formal devolution	6	1	2
Weak, formal devolution	2	4	7
No formal devolution	5	1	1

two sites with high impact, four sites with medium impact, and seven sites with low impact (see Appendix 1-1). In contrast, the strong, formal devolution category had six high-impact sites; one, medium; and two, low. Sites characterized by no formal devolution included five with high impact; one, medium; and one, low. Contrary to our initial hypotheses, then, local people are able to make ACM processes work just as effectively when there is no formal devolution of authority over forest resources as when there is strong, formal devolution. Our results are not statistically significant at the 1 percent level.[6]

I do not, however, find the figures or this conclusion compelling, from a causation perspective. First, the impacts are so diverse as to make comparisons difficult. Second, there is considerable variation in success within some of the multisite countries (Brazil, Cameroon, Malawi, Philippines)—characterized, of course, by the same devolution context. Third, three of the countries (Indonesia, Kyrgyzstan, and Nepal) each include four sites, meaning that a peculiarity (such as pressure from a dissertation adviser for the ACM facilitator to abandon participatory action research) can have a greater weight than might be warranted, when compared with single-site countries (Bolivia, Madagascar, Zimbabwe) or two-site countries (Ghana, Malawi, Philippines). And fourth, we have found successful ACM efforts at all levels of formal devolution.

My own view—and argument—is that the problem lies not with the data or the analysis, but with the question—phrased as it is within a unilinear, causal framework that seeks to link the devolution variable with the outcome, ACM's overall success. Without in-depth knowledge of the sites, the typical researcher operating within a reductionist framework is often unable to recognize the absurdity of some of the conclusions of such analyses.

I have systematically made these simple analyses (condition *x* correlated—or not—with condition *y*) for each dimension we identified as potentially important, as a demonstration of how inadequate this kind of research paradigm is for complex adaptive systems. I do this not from any desire to be overly critical or to deny the utility of reductionist approaches in more stable and simpler circumstances, but rather to make a strong case for changing our collective approaches to addressing development and conservation issues in complex, dynamic systems like forests and human societies. We need to look critically at our approaches to learn from the experience, then fashion better approaches for the future.

CHAPTER SEVEN

Forest Type and Population Pressure

*I*N THIS CHAPTER, I DESCRIBE THE kinds of forest contexts where we have worked and their implications for adaptive collaborative management (ACM).

Forest Type

Broadly speaking, forests can differ by type (humid or dry tropical, temperate, or subtropical montane), by landscape (a forest-rich or forest-poor area, cf. Salim et al. 2001; Porro et al. 2001; Tchikangwa et al. 2001),[1] and by commercial value. These characterizations are shown in Table 7-1. In the following text, the sites are divided into the four main forest types represented in our data set: humid tropical, dry tropical, temperate, and subtropical montane.

Humid Tropical Forests

Nearly two-thirds of our sites were humid tropical forests: Bolivia (Guarayos), Brazil (Gurupá, Muaná, Porto Dias, and Tailândia), Cameroon (Akok, Campo Ma'an, Dimako, Lomié, and Ottotomo), Ghana (Adwenaase and Namtee), Indonesia (Baru Pelepat, Bulungan, Rantau Buta, and Rantau Layung), and the Philippines (Lantapan and Palawan).

These 18 sites share four forest-related similarities:

1. significant pressure to use forestlands for other purposes (e.g., subsistence agriculture, cash cropping, timber extraction, plantation agriculture, mining) and related, widespread concern about forests;
2. impressive biodiversity—a defining feature of tropical rainforests;
3. abundance of rivers; and

4. uniformly hot, wet conditions, with rainfall (where measured) ranging from 1250 to 4000 mm/year.

We also noted six kinds of differences:

1. 10 forest–rich sites, 3 forest-poor, and 5 intermediate;
2. a correlation between accessibility and poor forest conditions for forest-poor sites (forest-rich sites are characterized by the whole range of accessibility, and access tends to improve as one moves from forest-rich to forest-poor contexts);

Table 7-1. *ACM Sites Classified by Forest Type*

Forest	Natural forests Forest-rich	Intermediate	Forest-poor	Plantation	Commercial value	Accessibility[a]
Humid tropical						
Bulungan, Indonesia	X				H	L
Campo Ma'an, Cameroon	X				H	H
Canta-Galo in Gurupá, Brazil	X				H	L
Dimako, Cameroon	X				H	H
Guarayos, Bolivia	X				H	M
Lomié, Cameroon	X				H	H
Ottotomo, Cameroon	X				H	H
Porto Dias, Brazil	X				H	L
Rantau Buta, Indonesia	X				H	L
Rantau Layung, Indonesia	X				H	L
Adwenaase, Ghana		X			M	H
Akok, Cameroon		X			M	M
Baru Pelepat, Indonesia		X			H	M
Jaratuba/Recreio/Nova Jerusalem in Muaná, Brazil		X			M	L
San Rafael–Tanabag–Concepción in Palawan, Philippines		X			M	H
Basac, Philippines			X		L	H
Namtee, Ghana			X		M	H
São João Batista/Nova Jericó in Tailândia, Brazil			X		M	M
Dry tropical						
Mafungautsi, Zimbabwe		X			L	H
Chimaliro, Malawi		X			L	H
Ntonya Hill, Malawi				X	H	H
Temperate						
Ortok, Kyrgyzstan	X				L	M
Arstanbap-Ata, Kyrgyzstan		X			L	H
Achy, Kyrgyzstan		X			L	H
Uzgen, Kyrgyzstan		X			L	H
Subtropical montane						
Bamdibhirkhoria, Nepal			X		L	M
Deurali-Baghedanda, Nepal			X		L	M
Manakamana, Nepal			X		L	L
Andheribhajana, Nepal			X		L	L

[a]This estimate is based primarily on access by road. L = no road, M = bad road, H = good road.

3. El Niño–related fire damage on several sites (Jambi and Pasir in Indonesia; Lantapan and Palawan in the Philippines);
4. widespread recognition of valuable biodiversity in comparatively intact forests (though in all cases, it takes a backseat to livelihood concerns, for local people);
5. Although shifting cultivation, nontimber forest product collection and hunting are important for local people's livelihoods in all humid tropical forests, hunting has greater significance as an issue in the South American and African sites than in Asia, and in forest-rich over forest-poor sites. In the Philippine sites, shifting cultivation is important primarily for indigenous groups, not so for migrant populations.
6. In several sites (Bulungan and Jambi in Indonesia and Campo Ma'an, Dimako and Lomié in Cameroon), there are also hunter-gatherer groups, who are even more dependent on the forest than the dominant population.

Dry Tropical Forests

Three sites, Chimaliro and Ntonya Hill in Malawi, and Mafungautsi in Zimbabwe, represent this forest type and were originally miombo woodlands; they have the following characteristics:

1. more seasonal temperature variation than in the humid forests (ranging as low as 5°C in Mafungautsi);
2. varying rainfall, from a minimum of 600 mm/year (Mafungautsi) to a maximum of 1200 mm/year (Chimaliro);
3. fire as a chronic problem;
4. apparently greater importance of soil conditions to human uses of the forest;
5. pressures on the forest primarily from human subsistence needs and population growth; and
6. significantly greater importance of firewood than in the humid tropics.

Temperate Forests

Four sites are temperate forests, all in Kyrgyzstan: Achy, Arstanbap-Ata, Ortok, and Uzgen. The differences between these sites and the tropical forests are striking, but a few also represent commonalities among this subset:

1. local communities' lack of strong traditional, cultural links to the forest (the people of Central Asia come from the steppes);
2. strong government presence under Soviet rule, which ensured day-to-day forest management through official channels (planning, implementation, sanctions, etc.) and shaped the relationship between the people and forests; and
3. lifestyles determined by the marked seasonality of temperate zones, which necessitates fuel for winter and encourages animal husbandry with some degree of transhumance related to pasture availability.

Looking at forest-related differences among the sites, I note the following:

1. varying abundance of the valued walnut trees: 47 percent of the forested area in Ortok, 40 percent in Arstanbap-Ata, 25 percent in Uzgen, and 11 percent in

Achy (Goslesagentsvo and LES-IC 1997 and State Forest Service 2002, quoted in Schmidt 2003b);
2. varying abundance of apple trees: 23 percent of the forested area in Ortok, 21 percent in Achy, 16 percent in Arstanbap-Ata, and 8 percent in Uzgen;
3. Achy, Arstanbap-Ata, and Uzgen forests have crown cover of 40–50 percent. Ortok's forest is better, with 8 percent of its forest characterized by >80 percent crown cover, 64 percent with a 50–80 percent crown cover, and the remainder (28 percent), 10–50 percent; and
4. a longer history of collaborative forest management in Ortok and Uzgen, through Swiss-supported experiments in 49-year forest leasing.

Subtropical Montane Forests

The final four sites, in Nepal (Andheribhajana, Bamdibhirkhoria, Deurali-Baghedanda, and Manakamana) are in subtropical montane forests. Beginning in the late 1970s, the government began to increase community control of forests, and an improving trend has been noted by many observers (e.g., Gilmour and Fisher 1991; Malla 2000, 2001), including our own teams. Commonalities among the sites include the following:

1. formal forest user groups (established between 1993 and 1995) that have the right to manage their forests, with considerable involvement from the Ministry of Forestry;
2. high elevation compared with tropical ACM sites (though intermediate, by Nepali standards);
3. seasonal variations in temperature that make fuelwood important for heating; and
4. small forest areas (ranging from 48 to 181 ha).

The most striking biophysical difference is that the two eastern sites (in Sangkuwasabha District) are considerably more remote than those in the west (Kaski District).

Implications of Forest Type

We see little evidence that forest type has any effect on the success of adaptive collaborative management. Turning to the quantitative data in Table 7-2, we find that the humid tropical forests included eight sites with high levels of success and five sites with low levels of success; five sites were intermediate. The three dry tropical forests were split among the levels. All temperate-zone forest sites, however, have so far attained only a low level of success, and the subtropical montane forests experienced uniformly high levels of success.

Such a finding highlights an important danger of reductionist science. It is highly improbable that forest type, per se, has had any significant effect on the actual ACM process. Although all temperate-zone sites have "failed" and all subtropical montane forests have "succeeded," it is clearly erroneous to attribute the difference in outcome to the forest type—reminiscent of the high correlation between number of skillets and the incidence of wife beating in the eastern United States: a spurious statistical relationship.

Table 7-2. *Forest Type and ACM Impacts*

Forest type	Overall success		
	High	Medium	Low
Humid tropical	8	5	5
Dry tropical	1	1	1
Temperate	0	0	4
Subtropical montane	4	0	0

The people in the Nepal sites (the subtropical montane forests) responded enthusiastically to the ACM teams' efforts to catalyze collaboration and social learning, with significant impacts on local people and potential impacts on their forests. ACM activities were stimulated by a large, committed, well-funded group of researchers with abundant access to pertinent expertise and interaction with the broader ACM team. In Kyrgyzstan (the temperate forests), the ACM process came to a grinding halt just as participatory action research, the heart and soul of the process, was to begin. Although research continues there, maintaining some of the goals that ACM and the collaborative forest management project share, local communities have not mobilized to address their concerns.

Tropical forests, both humid and dry, include both successful and apparently unsuccessful ACM experiences.[2] Even variations in accessibility—which would make it difficult for the facilitators to spend time in the communities, or even, in western Brazil, for community members to meet each other—seems to have had no effect. In Porto Dias, Brazil, for instance, Magna Cunha dos Santos had to walk long distances along muddy paths between isolated rubber tapper households, in a world where strangers were not always welcome to spend the night. In eastern Nepal, Kalpana Sharma, another young woman, had to walk six hours to reach her site. Such circumstances might reduce a facilitator's ability to serve as a link between communities and outsiders and to facilitate learning processes among community members, yet this was not the case.

Population Pressure

Growing populations can put added pressure on the forest and may affect people's interest in working and learning together. Information about population is not always readily available in remote, forested sites, however, and so we have relied on estimates. The resulting population density measures differ depending on the scale at which one measures the area. On several sites (e.g., Kyrgyzstan and Nepal), densities were computed based on the size of the official forest; in others (e.g., Rantau Layung in Indonesia), they were computed based on the formally recognized village territory; in still others (e.g., Campo Ma'an in Cameroon), densities were computed in two ways—taking the entire management unit into consideration, in the one case, and relegating the people to the area officially granted to them by the government, in the other. All of these approaches are valid, but the variation complicates cross-site comparisons. For such reasons, we supply various indicators of population pressure. Table 7-3 summarizes the demographic data available for each site.

Table 7-3. ACM Sites Categorized by Population Density[a]

ACM sites	Persons per km²	Person per ha	Total forest area (ha)	Total population	Families	Average family size	Growth rate	Sex ratio
High population density								
Adwenaase, Ghana		42.90	215	9,228				
Bamdibhirkhoria, Nepal		15.04	48	722	134	4.90		1.14
Andheribhajana, Nepal		9.03	113	1,020	185			1.02
Manakamana, Nepal		6.70	132	879	164			
Deurali-Baghedanda, Nepal		4.77	181	864	142			
Arstanbap-Ata, Kyrgyzstan		1.67	7,489	11,724			3.30	1.00
Achy, Kyrgyzstan		1.67	6,996	12,189			3.30	1.00
Uzgen, Kyrgyzstan		1.00	21,777	22,000	2,300		declining	1.00
Ortok, Kyrgyzstan		0.56	10,282	5,621	150		2.50	1.00
Chimaliro, Malawi	74		210					
Namtee, Ghana	62		190	11,500				
Mafungautsi, Zimbabwe	35		8,210					
Ndarire, Zimbabwe				1,000				
Batanai, Zimbabwe				1,800				
Gababe, Zimbabwe				2,000				
Ranomafana, Madagascar			41,600	26,000			>2.25	
Ntonya Hill, Malawi			370					
Intermediate population density								
Basac, Philippines	1.43	2,800	4,000	700		4.18		
San Rafael–Tanabag–Concepción, Philippines	0.58	5,006	2,900					
Campo Ma'an, Cameroon	29		771,000	60,000			1.7–5.3	
Akok, Cameroon	10							

Location	Village						
Muaná, Pará, Brazil	Jaratuba	5		250		5.00	
Muaná, Pará, Brazil	Recreio	5		500		5.00	
Muaná, Pará, Brazil	Nova Jerusalem	5		150		5.00	
Tailândia, Pará, Brazil	São João Batista	3		100		5.00	
Tailândia, Pará, Brazil	Nova Jericó	3		150		5.00	
Gurupá, Pará, Brazil	Canta-Galo	2		250		5.00	
	Low population density						
Guarayo, Bolivia			2,205,370				
Guarayo, Bolivia	Cururú	7.76	26,421	205	24	8.50	1.15
Dimako, Cameroon		1.08	16,240	15,000			1.28
Guarayo, Bolivia	Selvatierra	0.01	30,000	333	59	5.60	
Ottotomo, Cameroon		2.5	2,950	1,239			
Porto Dias, Acre, Brazil		1.2	22,345		100		0.49
Lumut Mountain, Indonesia	Rantau Layung	1.1	18,913	213	52		1.12
Lumut Mountain, Indonesia	Rantau Buta	0.5	16,546	82	20		1.12
Lumut Mountain, Indonesia			35,350				
Bulungan, Indonesia		0.76	321,000	5,000		37,747	
Lomié, Cameroon	Ngola			715	150		>2%
Lomié, Cameroon	Kongo			378			
Guarayo, Bolivia	Santa Maria			365	93	3.90	1.25
Dimako, Cameroon	Petit Pol			650			0.82
Dimako, Cameroon	Mayos (Baka)			450			0.92
Baru Pelepat, Indonesia				557		5.00	0.94

[a]The figures in this table are not always truly comparable; please see accompanying text.

Fourteen sites have comparatively high population pressure: Ranomafana in Madagascar, Namtee and Adwenaase Forest Reserves in Ghana, Chimaliro Reserve and Ntonya Hill Plantation in Malawi, Mafungautsi Forest Reserve in Zimbabwe, the four Nepali community forests (Bamdibhirkhoria and Deurali-Baghedanda in the Kaski District and Manakamana and Andheribhajana in the Sangkhuwasaba District); and the four Kyrgyz state forest farms, or *leshozes* (Achy, Arstanbap-Ata, Ortok, and Uzgen).

Seven sites were characterized by an intermediate level of population pressure: two Cameroon sites (Akok and Campo Ma'an National Park); the two Philippine sites (San Rafael–Tanabag–Concepción, Palawan, and Basac); and the Brazilian sites in Pará (the communities of São João Batista and Nova Jericó in Tailândia; Jaratuba, Recreio, and Nova Jerusalém in Muaná; and Canta-Galo in Gurupá).

Sites characterized by very low population pressure number nine: three Cameroon sites (Ottotomo, Lomié, and Dimako); all the Indonesian sites (Rantau Layung and Rantau Buta, Pasir, in East Kalimantan, Baru Pelepat, in Jambi, and Bulungan Research Forest, in East Kalimantan); Porto Dias, in Acre, Brazil; and Guarayos, Bolivia.

Implications of Population

Our central question was, Would communities find ACM more useful if population was sparse or dense? Would people who were intimately connected with their forests be more likely to work together to maintain the resource than people who had to compete among themselves for scarce forest products and landscapes, or vice versa? My personal expectation was that population density would make little difference—that people would be motivated to protect or exploit their environment for a whole host of reasons, which could include population pressure.

Of the densely populated sites, 5 were highly successful, 1 was so-so, and 13 were comparatively unsuccessful (see Table 7-4). Among the sites with intermediate levels of population, 2 performed very well, 3 were intermediate, and 2 fell in the low-impact category. The low-density sites included 6 high performers and 2 medium performers.

Most of the sites where ACM was not pursued wholeheartedly were in fact in densely populated areas (Ghana, Kyrgyzstan, Madagascar, and Malawi). Population density was significantly and positively correlated with poor ACM performance. Looking at Kyrgyzstan's population densities (which range from 1 to 1.8 persons/ha), Schmidt (2003b) reports no relationship between implementing action research and population density, though he does acknowledge that problems with the government's formal collaborative forest management program were greater in the densely populated areas. Within the intermediate category, two Cameroon sites (Campo Ma'an and Akok) have had some problems. Interestingly, Makak, a Cameroon site in a comparatively densely populated area, was abandoned early on because of conflicts.

One possible interpretation is that competition among community members in densely populated areas pulls them apart, whereas the equally fierce competition

Table 7-4. *Population Pressure and ACM Impacts*

	Overall success		
Population pressure	High	Medium	Low
High	5	1	13
Medium	2	3	2
Low	6	2	0

that exists in most sparsely populated areas is between communities and outsiders (companies and conservation projects), thus perhaps drawing the community together.

The Nepal and Zimbabwe sites were densely populated but showed impressive results. They had the benefit of large research and facilitation teams, who worked well together, led by individuals who thoroughly understood the approach.

I believe that in the densely populated sites, governments are often very apparent. Contact between the communities and their governments is longstanding and frequent, and relationships tend to be top-down, inflexible, and hostile. Such a context renders the ACM approach more difficult, at least at the beginning. That such constraints can be overcome is shown by the successes in Nepal and Zimbabwe.

CHAPTER EIGHT

Management Goals

O NE OF THE REQUIREMENTS FOR SELECTION as an adaptive collaborative management (ACM) research site was the involvement of one or more local communities in some form of forest management. Local people manage forests in all the sites for their own purposes. Although local management is a consistent component on all sites, in some cases, other stakeholders support the management of forests for timber production; in some, external stakeholders emphasize management for conservation purposes; and in still others, outsiders focus on management for both.

The ACM sites exemplify global diversity in forest management strategies but can be categorized as four types of management: large-scale timber, small-scale timber, large-scale conservation, and multipurpose forest reserves. One can imagine these types as continua; the boundaries are not tight. Multipurpose management actually exists on all the sites, within the context of community management. But within all the categories, community management is moderated and influenced by the management goals of other parties. How these outside influences affect local forest management and local communities is the subject of this chapter.

Large-Scale Timber Management

Large-scale external timber management demonstrates the interactions among systems operating at different scales. It also exemplifies the the links between local communities and outside forces.

The sites where large-scale timber management is a dominant force are the following:

• Akok, Campo Ma'an, Dimako, and Lomié, in Cameroon;

- Lumut Mountain (Rantau Buta and Rantau Layung , Pasir) and Bulungan, in East Kalimantan, and Baru Pelepat, in Sumatra, Indonesia;
- Nova Jericó and São João Batistain Tailândia, Pará, Brazil; and
- Guarayos, Bolivia.

Guarayos has a history of large-scale timber management, but its people are currently involved in small-scale timber production; hence this site is included below as well.

Commonalities in Large-Scale Timber Management

The sites that exemplify large-scale timber management share 11 features.

Corruption. The perceived severity of the problem is high in Bolivia, Cameroon, and Indonesia, all of which have been listed by Transparency International as among the six most corrupt nations in the world (with scores ranging from 1.9 to 2.0 on a 10-point scale, with 10 being the least corrupt); Brazil has a score of 4 (Transparency International 2002), but Adario (2002) identifies corruption as a central problem in forest management overall in Brazil as well, as do Barraclough and Ghimire (2000) and Pokorny (2002b).

 At the community level, accusations of corruption have been made in ACM sites in Indonesia and Cameroon, but not yet in the Bolivian or Brazilian sites [though Pacheco (2001) reports on such instances elsewhere in Bolivia].

Complex Local Management Systems. Forests have flourished under traditional systems, but local forest-dwelling communities tend to be seen as primitive and are not widely recognized as legitimate or useful within any of the four countries (Clay 1988; Dove 1988; Schmink and Wood 1992; Cronkleton 2001c; Oyono forthcoming). Local systems are being undermined by both misperceptions and national laws that are internally inconsistent, inconsistent with local customs, and applied inconsistently. "Authoritative knowledge" (Jordan 1997) is definitely in the hands of outsiders in all the Bolivian, Cameroonian, and Indonesian sites. In Brazil, there are competing views, with increasing awareness of the legitimacy of such systems in some sectors.

Influx of Farmers. Forest areas long in the hands of forest-dependent groups are being settled, "invaded," by more agriculturally oriented groups. The Indonesians refer to this process as transmigration; the Brazilians, colonization; and the Cameroonians, settlement. In Bolivia, formal settlement programs are proceeding slowly because of financial constraints (Pacheco 2001), but migrants from the highlands and immigrants from other countries have been moving into forested areas on their own, with similar impacts on those living there. The extent of the colonization schemes in the Brazilian Amazon is well documented.

Uncertainty about Land Tenure. In Indonesia and Brazil, government-sponsored transmigrants and colonists, respectively, theoretically have rights to plots of land (in Indonesia, ranging from 2 to 5 ha; in Brazil, more typically 50 to 100 ha). However, in both countries, indigenous communities may have longstanding claims to that land (Colfer and Resosudarmo 2002; Hall 1997), and large areas may be ceded by the government to timber concessions or other large operations.[1] In Cameroon as well, local communities have no legal rights to forestland (van den Berg and Biesbrouck 2000), and land they have considered their own is routinely handed over to timber companies or conservation groups. In Bolivia, legislation providing for communal indigenous territory *(terras comunales de origen)* is intended to address the issue of indigenous land rights, but the difficulties of sorting out the different claims mean that land tenure issues remain volatile (Kaimowitz et al. 2000).

Little Devolution of Power. Official efforts to devolve some power are under way in all these countries, with varying degrees of success, as described in a previous section. Bolivia and Cameroon have weak, formal devolution; and Indonesia and Pará, Brazil, no formal devolution.

Complex Bureaucratic Hurdles. As of June 2001, only 18 communities in Bolivia had managed to complete the bureaucratic requirements that would allow them to begin harvesting timber on their indigenous territories or via the local social associations, *Asociaciones Sociales de Lugar* (Cronkleton 2001c).[2] By September 2001, only 19 community forest plans in Cameroon had been approved (Diaw et al. 2002). Mvondo and Sangkwa (2002) document a complex routine required to obtain status as a council forest in Cameroon—another mechanism designed to decentralize forest management. Indonesia conducted a series of very small scale experiments (outside the law) in involving communities in forest management during the 1990s (Wrangham 2002) and has expanded the scale only since Soeharto's fall. The 1999 efforts to promulgate forestry cooperatives were followed by a range of efforts at the district *(kabupaten)* level that involved granting small concessions to communities (often coordinated by timber or plantation companies more accustomed to governmental paperwork and informal payment procedures; see Box 8-1)—a process sometimes called the decentralization of corruption (see Dudley 2000). Pokorny (2002b) reports that the bureaucratic impediments to getting title to land, for instance—which the government warns is a 10-year process—are a major constraint to sustainable forest management in Brazil.

Local Involvement in Conventional Timber Management. Indonesia conducted a small-scale, low-profile experiment in West Kalimantan in the 1990s in which it provided a timber concession to a group of communities. More recently, in a chaotic environment, local communities are involving themselves in unmanaged timber extraction through small-scale concessions (called IPPK, *izin pemanfaatan dan pengelolaan kayu*, permit for timber use and management) and through contracting arrangements organized by the military, buyers, and other more sophisticated stakeholders. Bolivia, working on a bigger scale through the U.S. AID project, BOLFOR, is helping communities (including members of the Guarayos indigenous territory) learn to manage timber, extract it in a low-impact way, and transport, process, and sell it. Pokorny (2002b) reports a disappointing experience in Tailândia, in which the local timber company tried to work with local communities in shared management; their efforts were frustrated by high turnover in the local population. In Cameroon, community forests are typically focused on commercial timber exploitation as the most immediately and obviously profitable use. There are also some similarities between the Indonesian IPPK permits and the

Box 8-1. Collaboration between Communities and Timber Companies in Pasir, Indonesia

IPPKs (*izin pemungutan dan pemanfaatan kayu*, or wood use harvesting permits) typically involve partnerships between a logging company and a community. Rantau Layung community members were dissatisfied with the Teguh Maronda Prima logging company (TMP) concerning promises of compensation fees, construction of facilities and provision of jobs for community members, promises that were denied by the company. They also suspected their own headman of collusion …. Furthermore, the people did not know where the areas designated for IPPK cutting were. Suspicion and confusion arose among the community members. They had never intentionally applied for such certificates.

To solve this problem, the community held a village meeting on April 7, 2002. At the end of the meeting, the people decided to revise their partnership contract with the company. For efficiency's sake, the community selected four persons to be stipulated as IPPK holders, to represent the community and take care of the problem.

When the community representatives requested a new agreement, the company initially refused. A meeting finally took place on May 22, 2002, facilitated by the head of Batu Sopang Sub-District. However, the community was not satisfied with the results of the meeting ….

After the meeting, the four village representatives visited the TMP office in Balikpapan to clarify the location map for the permit and the timber to be harvested. The company provided the representatives with the IPPK location map, but they refused to give any data on the volume of timber that had been cut. The company argued that they had not yet loaded the timber, though tree felling had been conducted.

When the representatives returned to their village, they held a meeting. The people realized that, based on the map, the company had actually made deviations from their joint plan. After the meeting, the people informed the Forest Service Office and asked clarification whether the office had already issued the documents the company needed to transport the timber. The head of the Forest Service Office was shocked when he heard what the company had been doing. He agreed to reprimand the company. However, the office had already given the transport document to the company. He promised that his office would not issue any new transport documents until the problem between the company and the community was resolved. Hearing this, the people were satisfied. However, by the end of this study, it was still not clear how the conflict had been resolved. During an interview with the Forest Service Office and the regional technical implementation unit, it was revealed that TMP was known for its "mischief," but the office found it difficult to take measures because the company had a logging permit issued by the Minister of Forestry, whose authority surpassed theirs (adapted from Adnan et al. 2002) ….

Turning to the village situation, which appears to have calmed, two of the four village representatives negotiating about the partnership between the village and the timber company inexplicably acquired new motorcycles worth Rp. 14 million at the same time—probably gifts from the timber company.

—Stepi Hakim

Cameroonian *vente de coupe* (sales of standing volume): in both, timber companies make arrangements with local government officials and village elites to harvest timber under the guise of community forests (Djeumo 2001). Efforts to ensure sustainability are most obvious in Bolivia, least in Cameroon and Indonesia.

Illegal Logging. In Indonesia (including the ACM sites), illegal logging is rampant (Dudley 2002; EIA and Telapak 2002a, 2002b), though the confusion about the division of rights and responsibilities between the central and district levels of government means that there is uncertainty about what is technically legal. Much illegal logging has been rendered apparently legal by district-level regulations like IPPK permits and HPHH (*hak pemungutan hasil hutan,* forest product harvesting rights).Additionally, logging by concessionaires has also traditionally included many illegal elements, since enforcement of forestry regulations has always been lax. In Pasir, efforts to counter illegal logging had become the prime focus of the ACM research by January 2002 (Yuliani et al. 2002).

Cameroon has also been marked by lax forestry regulations and a lack of enforcement of the regulations that do govern forestry operations. Auzel et al. (2001) note infractions ranging from logging outside the boundaries of *ventes de coupe* to "anarchic logging of forest management units."They also note the approximately 700,000 m³ of logs sold through the unregulated informal sector each year (see Fomété 2001, and Oyono et al. n.d.).

Pacheco (2001) calls illegal logging a serious problem in Bolivia.The Bolivian government has attempted to deal with such illegality by basically legalizing it.The *agrupacione sociale de lugar* (local social association) was created to allow illegal loggers to organize and conduct their logging in a more sustainable fashion (Alvira 2002), but illegal logging apparently continues in Guarayos (Bolaños 2001).

In Cameroon, Indonesia, and Bolivia, powerful people are often involved in organizing illegal logging, making it difficult to control (Etoungou 2002; Dudley 2002; Pacheco 2002). Adario (2002) maintains that 80 percent of the logging in Brazil is illegal, with the worst problems occurring in an "arc of deforestation," especially Mato Grosso, Pará, and Rondonia (see also Simula and Burger 2003). Pokorny et al. (2001b) reported illegal logging in all three Pará areas where the ACM team worked. In other parts of Brazil, illegal logging is reportedly now being addressed (Adario 2002).

Forest Degradation. The rate of deforestation in comparatively forest-rich landscapes is dramatic, though accurate measurement is lacking. Karsenty (2000) reports 450 authorized loggers in Cameroon, a small country. The confusion and chaos in Indonesia preclude any reasonable estimates, but the forest is being cleared at a rate that everyone agrees is alarming. Deforestation is less of a problem in Bolivia at this time. Gomes and Sassagawa (2002) report statistics from the *Instituto Nacional de Pesquisas Amazonica* (National Institute of Amazonian Research) that show deforestation rates in the Brazilian Amazon going down to 13,227 km² in 1997 but climbing back to an estimated 19,836 km² between August 1999 and August 2000; these data, based on satellite imagery, do not include deforestation of areas under 6.25 ha in size (meaning much subsistence agriculture). For Brazil, Adario (2002) reports that in 1970, only 1 percent of the Amazon basin was deforested; by 2000, 14.9 percent (representing an area the size of France) had been cleared. In all the countries, deforestation is proceeding more rapidly along roads.

Violent Conflict. In Indonesia, violence among forest stakeholders was kept in check under the Soeharto regime, whose military enforced the rights granted to timber concessionaires and plantation owners (Peluso and Harwell 2002). Now, under conditions of greater uncertainty, local communities are more likely violently to resist incursions by still-powerful external stakeholders (Anau et al. 2005). Cameroonian forests are marked by many small-scale incidents. Van den Berg and Biesbrouck (2000), for instance, report road blockages in eastern Cameroon; Fomété and Vermaat (2001) describe a violent conflict that arose when a community was not paid the fees promised by a logging company. In Bolivia, organized resistance to governmental expropriation of local resources persisted in the forested lowlands (Kaimowitz et al. 2000; cf. political upheaval in October 2003). Brazil is famous for its recurrent violent confrontations (Schmink and Wood 1992; Hall 1997; Hecht and Cockburn 1989), confirmed for the present, in Pará, by Pokorny (2002b).[3]

Economic Problems. All four countries had significant economic problems linked to structural adjustment programs required in connection with World Bank loans.

Differences in Large-Scale Timber Management

The similarities that emerge among the timber-dominated sites are more striking than the differences.

Cash Crops. Major efforts to introduce oil palm on a large scale, involving government and private industry, have been important in Indonesia (Casson 2002), Cameroon (Acworth et al. 2001), and Brazil (Porro 2002), and linked to deforestation. Cacao, on the other hand, has been central in Cameroon (Sonwa et al. 2001), less so in Indonesia, Brazil, or Bolivia. In Cameroon, the introduction of cacao led to what some analysts consider the emancipation of young men, who could now become economically and socially independent (van den Berg and Biesbrouck 2000).

Role of the State. Rural dwellers in Cameroon consider the state a much more significant actor (Diaw 2001; Diaw et al. 2002) than do rural dwellers in Indonesia, Brazil, and Bolivia (cf. forest-rich sites, in Tchikangwa et al. 2001).

Expectations. There is greater pessimism about the future in Cameroon than in Brazil or Indonesia, as observed in previous research (Porro et al. 2001) and confirmed in ongoing work (Pokorny 2002b; Tiani 2000).

Ethnic and Linguistic Differences. Ethnic and linguistic differences are highlighted in Cameroon and Indonesia; they are less obvious in Bolivia and Brazil. Schmink and Wood's (1992) discussions of social groupings in Brazil were reminiscent of the once-popular term applied to the United States—a melting pot, in which ethnic affiliations become muted, partially giving way to occupational and political groupings (with the exception of indigenous communities). This seems to be the case in Bolivia as well. In both countries, people often classify themselves more by occupation (rubber tapper, farmer) or by area of origin (e.g., the *cambas-colla*, or lowlander-highlander, distinction in Bolivia), than by ethnicity.

Size of Agricultural Holdings. In Cameroon and Indonesia, land holdings are generally 1- to 5-ha plots, though inclusion of forest fallows would imply larger holdings (up to 45 ha, depending on population density and other factors; Colfer with Dudley 1993). In Bolivia, 50 ha is a normal holding. In Brazil, settlers typically get 100 ha; rubber tappers may even have 600.

Economic and Subsistence Bases. In Bolivia and Brazil, large numbers of immigrants are involved in agriculture and ranching. In Cameroon and Indonesia, immigrants typically do agriculture and logging. There is no significant ranching in the forested areas of Cameroon or Indonesia.

Forest Fires. Fire has been a huge problem in Indonesia (Applegate et al. 2002), Bolivia and Pará; not so in Acre (Nepstad et al. 1999) or Cameroon.

Small-Scale Timber Management

The 11 sites in this category include 5 in Asia (all 4 in Indonesia, plus San Rafael–Tanabag–Concepción in Palawan, the Philippines), 2 in Africa (Lomié and Dimako, in Cameroon), and 4 in Latin America (Porto Dias in Acre, and Tailândia and Gurupá in Pará, Brazil; and Guarayos in Bolivia). Again, the spillover effects of large-scale timber management on sites in Indonesia, Cameroon, Brazil, and Bolivia cannot be ignored, despite recent efforts to accommodate small-scale operators.

Legalization of Small-Scale Logging

All 11 sites have seen recent efforts to legalize small-scale logging, partially in response to the adverse effects of large-scale timber harvesting. In 6 sites (San Rafael–Tanabag–Concepción, Lomié, Dimako, Porto Dias, Gurupá, and Guarayos), efforts to construct and abide by formal management plans are formalized by government and strengthened by externally financed aid, via NGOs (needed because of the labyrinthine bureaucracy and the lack of bureaucratic sophistication of the citizenry). In the others, the legalization efforts are more questionable (see Box 8-2). In all 11 sites, illegal timber harvesting remains a problem to varying degrees.

Ethnic or Local-Settler Differences

In San Rafael–Tanabag–Concepción, Palawan, the ACM team began work with the dominant migrants, but there are also smaller numbers of indigenous neighbors. In Lomié and Dimako, there are significant internal ethnic differences, as well as differences between indigenous communities and immigrants. In Porto Dias, the ACM team began work with the long-resident rubber tappers but realized it also needed to deal with more recent, and more numerous, colonists nearby. Tailândia is a hodgepodge of constantly shifting inhabitants. In Guarayos, the dominant group is indigenous, but there are significant settlements of immigrants, and the ACM team is trying to work with both. Rantau Layung and Rantau Buta in Indonesia and Gurupá in Pará are exceptional in being comparatively homogeneous.

Box 8-2. Small-Scale Timber Extraction in Tailândia

Ownership

A high level of conflict characterizes forest ownership in the Brazilian Amazon. By law, all land opened up by new roads becomes state property. Areas as large as tens of thousands of hectares are given or sold cheaply to powerful and influential landlords, often from the south of Brazil. Through settlement programs, plots of 50 to 100 ha are distributed to settlers, mainly from the northeast. Invasion of land by illegal settlers, who cultivate the land without land title, is widespread. Abandonment, invasion, and legal and illegal trade cause considerable upheaval. Small settlers normally cultivate rice, maize, and manioc using slash-and-burn techniques, while big landowners are often involved in cattle ranching.

Exploitation

The process of timber exploitation is complex. Small settlers with land title can sell their wood legally, with permission from IBAMA (*Instituto Brasileiro de Meio Ambiente e dos Recursos Naturais Renováveis,* Brazilian Institute for the Environment and Renewable Resources). For exploitation of bigger areas, mainly 500 to 5,000 ha, forest management projects have to be assessed as sustainable by the government. In this case the landowner sells four- to six-year exploitation concessions. A high percentage of such harvests are carried out illegally, without permission of the official landowner or IBAMA. The harvest of trees from smaller landowners, legal or illegal, is often carried out by small enterprises. They are often equipped with only a chainsaw and truck (often self-constructed) and buy trees for about U.S.$5–15 per stem. This sort of exploitation is generally more important in young frontier areas. Later, because of decreasing stocks, restricted transport capacities and increasing distances, the exploitation by bigger enterprises, equipped with skidders, tractors, and trucks becomes more important. Very often the sawmills themselves are involved in exploitation. Even large enterprises sometimes work without official management projects. Harvests are limited to the dry season (June to December). On average about 15 to 25m³/ha is harvested. If the transport distance exceeds 100 km, the enterprises tend to move to other areas.

Processing

Tailândia has about 50 sawmills, which have been processing timber for more than 10 years. Two-thirds of the sawmills obtain some timber from their own management projects, but more than 75 percent of the timber is purchased. The sawmills consume between 20 and 50 m³ of roundwood per day and employ 50 to 150 people in the dry season. In the region about 40,000 m³ of timber is processed monthly. The processing is not efficient: only 20 to 30 percent of the exploited volume is used. Almost the entire production is restricted to boards. There are no kilns for drying timber. The majority of the sawmills do not pay taxes. Because of chronic financial problems and strong competition, the fluctuation in numbers of active sawmills and ownership is extremely high.

Commercialization

Nearly all timber is sold commercially in Brazil. The most important markets are the northeast (Bahía) and the southeast around Rio de Janeiro. Only 10 percent of the sawmills are involved in the export of timber.

—Adapted from Pokorny et al. (1999)

Different Attitudes Toward Conservation

Palawan was subjected to a considerable amount of logging in the 1970s and 1980s (Lorenzo 2001c), but strong support for conservation has developed in recent years, including a logging ban that is still in effect ("salvage logging," in which the community is actively involved, is exempt). Acre manifests a similar "greening of the discourse" (Schmink and Wood 1992), without being subjected to previous logging. In Guarayos, large-scale logging has been under way for some time, and the most valuable trees have already been taken (though timber of commercial value remains). This site seems intermediate between those overtly concerned about conservation and those that are not. In the remaining sites with small-scale timber management, there is little overt concern about conservation.

Presence of Large-Scale Logging

The Indonesian sites and Dimako and Lomié in Cameroon are surrounded by industrial timber operations. The commercial operations are interested in "collaborating with" (and taking advantage of) the communities' fledgling forestry activities and demonstrate little concern for sustainability or conservation. In these sites, people are concerned for their children's futures and recognize their own dependence on the forests. But the fact that others continue to benefit from local resources, combined with insecurity about future access and availability, encourage them to be opportunistic.

Artisanal Versus Commercial Models

Porto Dias, Lomié, and Dimako have seen disagreements about the style of logging. In the African sites, the government and NGOs lean toward artisanal logging (as more environmentally friendly); locals favor the commercial. Conversely, in Porto Dias, Brazil, the local people are inclined toward artisanal logging, and the government and the important local NGO, in pursuit of economies of scale, wants more commercial endeavors. In Palawan (and among some NGOs in Acre), controversy remains about whether any logging should take place at all. In Indonesia and parts of Pará, there's simply a free-for-all.

Large-Scale Conservation (Parks)

Five ACM sites are formal conservation areas: Kerinci Seblat National Park, in Sumatra, Indonesia; Kayan Mentarang National Park in Bulungan, East Kalimantan, Indonesia; Kitanglad Natural Park in Lantapan, Mindanao, Philippines; Campo Ma'an National Park in southern Cameroon; and Ranomafana National Park in Madagascar. These parks were created based on the U.S. model of a national park. White et al. (2001, 241) note, "... the protected area approach derives primarily from the United States where it coevolved with democracy, increasing government transparency, accountability, and an ever-stronger civil society." Such conditions may not obtain in the tropics (and they may be declining in the United States!). The research approach taken in Kayan Mentarang and Madagascar, having started before the ACM program, differs somewhat from the other sites.

Commonalities among Parks

I have identified six significant commonalities in conditions in these parks.

Structure. All five ACM parks have core and buffer zones and are formalized units managed for conservation at the national level. The Campo Ma'an National Park is managed under Cameroon's Ministry of Water and Forests, with considerable external assistance (Nguiébouri et al. 2002) and a distinctly top-down approach (Diaw et al. 2002). Kerinci Seblat was nominated in 1982 by Indonesia's Minister of Forestry as a national park, with official boundary demarcation/gazettement occurring in 1995, when it became part of the country's system of protected areas (Kusumanto 2000). Kayan Mentarang was established in 1996 (Wollenberg et al. 2001b). The Indonesian protected area system has also been top-down. Mount Kitanglad is a part of the Philippine National Integrated Protected Area System, based on a 1992 law (Arda-Minas 2002b). Its evolution has involved greater efforts to involve local communities in decisionmaking through its multistakeholder body, called Mount Kitanglad Protected Area Management Board. Ranomafana was gazetted in 1996 (Ferraro 2002). It is managed by the *Association Nationale de Gestion des Aires Protegée* (National Association for the Management of Protected Areas), a parastatal organization, initially mandated to establish and choose pilot integrated conservation and development projects, of which Ranomafana is one (Wright 1997).

Danger of Degradation. Earlier logging and the fires that took place during the 1983 El Niño adversely affected Mount Kitanglad in the Philippines. Indeed, the level of degradation that has occurred has prompted concern among local inhabitants and outside observers alike (Valmores 2002). More accessible areas of Kerinci Seblat in Indonesia have been "invaded" by coffee, clove, and rubber farmers, and this trend could expand to include the ACM site, Baru Pelepat, and other areas. Similarly, Ranomafana is surrounded by land-poor people whose extensive, fire-based agricultural practices are considered inconsistent with conservation objectives. Managers are so fearful of establishing precedents that might reduce their level of control that they simply refused to negotiate the contracts proposed for comanagement of park resources (Rakotoson et al. 1999; see also Peters 1994). In Campo Ma'an (Cameroon) and Kayan Mentarang (Indonesia), concerns are less related to extant degradation than to potential degradation of biodiversity values and the threats posed by logging, plantation establishment, and to a lesser extent, swidden agriculture.

Recent Establishment. Kerinci Seblat was established in 1995; Kayan Mentarang, Mount Kitanglad, and Ranomafana in 1996; and Campo Ma'an in 1999.

Involvement of Major International Donors. The Campo Ma'an park was established and funded by the World Bank (with contributions from other sources) to compensate for the environmental damage done during construction of a major pipeline from Tchad through Cameroon (Nguiébouri et al. 2002). Kerinci Seblat National Park has likewise received large amounts of funding from the World Bank since the early 1990s. The Mount Kitanglad Natural Park is part of the Conservation of Priority Protected Areas Project, a World Bank–funded project through its umbrella Global Facility Biodiversity Conservation Project (Valmores 2002). Ranomafana aid has come from U.S. AID, which awarded $3.237 million to improving the management of natural resources in and around the park via Duke University and the government of Madagascar in 1991, augmented by the government ($29,000) and other participating organizations ($654,000) (Ferraro 2002). If, as Ferraro (2002) suggests, 5 percent of this money was spent on administration, $3.724 million would have been available to mitigate the negative effects of the park's establishment on local communities. Instead, Peters (1998) says, "less than two percent ... went towards 'village projects'. A full 18 percent of the total went towards U.S. university 'overhead', while nearly 37 percent went to expatriate personnel."[4] Kayan Mentarang has had multiple donors: WWF, CIFOR, ITTO, MacArthur and Ford Foundations, and the U.K. Department for International Development have all contributed significant amounts of funding at various times over the past 10 years.

Tension between Local, Indigenous Inhabitants and Settlers from Other Areas. Campo Ma'an is inhabited by a mix of indigenous groups, including settled Bantu and hunting and gathering Bakola (pygmies), as well as a wide range of immigrants, most of whom work for timber and plantation companies in the area (Nguiébouri et al. 2002). Kerinci Seblat's ACM site is composed of indigenous Minangkabau residents, hunter-gatherers (*Orang Rimba*), and immigrants from other areas in Jambi and from Java. About a third of the population is now immigrants (Kusumanto 2001a). The ACM site of Basac, in the buffer zone of Mount Kitanglad, is 95 percent indigenous Talaandig people, with small numbers of immigrants (Arda-Minas 2002b). In some nearby neighborhoods, recently arrived Igorot have brought much more intensive agricultural systems (Valmores 2002). In Ranomafana, Madagascar, the Tanala, a forest people, are interspersed with the in-migrating Beteseleo (Buck 2002).

Illegal logging. Timber theft occurs in all sites.

Differences among the Parks

Ten differences in site conditions are important to highlight.

Recognition of Local Forest Management. In Campo Ma'an, Kerinci Seblat, and Kayan Mentarang, there is little formal recognition of the value of indigenous natural resource management systems, which are considered backward and primitive. But in the Philippines, where "green discourse" has taken hold, in a form similar to that in Acre, Brazil, indigenous systems are increasingly recognized for their potential contribution to management.

Livelihood Security. Local stakeholders in Campo Ma'an and Ranomafana fear the loss of their livelihoods. Portions of their traditional lands were appropriated for the parks, new rules limit their access to important forest products, and in Ranomafana, shifting cultivation—essential to traditional livelihoods—is prohibited. The Tanala of Ranomafana have had bitter experiences with outsiders, including a brutal quashing of a rebellion in 1947 (Freudenberger 1999; Rakotoson et al. 1999). In the Indonesian park areas, the enforcement of regulations is less effective, apparently resulting in less fear about livelihood security there.

Level of Conflict. In Campo Ma'an, Cameroon, the serious conflict relates to the perception that important decisions are made elsewhere with little regard for local needs and rights. People are particularly angry with the World Bank (Tiani et al. 2002, 2004). In Mount Kitanglad, Philippines, the conflict is about which stakeholders have stronger rights and responsibilities in management (Valmores 2002). Conflicts in Ranomafana derive from longstanding fears of outsiders, as well as the actual removal of some communities from the park and the economic opportunity costs to residents (cf. Ferraro 2002; Buck 2002; Peters 1994). Conflict levels in Kayan Mentarang National Park were not particularly high until recently, when political turmoil in Indonesia began to affect park dwellers. The remoteness and difficulty of access once made interventions by outsiders infrequent, but now loggers from Malaysia pass through in search of illegal timber. Although the Kerinci Seblat site in Jambi is marked by conflict, very little of it relates to the neighboring national park.

Rule of Law. Legal systems seem a bit more firmly established in the Philippines than in Indonesia or Cameroon (on Transparency International's 2002 corruption index, for instance, Indonesia is 88 out of 91, with a score of 1.9 on a 10-point scale; Cameroon ranks 84, with a score of 2.0; and the Philippines is ranked 65, with a score of 2.9; Madagascar is not listed). Madagascar's more limited natural resources of commercial value have meant that the scale of corruption is considerably smaller than in the other sites (although two Ranomafana Park rangers were removed for corruption between 1995 and 2000; Buck 2003).

Forest Degradation. The forest is further along in what might be termed a degradation cycle in the Philippines and Madagascar than in the ACM sites in Cameroon or Indonesia. Fires have been a serious problem in both Lantapan, Philippines (Valmores 2001), and Ranomafana. Although El Niño fires of 1997–98 were also important in Indonesia, the ACM sites were only marginally affected.

Extent of Logging. This has been an important issue in Campo Ma'an, Kayan Mentarang, and Kerinci Seblat national parks, whereas in Mount Kitanglad and Ranomafana, most logging ceased some time ago.

Access. Local communities' access to markets, education, and medical care varies by site (see Table 7-1).

Population Growth. The rate of growth is 5.3 percent in Campo Ma'an (Nguié-bouri et al. 2002), and 4.2 percent in the watershed of Mount Kitanglad (Hartanto 2001). Madagascar is growing more slowly, at a 3 percent rate (Pollini 2000). Our researchers did not report population growth rates in the sites characterized by low population density, including those in Indonesia, reflecting their perception of relative unimportance (see Table 7-3).

Conservation awareness. There is now widespread interest in conservation in civil society in the Philippines—a feature far less evident in Cameroon, Madagascar, or Indonesia.

Level of education. Although education attainment was generally low, there are a few politically sophisticated indigenous people in Mount Kitanglad (Burton 2001b). No indigenous leaders adept at dealing with outsiders have come to the fore in the other sites. Rural Filipinos appear, from our experience, more adept and comfortable dealing with their own governmental bureaucracy than do rural Cameroonians, Malagasy, or Indonesians.

The inflexible model on which these parks have been built has consistently caused problems for ACM researchers and communities. Researchers in Campo Ma'an and Ranomafana had persistent conflicts with park managers in their efforts to catalyze community (and broader) action. In the latter case, these conflicts basically stymied the effort until the funding source dried up—though the team began to make progress by shifting from a focus on partnerships with the park to partnerships between indigenous healers and "modern" medical practitioners, relating to medicinal plants (cf. Rakotoson et al. 2000; Buck 2003). In Baru Pelepat and Kayan Mentarang, park management was comparatively removed from the site of our ACM activity; enforcement of regulations is notoriously weak in Indonesia. The situation in Mount Kitanglad is still unfolding.

Box 8-3. Policy and Legal Elements of Community Forestry in Nepal

The current policy on community-based forest management focuses on the following:

- handing over accessible forests to forest user groups irrespective of political or administrative boundaries;
- ensuring that the real or traditional users of the forest make up the group;
- sharing of all benefits from community forests among the users;
- using the group's fund in community development work, including forest development;
- changing the role of forestry staff from custodial to facilitators; and
- building capacity of the forest user groups and district forestry staff through training.

The legal processes for handing over a community forest to the user groups are as follows:

- identifying and forming a forest user group within a community;
- preparing the group's constitution and registration at the district forestry office;
- preparing the operational plan of the forest;
- applying to the district forestry office for handing over the management responsibility of the forest; and
- approving the operational plan and handing over of the community forest to the user group.

—New Era Team (2001a)

Multipurpose Forest Reserves

With the exception of Muaná in eastern Brazil, 15 sites are multipurpose forest reserves—for both conservation and production; they have formal status and are recognized by the government. Here, governmental agencies are the most significant external entities involved in forest management (see Box 8-3), and the reserves do not fit into a standard, international mold. The 15 sites are as follows:

- Ottotomo Forest Reserve, Cameroon;
- Mafungautsi Forest Reserve, Zimbabwe;
- Chimaliro Forest Reserve, Malawi;
- Ntonya Hill Plantation, Malawi;
- Adwenaase and Namtee Community Forests, Ghana;
- Achy, Arstanbap-Ata, Ortok, and Uzgen state forest farms (*leshoz*), Kyrgyzstan;
- Manakamana and Andheribhajana Community Forests of Sankhuwasabha District, and Bamdibhirkhoria and Deurali–Baghedanda Community Forests of Kaski District, Nepal; and
- Muaná, Pará, Brazil.

Muaná is included here because the people's forest management is indeed multipurpose, despite the absence of any formal reserves. With the exception of Muaná, these sites share four characteristics:

1. formally declared and comparatively clear boundaries;
2. legal instruments for regulating forest use by communities;
3. significant, though not extreme, conflict between local communities and forestry officials; and
4. some illegal logging (though this is not a serious problem).

Management Goals and ACM Success

Contrary to our fears, the presence of large-scale timber operators does not seem to have adversely affected local communities' motivation and capacity to act together to achieve their goals, nor does the presence of industrial forestry guarantee success. Five sites where large-scale timber management was important did very well, three did moderately well, and two were fairly unsuccessful (Table 8-1). Careful

Table 8-1. *Management Goals and ACM Impacts*

Management type[a]	Overall success		
	High	Medium	Low
Large-scale timber	5	3	2
Small-scale timber	7	3	1
Large-scale conservation	2	2	1
Multipurpose	7	1	7

[a]Some sites have multiple management types.

reading of the data from each site confirms the importance of contextual factors. The problems identified and the difficulties confronted in these sites are clear. But in most cases, the people in the ACM communities have been able to proceed in the manner anticipated, despite the presence of large-scale timber producers.

Nor did the presence of small-scale logging seem to reduce communities' abilities to use ACM effectively. In fact, this type of management had the best formal success rate, with seven high achievers, three medium, and only one "failure."

Looking at the five large-scale conservation sites from the quantitative and reductionist perspective, we have two with high overall success, two with medium, and one with low—again, insufficient to confirm that large-scale conservation areas interfere with or help people who are using an ACM approach to address problems.

Among the multipurpose forest reserves, we see a sharply bimodal distribution, with seven sites highly successful, seven sites minimally successful, and only one in the middle. As with forest type, we have the four Nepal sites at one end and the four Kyrgyz sites at the other.

This section has provided a telling (and perhaps depressing) overview of management conditions in many of the world's forests. Looking through the "management lens," we do indeed see different implications for the conduct of the ACM process under the influence of different external stakeholders with their different forest management goals. However, they are not the results we had anticipated. Several local communities have successfully identified their goals, made plans together, monitored their progress, and changed their subsequent plans accordingly, under the watchful eyes of powerful people involved in large-scale logging and large-scale conservation.

This is not to minimize the problems that the researchers encountered in these locales characterized by striking power imbalances. Diaw (2002, 12), for instance, in a recent report on our work in Cameroon, noted,

> When [supra-local policy] incentives are clearly set to achieve comanagement objectives (e.g., Ottotomo, Lomié), collaboration and mutual learning are easier to achieve with the help of appropriate facilitation. When those incentives are primarily set to achieve supra local or global conditionality (e.g., Campo Ma'an), collaboration and mutual learning are difficult to achieve, local handicaps to the management goal tend to increase, and independent facilitation may be looked at with mistrust ...

He stresses the difficulties our teams encountered in Cameroon, where Ottotomo is a forest reserve, Lomié has community forests engaged in small-scale logging, and Campo Ma'an has large-scale conservation and large-scale logging.

Our work in the national and natural parks also encountered barriers—in the form of comparatively clear, legal definitions of park boundaries and functions, created without much input from local communities, and supported and maintained by international wealth and enforcement power, sometimes even in conflict with national priorities. It was necessary in some cases to sidestep or postpone direct engagement with park management.

In some cases, the communities and ACM teams also had to postpone working with the powerful logging companies until they had established some legitimacy with other stakeholders. In other cases, the ACM process involved focusing on forms of collaboration and adaptation that did not directly involve the companies. And there were, as anticipated, efforts by some companies to subvert the collaborative process.

But in both large-scale logging and large-scale conservation settings, some teams were able to meet such efforts to subvert the process with perseverance and collaboration, and it was often possible to get some buy-in from logging companies and conservation projects.

Communities were able to conduct a useful ACM process in virtually all the large-scale logging, small-scale logging, and large-scale conservation sites. The results in the multipurpose reserves were much more patchy. ACM worked well in Nepal and most of Cameroon, not so well in Kyrgyzstan, Ghana, and part of Malawi. There was no statistically significant link between ACM success and these other management goals. One constraint in areas with multipurpose reserves was the bureaucracy. Where it was possible to mobilize that bureaucracy or fit into its goals, as in Nepal, ACM teams were able to make progress; in other sites, the bureaucratic context appears to have sapped people's energies and inhibited their creativity. The inertia of a bureaucracy proved a more potent foe for adaptive collaborative management than the raw power of timber companies and international conservation projects.

Although it appears that areas with a history of timber management, both large and small scale, may be more conducive to ACM than those managed for conservation, the differences are not striking or statistically significant. Here I want to return to my argument that this kind of analysis is likely to mislead us. The reductionist approach obscures differences attributable to site-specific peculiarities—researchers' motivation or length of time in the field, say. Precisely because this study was in fact very carefully conducted and analyzed, we *know* that some of the findings are suspect. We were able to triangulate on the subject we were studying, using a variety of methods, and thus we can avoid concluding some things that I don't believe are true.

Perhaps more important, I fault reductionist analyses because I think they give us little concrete insight to help us improve forest management and human well-being. Suppose we conclude from this study that people have more trouble working together and improving their efforts iteratively in conditions of high population density than in conditions of low population density. Should we then stop trying to use the ACM approach in high-density sites? I believe that we can make much better use of our research dollars and time if we look at each site holistically (and in concert with local people). We can then trace the plausible causal links among human and ecological factors, develop ways to help us (including local communities) address those links, and solve the problems creatively and in a focused, relevant manner.

One conclusion I can confidently assert is that the ACM process thrives on a direct and obvious challenge, working well under difficult, clearly inequitable, and even chaotic conditions. Additionally, the enthusiasm and commitment of ACM staff were consistently striking under those conditions.

But let us continue examining how communities have used ACM under varying conditions. In the next chapters, we look at the more microlevel issues we anticipated might affect communities' readiness to implement ACM.

CHAPTER NINE

Human Diversity

THIS CHAPTER LOOKS AT THE EFFECTS OF the human diversity variable on the process of adaptive collaborative management (ACM). We initially imagined diversity as primarily a local-level issue. However, it soon became clear that diversity had implications beyond the village level. A diversity of stakeholders can affect a forest, often through different management goals. It seemed important to examine the variations along this dimension to see whether such complexity made a more or less hospitable environment for the ACM process. We were unsure whether homogeneity or heterogeneity might be preferable. Would the ease of communication in a homogeneous group or the range of talents and contacts provided by a heterogeneous group make an ACM process more desirable, feasible, and effective? What were the implications of many versus few human differentiations?

Our results have suggested two parts to the diversity issue, one dealing with the diversity of external stakeholders (the number and importance of outsiders with whom communities interact), the other with diversity within communities, focusing on age, gender, ethnicity, caste, and wealth status.

External and Formal Diversity

At the beginning of our research process, the ACM facilitators made an initial estimate and analysis of their sites' community heterogeneity and stakeholders. Although one might expect these data to be easily comparable, levels of detail differed, knowledge improved over time, and conditions changed. The categorization below derives from the facilitators' data sets, complemented by information in other reports, informal discussions, and direct observation on some sites.

In many sites, external stakeholders wield power, authority, and resources. In the rich tropical rainforests of Indonesia or Cameroon, we find a proliferation of powerful and well-funded timber companies, government agencies, and conservation projects. In forest-poor countries, like Nepal or Zimbabwe, interactions tend to be primarily between communities and governments. The roles of NGOs also vary considerably from country to country and from community to community.

Here, I have categorized the sites by high, medium, and low levels of diversity. I am specifically discussing here both external stakeholders (like government agencies, timber companies, and conservation projects) and formal internal ones (like mothers' clubs, local unions, and churches or mosques). The basis for this categorization scheme is shown in Table 9-1. Because of cross-site variability in both importance of external stakeholders and the scale at which stakeholders were listed, it has not been possible simply to add up the total number of stakeholders; a fair amount of judgment has been necessary. In the subsequent listings, ACM facilitators and CIFOR are not listed as stakeholders, simply to avoid redundancy.

Nine sites have high external diversity: Akok, Campo Ma'an, Dimako, and Lomié in Cameroon; Ranomafana in Madagascar; Baru Pelepat in Indonesia; the two Philippine sites of San Rafael–Tanabag–Concepción, Palawan, and Basac, Lantapan in Mindanao; and Guarayos in Bolivia.

In several of these sites, the links between local communities and outsiders were identified early on as problematic. The numerous workshops held—in Campo Ma'an (Nguiébouri et al. 2001), Lomié (Oyono and Efoua 2001), and Dimako (Oyono and Assembe 2003) in Cameroon; in Baru Pelepat, Indonesia (Kusumanto 2001a); in Palawan, Philippines (Lorenzo and Hartanto 2001)—are an indication of the importance of addressing these issues.

In Campo Ma'an, community problems dealing with the government and the conservation project prompted the team to provide training to community members in legal matters. In Bolivia, the team identified the attitudes of formal forest managers and government officials as a problem for addressing local concerns. The team arranged training for these outsiders on how to deal more sensitively and equitably with rural populations. Topics included training in the ACM approach and in gender issues (Cronkleton 2001a, 2001b). In Akok, Cameroon, the ACM team members worked closely with researchers from the International Institute for Tropical Agriculture, a stakeholder with a more conventional approach to research. The ACM component appears to have dissolved in this interaction—another indication that the ACM team itself was subject to the same kinds of human constraints and complications as the communities with whom we worked.

In Basac, Philippines, the diversity of stakeholders was seen as directly providing opportunities for learning. Cross-site visits exposed community members to neighboring groups with skills needed locally and strengthened ties between communities (Arda-Minas 2001). We found that the presence of potential partners can be both an advantage and a risk. If one encounters a stone wall in efforts with one partner, one can turn to others, but having many partners can immobilize a group because of transaction costs, including conflicts, miscommunications, and differing goals.

Twelve sites were characterized by medium levels of external and formal diversity: Rantau Buta, Rantau Layung, and Bulungan in Indonesia; Ottotomo, in Cameroon; Ntonya Hill in Malawi; Tailândia and Gurupá in Pará and Porto Dias in Acre, Brazil; and Arstanbap-Ata, Achy, Uzgen, and Ortok in Kyrgyzstan.

Relations among stakeholders in the sites with medium levels of external diversity were typically more intense and focused in the conduct of ACM. The Ottotomo team worked very closely with the community, the *Office National de Développement des Forêts* (National Office of Forest Development), and the *Association Terre et Développement* (Land and Development Association), an NGO. The Porto Dias ACM team worked with the rubber tappers and *Centro dos Trabalhadores da Amazônia* (Center for Amazonian Workers), an NGO. Such focused partnerships characterized all the sites in this category, and under this scenario, good relations and mutual understanding can result in success. But the lack of choice in potential partners also spells a risk, if the few partners are insufficiently motivated to cooperate with one another.

Nine ACM sites were characterized by low external and formal diversity: the four Nepal sites (Manakamana, Andheribhajana, Bamdibhirkhoria, and Deurali-Baghedanda); Chimaliro, in Malawi; the two Ghanaian sites (Adwenaase and Namtee); Mafungautsi in Zimbabwe; and the Muaná sites in Brazil.

In sites with low levels of external diversity, relationships were between communities (or community-level organizations, such as forest user groups) and government forestry establishments (with the exception of Muaná, which had no such resident establishments). In Muaná, the community formed local researcher groups, which organized themselves to strengthen their capabilities and external ties for marketing local products. With regard to ACM processes, these sites ranged from the very successful Nepali and Zimbabwe sites to Adwenaase and Namtee in Ghana, where little seems to have occurred.

External and Formal Diversity and ACM Success

The wide range of variation among these sites reflects differences in the number, types and importance of external actors. We obtained the best results among the sites with the least diversity (see Table 9-2; statistically significant at the 1 percent level). Among that group, six sites had high levels of success, one had medium, and two had low. Among the sites with intermediate levels of diversity, five were high achievers, one was intermediate, and six were low. The high-diversity sites had the flattest distribution, with three high achievers, four medium, and two low. There are examples of obvious successes and at least one case of dramatic lack of progress in all three categories.

In interpreting these results, it is important to note that the four uniformly successful Nepal sites were among the six high achievers in sites characterized by low external diversity; and that the four Kyrgyz sites were all among the six low achievers in the sites with intermediate levels of external diversity.

Table 9-1. *ACM Sites Classified by Presence of Formal External Stakeholders*

Forest	Neighbor communities	Government Local	Government Higher	NGOs, church projects	Company (logging, plantation)	Research institutions[a]	International donors[b]	Conservation projects	Poachers[c]
Baru Pelepat, Indonesia	X	X	X	X	X	X	X	X	X
Bulungan, Indonesia	X	X	X	X	X	X	X	X	X
Campo Ma'an, Cameroon	X	X	X	X	X	X	X	X	X
Lomié, Cameroon	X	X	X	X	X	X	X	X	X
Dimako, Cameroon	X	X	X	X	X	X	X	X	X
Akok, Cameroon	X	X	X	O	X	X	X	O	X
Guarayo, Bolivia	X	X	X	X	X	X	O	O	O
Ntonya Hill, Malawi	X	X	X	X	X	X	X	O	O
Chimaliro, Malawi	X	X	X	X	X	X	O	O	O
Rantau Layung, Indonesia	X	X	X	X	X	X	O	O	X
Rantau Buta, Indonesia	X	X	X	X	X	X	X	O	X
Basac in Lantapan, Philippines	X	X	X	X	X	X	X	X	O
Arstanbap-Ata, Kyrgyzstan	X	X	X	O	O	X	O	O	O
Achy, Kyrgyzstan	X	X	X	O	O	X	X	X	X
Ortok, Kyrgyzstan	X	X	X	O	O	X	X	X	O
Uzgen, Kyrgyzstan	X	X	X	O	X	X	X	X	X
Mafungautsi, Zimbabwe	X	X	X	X	X	X	O	X	O
Ottotomo, Cameroon	X	X	X	X	O	O	O	O	O
Bamdibhirkhoria, Nepal	X	X	X	X	O	O	O	O	O
Deurali-Baghedanda, Nepal	X	X	X	X	O	O	O	O	O
Manakamana, Nepal	X	X	X	X	O	O	X	O	O
Andheribhajana, Nepal	X	X	X	X	O	O	X	O	O
San Rafael–Tanabag–Concepción in Palawan, Philippines	X	X	X	X	X	O	X	O	O
Porto Dias, Brazil	X	X	X	X	X	O	X	O	O
Adwenase, Ghana	X	X	X	O	X	O	X	O	O

Namtee, Ghana	X	X	X	O	X	O	X	O	O
São João Batista/Nova Jericó in Tailândia, Brazil	O	O	O	O	X	O	O	O	X
Jaratuba, Recreio, Nova Jerusalem in Muaná, Brazil	X	X	O	O	O	O	O	O	X
Canta-Galo in Gurupá, Brazil	X	X	O	X	X	?	X	O	X
Ranomafana, Madagascar	X	?	X	X	O	?	X	X	?

[a] Excludes CIFOR, since all sites have a CIFOR presence.
[b] Excludes CIFOR's ACM, since funds were not generally available for community use.
[c] Poachers identified as significant stakeholders (including illegal harvesting of all kinds).

Table 9-2. *External Stakeholders and ACM Impacts*

Number of external stakeholders	Overall success		
	High	Medium	Low
High	3	4	2
Medium	5	1	6
Low	6	1	2

Internal Diversity

Diversity also occurred within the ACM communities. The guidelines to help researchers characterize their sites on this heterogeneity dimension are provided below (Box 9-1).

Although gender and age differentiations exist in all societies, they take radically different forms—and the ACM sites represent the range of global differences along these common lines. Here I have categorized the sites on the basis of gender, age, and a combined category consisting of ethnicity, caste, and occupation. We began the ACM research with a resolve to address such internal community differences in our work because of our interest in empowerment and equity issues. This classification helps set the stage for the discussions of conflict, social capital, and ACM processes that follow.

Box 9-1. Guidelines for Categorizing Communities by Heterogeneity

Low:
- People in the area are almost all of one ethnic group, caste, or class.
- People in the area share the same subsistence or economic base.
- Gender roles are flexible, with much overlap and shared responsibility.

Medium:
- Although one group has traditionally lived in the area, another group has recently begun moving in.
- Although most people share the same profession, some specialization is beginning to occur.
- Men and women share some responsibilities, but most are separate.

High:
- There are distinct ethnic or other social groups within the area, such as indigenous communities of various ethnic groups, combined with settlers from distant areas (e.g., parts of East Kalimantan in Indonesia, and the Trans-Amazon in Brazil).
- People practice a variety of means of livelihoods (e.g., logging, mining, plantation workers, subsistence farmers, fisherfolk).
- Gender roles are strictly divided, with little shared responsibility for specific tasks.

—Carol J. Pierce Colfer (2000)

Gender

In all societies men and women have different ways of interacting with their environment and with each other. Specifically in forest communities, women and men have different amounts and kinds of access to forest resources, they have different kinds and degrees of voice in forest-related decisionmaking, and they derive different levels and kinds of benefits from forests: in short, they have "asymmetrical entitlements" (Rocheleau et al. 1996).

In Table 9-3, I have ranked each site from 1 (low) to 5 (high) along four dimensions:

1. clear division of physical space for women and men;
2. strict division of labor between the sexes;
3. strong ideology of male dominance; and
4. hostility to women's involvement in public arenas.

Nepal and Kyrgyzstan have the strictest gender differentiation, followed by the African sites, most of which are in the intermediate category, along with the somewhat less differentiated Latin American sites. The Philippine and Indonesian sites have the least gender differentiation. This distribution also fits with what one might expect from the literature about communities in these parts of the world. High gender differentiation occurs in 9 sites; intermediate differentiation in 15 sites; and low levels of differentiation occur in 6 sites.

Attention to gender in the ACM sites spanned the range of gender differentiation, as did ACM success. Zimbabwe, Nepal, and Kyrgyzstan exhibited the most gender differentiation, and the issue was addressed in all nine of these sites. The Nepal and Zimbabwe sites made good progress in strengthening women's voices in planning, decisionmaking, implementation, and forest product distribution; the advances in the Kyrgyz sites are much more modest, though there was some increase in women's participation in meetings and access to the desirable forest leases (Fisher 2003b).

Sites with intermediate gender differentiation that explicitly addressed the issue included Chimaliro in Malawi, Campo Ma'an, and to a lesser extent, Ottotomo in Cameroon and Guarayos in Bolivia. Among the sites with low gender differentiation, only the Basac (Lantapan) site in the Philippines explicitly addressed gender issues. The sites with low gender differentiation (all in Indonesia and the Philippines) had considerable success at implementing ACM; some of the medium and high gender differentiation sites saw low levels of success.

The implications for ACM success are shown in Table 9-4: Sites with minimal gender differentiation did quite well: five attained high success, one attained middling success, and none did poorly. High gender differentiation resulted in a bimodal distribution: five did well, four did poorly, with none in the middle (again reflecting the dominance of Nepal in the first category and Kyrgyzstan in the last). Sites with intermediate levels of gender differentiation showed a flat distribution: four were successful, five showed intermediate impacts, and six did poorly. According to this kind of analysis, success is possible with all levels of gender differentiation, and somewhat more likely with low gender differentiation.

Table 9-3. *Elements of Gender Differentiation in ACM Sites*

ACM sites	Clear division of space	Strict division of labor	Strong male dominance	Hostility to women in public arenas
High				
Manakamana, Nepal	5	5	5	5
Deurali-Baghedanda, Nepal	5	5	5	5
Bamdibhirkhoria, Nepal	5	5	5	5
Andheribhajana, Nepal	5	5	5	5
Astanbap-Ata, Kyrgyzstan	5	5	4	4
Achy, Kyrgyzstan	5	5	4	4
Ortok, Kyrgyzstan	5	5	4	4
Uzgen, Kyrgyzstan	5	5	4	4
Mafungautsi, Zimbabwe	5	5	4	4
Medium				
Lomié, Cameroon	5	5	4	3
Dimako, Cameroon	5	5	4	3
Chimaliro, Malawi	5	5	4	3
Campo Ma'an, Cameroon	5	5	4	3
Akok, Cameroon	5	5	4	3
Adwenaase, Ghana	5	5	4	3
Ottotomo, Cameroon	5	5	4	3
Ntonya Hill, Malawi	5	5	4	3
Namtee, Ghana	5	5	4	3
Porto Dias, Acre, Brazil	4	4	3	3
Tailândia (2 villages), Pará, Brazil	4	4	3	3
Muaná (3 villages), Pará, Brazil	4	4	3	3
Gurupá (1 village), Pará, Brazil	4	4	3	3
Guarayos, Bolivia	4	4	3	3
Ranomafana, Madagascar	3	4	4	3
Low				
San Rafael–Tanabag–Concepción, Philippines	2	2	2	1
Basac, Lantapan, Philippines	2	2	2	1
Rantau Layung, Indonesia	2	1	2	1
Rantau Buta, Indonesia	2	1	2	1
Baru Pelepat, Indonesia	2	1	2	1
Bulungan, Indonesia	1	1	1	1

Note: 1 = low, 5 = high.

Table 9-4. *Gender Differentiation and ACM Impacts*

Gender differentiation	Overall success		
	High	Medium	Low
High	5	0	4
Medium	4	5	6
Low	5	1	0

Age

Turning to the question of age—which we did not initially consider—we detect striking differences in attitudes, specifically in the relationships between young and old. Although in none of the sites did people group for collective action based primarily on age, there were interesting and potentially important contextual differences from country to country. Since the importance of age emerged in the course of the research, the following analysis is qualitative and derives from my reading of the case material and e-mail communications on each site (except Nepal, where no information is available).

In all the Cameroon sites there is marked hostility between young adults and the old. In the past, the young were dependent on the old: parents' financial contributions were needed for young people to marry, and their blessings remained important in ensuring a successful marriage (Akwah 2002). Today, however, the young may embrace changes that many of the old fear (Russell and Tchamou 2001; Tiani 2001), and in some areas, some young people fear that the old will curse them. In our studies on intergenerational access to resources, Cameroon stood out as unusual in the degree to which the current generation was thought to be using up resources that would no longer be available to their young (Porro et al. 2001; Mala et al. 2002). Cameroon had the full range from high-impact to low-impact sites.

Similarly, in Ghana there is recurrent conflict over the sale of community timber resources by the elders (who retain the proceeds). "While the elders claim to be in charge because of their social positions, the youth also claim to be in charge because of their commitment to the establishment of institutional structures for the management of the community forests" (Blay 2002c). Relations between the young and the old in Madagascar, at least among the Tanala, also seem to be marked by considerable antipathy. Freudenberger recounts a 1990 revolution of the young against the old in a number of communities in the forest corridor of which Ranomafana National Park is a part (Freudenberger 1999). The youths' primary complaint was the harshness of traditional justice, including the possibility of complete social ostracism (becoming a "black banana") and the imposition of fines higher than those imposed by the state. Rakotoson et al. (1999) describe the unwillingness of elders to listen to the young people who dared to speak up in public meetings in the village of Sahavoemba. Stakeholders on the two Ghana sites and the Madagascar site did not adopt the ACM process as hoped, and results were unimpressive.

Relations between old and young in Zimbabwe seem less conflict-ridden than in Cameroon. Traditional roles—the dependency of the young and the responsibility of the old—are combining with the dynamics of change (Matose 2001; Chitiga and Nemarundwe 2003). The young are drawn to new approaches and question the viewpoints and perspectives of their elders. But HIV/AIDS, which has decimated the young population, has also changed these relationships, often making young adults dependent again on parents at a time when they would have been emancipated. The recruitment of young men to back up government policy in this time of political unrest has also created an unusual situation in which the young are widely feared as enforcers (Sithole 2002a). ACM went very well in Zimbabwe.

In Kyrgyzstan, the young are also financially dependent on the older generation, though intergenerational hostility is not as palpable as in the African sites. As they grow older, they obtain the resources (often animals, land, and labor) needed to prosper. In their youth and young adulthood, they remain part of their parents' household and subject to their decisions. Parents are expected to provide a home for each married son and find him employment. The young find more opportunities to make money in forest resources (which are peripheral) than in the central economic activities of the area. A fair number of young Kyrgyz and Uzbek men and women leave to work in the cities or in other countries, mainly Russia and to a lesser extent Kazakhstan (Marti 2000; Schmidt 2002a). The ACM process was cut short in Kyrgyzstan.

In the Latin American sites, the young are expected to make their own way once married, and marriage choices are made by the young couple (with varying amounts of advice from the parents). Parents in the western Amazon make significant sacrifices to ensure, for instance, that their children obtain an education (Campbell et al. 2005). Cronkleton (2005) notes differences in perceptions between older and younger community members about how the benefits from forest management might be used in the Bolivian site, but serious conflict has not been reported. Whereas in the African sites, local people discussing age emphasize the problems related to relations between young and old adults, in the Latin American sites, the responsibilities of adult parents for their young children are emphasized (Campbell et al. 2005; Bolaños and Schmink 2005). Guarayos in Bolivia and Porto Dias in Acre, Brazil, did very well; Muaná in Pará, Brazil, had intermediate success; and Tâilandía in Pará did poorly.

In Borneo and Sumatra, Indonesia, although there are differences in perspective between young and old (cf. Colfer 1981; Hakim et al. 2001b), in general the relationship is amicable. The young are expected to respect the old but also to take responsibility for decisionmaking as they grow into mature adulthood; parents expect both to make significant sacrifices for their children and to relinquish control as their own labor contribution decreases. The idea that the old might harm their own young—for example, by cursing them—is inconceivable.[1] All four of these ACM sites did well.

The African sites had the most pronounced differentiation based on age and the full range of ACM outcomes. In eastern Cameroon, Oyono et al. (n.d.) explicitly link the belief among the young that previous generations have wasted their environmental heritage to a lack of environmental concern for the future—indeed, he argues that the young would prefer obliterating the forest. Previous research has shown the negative perceptions of Cameroonians about what their forests will be able to provide for the coming generation (Porro et al. 2001; Tiani 2000). One might anticipate that such perceptions could interfere with the ACM process, and with its focus on collective action pertaining to the environment. But many of the African communities have made excellent progress with ACM, including attention to environmental issues. As with gender, those sites characterized by little age-related differentiation (South America and Southeast Asia) did not show any obvious pattern of problems with the ACM process.

Ethnicity, Caste, and Occupation

Ethnicity is a critical dimension in all ACM sites in Africa and Asia. Cameroon, for instance, has 286 languages, each representing ethnic differences (Summer Institute of Linguistics 2003). Within the data set discussed here, caste is relevant only in Nepal. Occupation is a dominant social grouping mechanism in the Latin American sites.

Here, I characterize each site in terms of its internal diversity (Table 9-5); and then examine the variation in ACM success based on diversity. Information about diversity that can be of direct help to those trying to manage natural resources more effectively will be available in such works as Fisher et al. (forthcoming), Matose and Prabhu (forthcoming), and Diaw et al. (forthcoming), for Asia, Cameroon, and Zimbabwe, respectively (see the Appendix of Cases for more contextual information).

The four sites in Nepal—Bamdibhirkhoria, Deurali-Baghedanda, Manakamana, and Adheribhajana—stand out as the most internally diverse in our sample. Historically and culturally influenced by the hierarchical world view of Hinduism, each community has its own hierarchy of castes, each with its own occupation. Additionally, there are other ethnic groups (see Dangol et al. 2001b; New Era Team 2001; and McDougall et al. 2002a). The prominence of diversity as a contextual feature prompted considerable attention to it in the ACM work. Box 9-2 describes some of the measures taken to address this issue by Nepali communities as part of their ACM efforts.

Ten sites were characterized by intermediate levels of internal diversity. Five large sites in fact included both high and medium levels of internal diversity. Since our work focused more on communities with medium amounts of internal diversity, I have categorized these five sites—Campo Ma'an, Lomié, and Dimako in Cameroon, Bulungan in Indonesia, and Guarayos in Bolivia—accordingly. Other sites in this intermediate category were Baru Pelepat, in Jambi, Indonesia; San Rafael-Tanabag-Concepción, in Palawan, Philippines; Arstanbap-Ata and Achy in Kyrgyzstan; and Mafungautsi in Zimbabwe.

Table 9-5. *ACM Sites Classified by Heterogeneity (Diversity of Ethnicity, Caste, and Occupation)*

High diversity	Medium diversity	Low diversity	
Bamdibhirkhoria	Bulungan	Rantau Layung	Namtee
			Ranomafana
Deurali-Baghedanda	Campo Ma'an	Rantau Buta	Ntonya Hill
Manakamana	San Rafael–Tanabag–Concepción	Basac	Chimaliro
Andheribhajana	Achy	Ortok	Porto Dias
	Astanbap-Ata	Uzgen	Muaná (3 villages)
	Lomié	Ottotomo	Tailândia (2 villages)
	Dimako	Akok	Gurupá (1 village)
	Mafungautsi	Adwenaase	
	Baru Pelepat		
	Guarayo		

Box 9-2. Dealing Constructively with Diversity in Nepal

In the Nepali sites, the issue of diversity—or more specifically, inequality—was raised explicitly by community members and colleagues within the forestry bureaucracy. There was widespread recognition that elite community members generally dominated the forest user group and its committee. The desire for greater equity emerged in informal discussions and in the local workshops that initiated the participatory action research process on each site. As workshop participants developed shared local visions for the community forest and considered criteria and indicators, equity in making decisions and sharing benefits was identified as important in almost all the groups. The groups developed their plans based on priorities identified in this process.

The decisionmaking processes evolved over the course of the research. One pattern that we observed in all the sites was a shift in decisionmaking from the centralized forest user group committee to the hamlet (tole) level. The hamlet groups were smaller and generally more homogeneous in terms of caste and ethnicity. It appeared that men and women felt freer to express their views in these smaller and more like-minded groups. The issues raised at the hamlet level were then fed to the centralized committee for further discussion and ratification.

A second pattern that emerged in most forest user groups was an effort to elect a wider representation of caste, ethnicity, and gender to leadership positions, including the newly created hamlet committees and in the centralized committee.

An important strategy in two sites was the development of a mechanism to track participation and equity. The committees led processes of categorizing households into high, medium, or low privilege, then tracked who was participating in forest user group events and who was benefiting. This helped the groups move away from lip-service commitment to providing benefits and opportunities to marginalized users. Other important mechanisms to enhance equity included establishing working groups that would assess the need for various benefits in an interactive way at the hamlet level and draft plans for benefit sharing (rather than leaving it to the committee).

In all the sites, the research team and other stakeholders supported forest user group members in capacity-building activities, some targeted marginalized users, to support more equitable decisionmaking processes, such as facilitation skills.

—adapted from Cynthia McDougall

More than half our sites (16) were marked by low internal diversity: Rantau Layung and Rantau Buta in Indonesia; Basac in the Philippines; Ortok and Uzgen in Kyrgyzstan; Ottotomo and Akok in Cameroon; Adwenaase and Namtee in Ghana; Ranomafana in Madagascar; Chimaliro and Ntonya Hill in Malawi; Porto Dias in Acre, Brazil; and Muaná, Tailándia, and Gurupá in Pará, Brazil. Most of these communities consist primarily of one clearly dominant group.

Interestingly, class was not used as an analytical tool by the team. Using a social class framework, it would be hard to see any of these people as anything other than "lower class": they are generally financially poor, poorly educated or uneducated, and marginalized by the wider society (with possible exceptions in Nepal). Class as an analytical category is perhaps most congruent with, and more commonly used by others to describe, the situation in Brazil.

Table 9-6. *Community Heterogeneity and ACM Impacts*

| Community | Overall success | | |
heterogeneity	High	Medium	Low
High	4	0	0
Medium	5	3	2
Low	5	3	8

Quantitatively examining the links between community heterogeneity (ethnicity, caste, and occupation) and ACM outcomes, we find that the four most heterogeneous sites (all in Nepal) had uniformly high overall success (see Table 9-6). The communities marked by intermediate levels of diversity include five high-success sites, three medium-success sites, and two low-success sites. Five of the low-diversity sites had high success rates, three had medium success rates, and eight had low success rates. Success is possible at all levels of diversity but somewhat more likely in diverse sites; statistically, however, these results were not significant at the 1 percent level. I remain skeptical of analyses implying unilineal causation.

Conclusions on Human Diversity

This section has examined the levels of external diversity and community heterogeneity in the ACM sites. The practical implications of such diversity have often been overlooked in attempts to manage forests. Busy forest managers not infrequently ignore or trivialize the different use patterns, organizational traditions, and kinds of knowledge of users and stakeholders—to both the people's and the forests' disadvantage. These findings suggest that both the quantity and quality of diversity may be important factors to consider in creating sustainable management systems.

Diversity affects the conduct of ACM on any given site. Gender, age, and other kinds of differences are crucial elements to take into account in efforts to achieve sustainable forest management and human well-being because people in different categories tend to have different relations with the forest in terms of use, conceptual models, and rights. Diversity also has implications for equity. Subsets of the population in most forest villages are disadvantaged in their rights, use, and access to resources. The ACM teams' experience in working with the subgroups strengthens our conviction that this is an important issue to address and that ACM provides a practical tool to do so.

As with most of the other variables we have examined so far, diversity itself has little *predictable* effect on the conduct of ACM. Quantitative analysis without a thorough understanding of the context can mislead. Holling et al.'s observations in Box 9-3 pertaining to systems in general help us consider some of these issues in a more abstract, conceptual—perhaps generalizable—way.

The authors' discussion of biological functions within scales ("lumps") and between scales seems comparable to the distinction made in this chapter between internal and external human diversity, respectively. I have long believed that global

Box 9-3. Functions of Diversity in Systems

There are two types of such diversity, one concerning how diversity affects biological function within a range of self-similar scales—within a lump (Walker et al. 1999); and one concerning the way it affects biological function across scales—between lumps (Peterson et al. 1998). Both types of diversity contribute to the resilience and sustainability of the system.

… [W]ithin-scale and between-scale diversity produces an overlapping reinforcement of function that is remarkably robust. We call it imbricated [overlapping] redundancy.

—C.S. Holling et al. (2002c)

cultural diversity serves as an insurance mechanism for the human species—a thought related to theories about the function of biodiversity in maintaining ecosystems' resilience, sustainability, and robustness.

CHAPTER TEN

Conflict and Social Capital

W HEREAS CONFLICT AT THE LOCAL LEVEL was important for
our efforts to encourage collaboration locally, the impacts of national-level
conflicts were also keenly felt. And social capital, which we had initially conceptualized as something fairly local, turned out to have important implications pertaining to links with outside actors.

Level of Conflict

Preeminent in my own mind when I thought of conflict, as we began our research
for adaptive collaborative management (ACM), was conflict between communities
and powerful local stakeholders. Our previous fieldwork, focused on criteria and
indicators for timber management in tropical forests, had burned these kinds of
conflicts into my mind. Yet I also realized that microlevel, intracommunity conflict
was likely to play a part in our attempts to strengthen social capital and promote
collective action and social learning as well. The guidelines we used to assess conflict levels are listed in Box 10-1.

Large-Scale Conflicts

What I did *not* have in mind when we began our research was the kind of large-
scale, national-level conflict that has erupted in several of our host countries (see
Boxes 10-2 and 10-3). The ongoing problems in Zimbabwe, Madagascar, Indonesia, the Philippines, and Nepal have been widely documented in the international
press; problems in Brazil, Bolivia,[1] and Cameroon have been less dramatic but still
significant.

Box 10-1. Guidelines for Categorizing Level of Conflict

Low:

- small number of stakeholders and abundant resources.
- homogeneity of community, sharing similar interests.
- strong links and interdependence among community members.
- shared or complementary interests among stakeholders.
- fairly equal sex ratio.

Medium:

- small, but possibly growing, number of stakeholders with decreasing availability of resources.
- heterogeneity of communities, with longstanding coexistence.
- early fears that access to resources may be in danger in the future or that inequitable distribution of resources may become a problem.
- changing sex ratio, with inmigrants primarily of one gender.

High:

- large number of stakeholders competing for scarce and/or decreasing resources.
- heterogeneity between communities (e.g., indigenous vs. settler communities), particularly when some are newcomers.
- reality or perceptions of inequitable distribution of benefits among stakeholders.
- vulnerability of one or more stakeholders, vis-à-vis some other stakeholder(s).
- highly unequal sex ratio.

—Carol J. Pierce Colfer (2000)

As noted in Chapters 1, 2 and 3, I have linked my comparative analysis to some of the ideas discussed in *Panarchy* (Gunderson and Holling 2002). Gazing through the macrolevel conflict lens within their framework (see Figure 10-1), one might characterize the national political systems of five countries (Zimbabwe, Madagascar, Indonesia, Philippines, and Nepal) as in the release (or chaotic) stage of Holling and Gunderson's figure 8; three (Bolivia, Brazil, and Cameroon) are in the early stages of reorganization (or rebirth). Kyrgyzstan, having undergone the change from communism to its current state, is probably a bit further along in the reorganization phase, en route to exploitation; Ghana and Malawi appear to be in the conservation phase.

Our reductionist analysis shows striking—and unexpected—links between conflict at the national level and ACM success (see Table 10-1). Among the 12 sites in high-conflict countries, 10 had high levels of success, and only 1 each achieved medium and low levels of success! Among the countries with intermediate levels of conflict, there were 4 high achievers, 4 mediums, and 2 lows. Of the 8 sites in low-conflict countries, 1 achieved an intermediate level of success, and 7 were ranked low. This result is statistically significant at the 1 percent level. The "plausible causal connections" we need between correlations (Fisher et al. forthcoming) are not obvious, but the conclusion that ACM thrives in difficult conditions is hard to avoid.

Box 10-2. Conditions in Mafungautsi

Political instability

No meetings from December to end of March. After the elections at Batanai resource management committee, the ward councilor was not comfortable working with the ACM community partner, allegedly because he and his brothers had served as polling agents for the opposition party. At one point the community partner could not freely move around for fear of being victimized by members of the ruling party. The ACM team too had to flee the study area for their own safety, and when they resumed work after the elections, a lot of time was wasted moving from one government institution to another explaining their political innocence.

Hunger and economic hardships

As locals try to eke out a living in these trying times, there has been a massive out-migration of the able-bodied men to engage in menial jobs in South Africa or to be gold diggers at mine dumps in the country. In the case of the ACM beekeeping group at Batanai, about 80 percent of the group have left the area for gold panning, and this has put on hold the group's plans.

—Frank Matose et al. (2002, 43)

Box 10-3. Impacts in Manakamana

The researcher also experienced some miserable feelings from the national incidents while he was working in the field. The massacre of the royal family (king and queen and other members of the royal family), the attack of the Maoist rebels on the army camps and remote district headquarters, and the fatalities on both sides had really created psychological tensions in the field (anything could happen at any time). Once when the researcher was working in the field, Maoist rebels attacked the Achham district (a remote district in western Nepal) and killed 150 security personnel, dozens of government officials, and a few civilians, and this news broadcast had negative effects on security at the research site. The only link to the research site from Kathmandu was by air transport … With the fear of impending attack and the advice of the civil administration, the 60 police stationed at the airport left rather than die, leaving it and the research area in a security vacuum. Such incidents discouraged the researcher from working in the field. They might be attacked by the rebels who were against NGO activities. And they might be trapped in the fighting between the rebels and security forces. Fortunately, the researcher overcame this situation and completed the research work successfully.

Definitely, such incidents affected ACM activities in the field. Due to the royal massacre, the Community Forest User Group general assembly could not be held despite the huge gathering of users. Similarly, the search operation of the security forces and the declaration of a "state of emergency" in the country, suspending all the fundamental rights of the citizens, prevented the local users from gathering for workshops/meetings. As a result, the users also could not visit other user groups to collect money promised them for building construction. The users still had not called their general assembly in June 2002, so the action plan made with ACM through self-monitoring could not be approved.

—Laya Prasad Uprety, Narayan Sitaula, and Kalpana Sharma, New Era Team
See also McDougall et al. (2002b, 111–12).

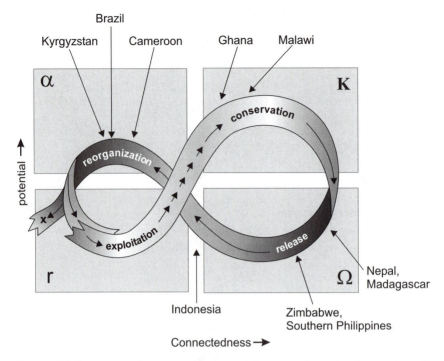

Figure 10-1. *Estimated Stages of National Systems, from the Perspective of Macro-Level Conflict, adapted from Holling and Gunderson's (2002: 364) Stylized Representation of Four Ecosystem Functions as They Change Over Time*

Local Conflicts

In the following discussion of the kinds of conflicts originally anticipated to be important in our research, the sites are divided into those with high, medium, and low levels of local conflict. There is no necessary correspondence between national-level conflicts and the level of conflict in the study communities or between those communities and other accessible stakeholders. The descriptions in the boxes will give the reader a better feel for the situation on the ground.

The 10 sites categorized as having high levels of conflict are those where conflicts are many, passionate, and likely to explode into violence: Lomié, Dimako, Campo Ma'an, and Akok in Cameroon[2]; Rantau Layung, Rantau Buta, Bulungan

Table 10-1. *National-Level Conflict and ACM Impacts*

National-level conflict	Overall success		
	High	Medium	Low
High	10	1	1
Medium	4	4	2
Low	0	1	7

(Box 10-4), and Baru Pelepat in Indonesia; Lantapan in the Philippines (Box 10-5); and Andheribhajana in Nepal.

The sites characterized by medium levels of conflict typically have many conflicts, but of a less intense and more readily solvable nature. These are Ottotomo in Cameroon; Adwenaase and Namtee in Ghana; Ntonya Hill and Chimaliro in Malawi; and Mafungautsi in Zimbabwe; Ranomafana in Madagascar; Porto Dias in Acre, Brazil; Guarayos in Bolivia; Manakamana, Bamdibhirkhoria, and Deurali-Baghedanda in Nepal; San Rafael–Tanabag–Concepción, Palawan in the Philippines; São João Batista, Nova Jericó, and Tailândia in Pará, Brazil; Achy, Arstanbap-Ata, and Uzgen in Kyrgyzstan.

Although we generally eschewed sites with low levels of conflict—on the theory that the issues were more pressing in areas marked by conflict—three sites had relatively low levels of conflict: Canta-Galo, Gurupá, in Pará, Brazil; Jaratuba, Recreio, and Nova Jerusalém in Muaná, Pará, Brazil; and Ortok in Kyrgyzstan.

Table 10-2 provides a systematic examination of the correlations between meso- and local-level conflicts and ACM success. The low-conflict sites show a completely even distribution, in terms of overall ACM success (one high, one medium, and one low). At the intermediate level of conflict, we have a perfectly bimodal distribution, with eight sites quite successful, eight sites quite unsuccessful, and one in the middle. For high-conflict settings, we have five with high levels of success, four with intermediate levels of success, and one with a low level of success. None of

---

Box 10-4. Conflicts in Malinau[3]

On the Malinau River, four villages—Sentaban, Setarap, Gong Solok and Langap—argue that they were first in the area and that the others asked permission [to settle] from them. This argument has had mixed results depending on the political strength of the villages. In the case of Sentaban and Setarap, neither had the political connections or strength of numbers to win their case, and Setulang has been able to maintain its claims. In the case of Langap, a small neighboring Punan village was pressured to give up its land rights completely, and another neighboring Punan village was forced to accept boundaries dictated by Langap. The more politically connected and larger Kenyah villages of Tanjung Nanga and Long Loreh negotiated boundaries that gave these neighbors reasonably large territories ....

In the village of Adiu, community members protested in November 2000 when their advance payment of Rp 50 million was not received from CV Wana Bhakti [logging company], after a month of operation. Chainsaw operators from Adiu had also not been paid. In October 2000, youth from Setarap protested because the company had employed only nonlocals, despite promises otherwise. They also protested that the distribution of the advance payment funds was neither transparent nor fair and accused the village leader of not being honest about the benefits he, himself, was getting from the deal. Reflecting a different problem related to land claims ... the contract between Sengayan and CV Putera Surip Widjaya states that the company has the right to an area of 35,000 ha, but the total area of the joint territory in the Long Loreh settlement (where Sengayan is located) is only 12,000 ha! In another case in Bila Bekayuk, a local man involved in negotiations with the company was not even sure of how much 1 hectare was.

—N. Anau, R. Iwan, G. Limberg, M. Moeliono, M. Sudana, and E. Wollenberg
See also Barr et al. (2001).

Box 10-5. Conflict in Lantapan

In 1997, Datu Migkitay of the Mt. Kitanglad Council of Elders applied to the Department of Natural Resources for a certificate of ancestral domain as a unified claim covering a total of 40,000 ha under eight different municipalities. This was presented to the Protected Area Management Board for Mount Kitanglad Natural Park. The majority of mayors sitting on the board, however, preferred that claims be put forward by the municipality (following the administrative boundaries)—apparently fearing the political power implied by a large, unified claim. The unified claim, therefore, was rejected. Notwithstanding this rejection, Datu Migkitay submitted his application to the National Commission on Indigenous Peoples. The commission was then a newly formed body, based on the Indigenous Peoples' Rights Act, to handle matters related to the indigenous communities and their rights. The act was later suspended, which stalled the issuance of certificates by the commission all over the Philippines.

In the meantime, while waiting for the implementation of the act, several communities applied for the community-based forest management program, which was also introduced into the buffer zone of the protected area. The people were encouraged to apply as a means of securing tenure over their farm lands located in the park's buffer zone. Furthermore, assistance was available for community forest holders. Their applications were approved by the Department of Natural Resources. In total, there are 13 community forest areas being established within Mount Kitanglad. This situation created conflicts in which the agency was accused of not supporting the unified claim and thereby contributing to the disintegration of the communities (Hartanto 2001; Burton 2001b).

The conflict between the tribal leader and the Department of Natural Resources on the claims issue was probably the most intense conflict in the area. Although the conflict occurred at a higher level (above the *barangay* [village] level), this conflict indirectly involved our collaborator, the Basac Upland Farmers Association, as it was the holder of the community-based forest management agreement and had declared its stand by not supporting the unified claim. Several People's Organization members received threats for their stand. This also deteriorated their relationship with the local Basac Council of Elders, who supported their leader's claim. The Basac elders felt that they were excluded from forest resource management with the granting of the agreement to the Bufai farmers' association.

—From Herlina Hartanto et al. (2002)

---

the three levels of conflict precluded successful ACM, and high conflict at this local level, as at the macro level, appears to be a somewhat congenial condition for ACM success. The results were not, however, statistically significant.

Looking at the findings, one sees that large-scale commercial logging is important in 8 of the 10 high-conflict sites: most of the Cameroon sites and all of the Indonesian ones. This is consistent with the views expressed by Peluso (2002, 1), who has argued that "violent conflict is more likely to occur under conditions of resource abundance rather than scarcity" (see also Dove 1993). Andheribhajana in Nepal and Basac, Lantapan, in the Philippines are the exceptions to this rule, having high conflict but no large-scale logging.

Conclusions on Level of Conflict

The ubiquity of conflicts in forested areas is clear, and none too surprising. What is surprising is the minimal degree to which conflict at the meso and micro levels

Table 10-2. *Local-Level Conflict and ACM*

Local conflict levels	Overall success		
	High	Medium	Low
High	5	4	1
Medium	8	1	8
Low	1	1	1

interfered with our efforts. Of course, in an attempt to stimulate or catalyze collaborative behavior, one would expect to encounter conflicting wishes and goals among different stakeholders, as well as oscillation in the degree of conflict over time. We had two contrary hypotheses early on: that conflict might inhibit our success, or it might strengthen local commitment to solving the problems and thus contribute to success. But the degree to which conflicts in fact emerged as *opportunities* for bringing about collaborative behavior was considerably greater than expected.

In support of the second hypothesis, existing conflicts often provided the motivation for community members to work with us—once we got established (one site each in the Philippines, Cameroon, and Indonesia was abandoned early on because of high conflict levels). In most cases, community members (and others) wanted to resolve their conflicts, and they saw the ACM teams as a potential means to such resolution, by facilitation and improved communication. The open-ended nature of our approach to collaborative action allowed us to follow up on these locally important issues and conflicts. Without that freedom, the conflicts themselves would likely have been serious hindrances—as was the case in Madagascar, where the team shifted its agenda from creating partnerships with park managers to creating partnerships among users of medicinal plants (again reminding us of the value of diversity of stakeholders).[4]

National-level conflicts present more intractable problems for work in communities and forests. Whereas facilitators confronted with a local conflict can work with community members on issues they identify together and make some immediate progress, influencing national-level politics is a more long-term goal, given the slow-moving nature of large-scale systems.

The idea that ACM thrives in difficult contexts seems to gain ground as the analysis proceeds. Our theory is that the kinds of processes that ACM invokes should stimulate a kind of governance that is less prone to violent upheaval. Strengthened self-confidence and attention to equity, collaborative identification of problems and solutions, systematic monitoring of community experiments, stronger links between communities and government and other stakeholders—all these represent components of more effective governance. Although such efforts cannot make conflict disappear, they are likely to avert some of it and provide effective mechanisms for dealing with it in a nonviolent and constructive—or "bounded"—way. Perhaps the motivation to seek bounded ways to deal with conflict bubbles up only when conflict is a significant problem.

Let me conclude with an observation by Andy White et al. (2001, 233): "True partnership rests upon the precepts of trust, reciprocity, transparency, and fairness, and upon near equal standing in terms of power to effect the outcome. The absence of perceived fairness inevitably breeds antagonism and resentment, true and false accusations of abuse, and, eventually, all possible forms of resistance and subterfuge."

Social Capital

I now turn to social capital—a topic that some might consider the mirror opposite of conflict. Others might see it as a backdrop against which conflict and cooperation play themselves out. Social capital is a feature of communities that both has the potential to affect the implementation of ACM and represents a useful indicator of success. In systems terms, social capital is part of a crucial feedback loop for ACM.

In 1998, when we were conceptualizing the dimensions that we anticipated would be important in understanding the success or failure of our approach, we were rather naïve about social capital.[5] We saw it as a characteristic of a community—putting us squarely in the Putnam (1995) camp, as opposed to the more individualistic conceptualization preferred by some authors. We imagined networks of people having an expectation of trust and trustworthiness among themselves and an ability to act together. We thought this might be a causal factor in communities' propensity for collective action and social learning. Only later did we fully realize that a viable interpretation of what we were trying to do through ACM was to build or strengthen social capital.

Some would argue that catalyzing collaborative action depends on strengthening social capital; the social element in "social learning" likewise implies a certain amount of social capital. Whittaker and Banwell (2002, 253) have criticized others for conflating cause and effect when talking about social capital. But in my view, social capital is part of a positive feedback loop that can contribute to strengthening itself, weakening itself, or remaining at the same level—thus becoming both cause and effect (Colfer 2002b; see also Brown and Asman 1996). For example, a community might decide to mobilize to protect its local resources and in the process draw on existing social capital (a cause). Working together involves collaboration (or collective action), which in turn strengthens social capital among the group (an effect). The process can work in reverse as well (when collective action is ineffective, trust is betrayed, or leadership is lacking), leading to a decrease in social capital. Social capital, like collaboration, is then both a cause and an effect. The measures of collaboration or collective action are closely linked to those that explicitly deal with social capital.

Box 10-6 reproduces the guidelines used by ACM team members in September 2000, as most teams were getting under way and assessing their sites' social capital. We simply identified communities as having low, medium, or high levels of social capital.

As we progressed with our research, it became clear that simple measurements were inadequate. One important reason brings us back to intracommunity variability: groups within communities, like *toles* (hamlets) in Nepal, might have very high levels of social capital, while the larger communities of which they are a part might have quite low levels. Similarly, adjacent communities (like the rubber tappers and the settlers in Porto Dias, Brazil, or the Bataks and the migrant communities involved in the cooperative in Palawan, Philippines) might have strong social capital within their respective communities but different levels between. Finally, the issue of social capital among different stakeholder groups cannot be ignored. I have included in the "high social capital" category sites where there seemed to be *some* high level of social capital—not necessarily a uniformly high level. The amount of information available from the sites is variable. Our purpose has not been to

---

Box 10-6. Guidelines for Categorizing Sites by Level of Social Capital

Low:

- The community rarely plans events or functions together and seems unable to work cooperatively.
- There are significant and obvious factions within the community that interfere with accomplishment of joint goals.
- Leaders are continually under attack from community members and seem unable to bring the community together productively.
- People express and demonstrate their distrust of each other readily to strangers.

Medium:

- There are factions within the community, but they function effectively to accomplish goals within each faction.
- The community occasionally accomplishes joint goals, but these joint activities may be marred by strife and discord.
- Although some distrust is evident among community members, they sometimes coalesce to confront outsiders whom they may distrust more.

High:

- The community regularly accomplishes joint goals in a smooth, efficient, and congenial manner.
- Although there may be different groups within the community, they do not compete in a destructive manner, but rather perform complementary tasks, widely valued by the community at large.
- Leaders are respected and valued for their leadership roles.
- Community members express and demonstrate trust in each other.

—Carol J. Pierce Colfer (2000)

---

contribute analytically to improving the concept of social capital per se, but rather simply to use the concept loosely to characterize our sites (but see Appendix 10-1 for some examples of relevant measurements on several sites). As with the section on conflict, I make extensive use of boxes, to convey better our uses of the term *social capital*.

Because sites sometimes included several villages, each with a different level of social capital, we produced more than 30 assessments of social capital. Seven sites were determined to have high levels of social capital: Mafungautsi in Zimbabwe; Ntonya Hill in Malawi; Porto Dias in Acre, Brazil; Canta-Galo in Gurupá, Pará, Brazil; Salvatierra and Cururú in Guarayos, Bolivia; and Uzgen and Ortok in Kyrgyzstan. Boxes 10-7 and 10-8 show the distinction between social capital within a community and that between community members and outsiders.

The 12 sites with intermediate levels of social capital were Ottotomo, Dimako, Lomié, and Akok in Cameroon; Adwenaase and Namtee in Ghana; Chimaliro in Malawi; Ranomafana in Madagascar; and San Rafael–Tanabag–Concepción in Palawan and Basac in Mindanao, in the Philippines; and Arstanbap-Ata and Achy in Kyrgyzstan. Here we highlight Cameroonian examples. Tiani (2000)

Box 10-7. Social Structures and Support Networks in Uzgen, Kyrgyzstan[6]

Social capital seems to be good in the village. This impression was increased through the observation of the mass participation of the community in the funeral that took place while we were in the village. Social capital is important for all households classified at different levels of wealth ranking. [A] family who became impoverished because of the wife's illness has been able to meet the cost of medical expenses not just through livestock sales, but also thanks to the help of relatives. The volume of this help is more significant than what was made through livestock sales. The husband's sister living in Uzgen gave them 6000 *som*,[7] a brother from Mirzake gave them 2000 *som*, yet another brother from Ak-Terek gave them 500 *som*. In addition to help from relatives, the family is also being helped by friends and neighbours. A friend of the husband has given the family wheat on credit, a cousin of the family's helps them by transporting their hay for free. The husband has been reciprocating these forms of support in many ways. He has helped his sister build her house and has helped his brothers by moving their beehives, cutting hay for them or collecting firewood.

Some of the respondents discussed how social networks are of crucial importance these days, especially since there has been a deterioration in the reliance on institutional networks compared with Soviet times. To have people you can trust and call on for help at any time is the best risk avoidance strategy against poverty and negative situations. One respondent has a neighbour to whom things are automatically entrusted when he is away. [The neighbour] helps look after his property; his wife and family can call upon him for any type of help. Another household described their network of relations in detail, highlighting the mutual support mechanisms. One extended family are all contributing to pay for the younger sister to study at the Medical Institute in Osh. Intergenerational mutual help is also very important. Young people help older generations with physical work, while older (and usually wealthier) people help them in kind.

Some households interviewed hold *sherine* (reciprocal entertainment [meals] between friends and neighbours in each others' houses) and *chornaya kassa* (a circulating pool of cash between a close circle of friends and neighbours, in which no one ultimately loses or gains), while others disapproved of these practices.

—U.K. Department for International Development (DFID) (2001)

Table 10-3. *Parameters, Stakeholders and Measures in Ottotomo, Cameroon*

Parameters	Stakeholders	Level
Cooperation	—	very weak
Institutional pluralism	—	weak
Shared learning	—	weak
Capacity building	ATD—Populations	average
	MINAGRI—Populations	average
Sharing of methodologies	ATD—Population	average
	ONADEF—Population	weak
Conflict resolution mechanisms	ONADEF—Population	weak

Source: Tiani 2000, 11–12.

Box 10-8. Community Relations with the State in Four *Leshozes*, Kyrgyzstan

Generally speaking it appears that there exists great distrust between the *leshoz* and people living in the *leshoz* area. During interviews and most of all during informal talks with female tenants and nontenants, women referred to the *leshozes* as being somehow "useless institutions that only hinder the people in general, because they extract money from them." … This female (but also male) opinion develops primarily due to corruption, which is present in most state institutions and, as such, also in the *leshozes*. Resulting from this unlawful behaviour of state bodies and institutions is another problem: with the perceived disintegration of law and order and the increasing corruption and misbehavior of official state institutions, people have lost their trust both in the workings of the courts and in the validity of written contracts. The performance of the state court and law enforcement bodies have got the most negative appraisal of the population among other state administration bodies— almost half of the population have no confidence in them—48.7 percent and 48.9 percent of the population, respectively.

—Research & Consulting Group (2000, 7)

Messerli (2000) finds it unsurprising, that "it is usual for people to place greater trust in verbal agreements made with individuals in authority, than with anything on paper. If the individual changes, the agreements usually have to be renegotiated, which raises considerable uncertainty over any supposedly long-term commitment."

—Jane Carter (2000, 14)

identifies important parameters and measures their relevance for major stakeholders in Table 10-3; and Mala discusses the efforts to create formal groups within Cameroonian villages in Box 10-9.

Twelve sites were considered to have low levels of social capital: Campo Ma'an in Cameroon; all four sites in Indonesia; all four sites in Nepal; Tailândia and the Muaná sites in Pará, Brazil; Santa María and Urubichá in Bolivia. This section highlights the Indonesian cases. Box 10-10 describes the historical antecedents of current levels of social capital within the large area of Bulungan Research Forest. Box 10-11 explains the process of improving social capital in the site in Jambi, Sumatra.

Conclusions on Social Capital

When we began looking at social capital, we imagined that we might be able to neatly place each site on a continuum from low to high social capital. Without any accepted standards for measurement across the wide range of contexts in which we were working, we made do with the broad guidelines outlined in Box 10-6. I was initially skeptical, and remain so, that one can come up with universally valid measures of social capital (cf. Krishna 2002), though Uphoff's and Krishna's thoughts on structural and cognitive elements are helpful (Uphoff and Wijayaratna 2000; Krishna and Uphoff 2001; Uphoff 2000).

Box 10-9. Social Capital within Formal Groups in Akok, Cameroon

According to members of the Akok community, one reason for failure of the management committee assigned to manage fees from timber harvesting was the quality of its membership. The president and treasurer were civil servants, respectively the *sous-prefet* and the Financial Officer of Ebolowa Municipality. The group's accountant was the mayor of Ebolowa town, and the other four members were from the local community. A bank check for 5,000,000 CFA (U.S.$7,937), which was issued in 2000 by the local logging company, has not yet been recovered by the community. Community members went many times to the offices of the *sous-prefet* and the financial officer of the municipality in search of the funds, but eventually they were told simply that the check had disappeared.

In Akok, there are also six farmers' organizations called Common Initiative Groups. These groups have created a Union of Common Initiative Groups. Some of these farmers' groups emerged within the implementation of the Sustainable Tree Crops programme, which is a private partnership approach between the chocolate industry, farmers' organizations and research institutions. These farmers' organizations have been particularly active in cocoa production and marketing. For the past two years they have succeeded in collective marketing of their cocoa, which has helped to double the price they received compared to individual marketing. This year, they plan to address forest farms (or *ngon*).

On the downside, there are still very weak connections among these activities, the development of appropriate technologies, their connections to landscape changes, and to the improvement of rural livelihoods. Technology development remains focused on the plot level, with little connection with the landscape. This is one of the issues being addressed in our research.

—William Mala

Box 10-10. Social Capital in Bulungan, East Kalimantan, Indonesia

Alliances among different groups in Malinau underlie how decisions occur and favors are allocated. These relationships are based on fluid, interlocking networks of ethnic affiliations, economic interdependencies, strategic kin relationships, and even historical alliances from the headhunting period. Historically, patron-client relationships were important between the "aristocratic lords" of the Kenyah or Merap and Punan (a hunter-gatherer group), who previously served these leaders in warfare and procured forest products for trade for them. But as Punan groups have moved towards greater economic and political independence in the last 15 years, they have increasingly rebelled against such associations. Meanwhile, Kenyah, Lundaye, and Tidung groups have been the most politically aggressive in recent years and now dominate Malinau's new local government. These groups have also sought, as have the Merap, to consolidate their claims to land. Punan groups, meanwhile, have had no representation in the *Kabupaten* (district) government, have had weak historical claims to lands and the weakest alliances with other ethnic groups. Individuals from all groups have maintained an opportunistic attitude towards building alliances and sought to strike new deals as they may. Unfortunately, only a relatively small group of leaders and their circles enjoy the benefits of these deals and exert any real influence over decisions. .

—N. Anau, R. Iwan, G. Limberg, M. Moeliono, M. Sudana, and E. Wollenberg
See also CIFOR Bulungan Research Forest ACM Team (2001, 2)

> ## Box 10-11. Improvement in Social Capital in Jambi and Pasir, Indonesia
>
> Too often, the villagers in Pasir and Jambi ACM sites had witnessed past development projects fail to bring improvements into their lives, despite having tremendous human and financial resources at their disposal. Most stakeholders attributed this failure to weaknesses in social relationships between individuals, social groups, and communities contributing to low social capital, social institutions, and leadership. Social cohesiveness was not only affected by external initiatives (e.g., a government resettlement project), but also by divisiveness among community stakeholders ...
>
> An evident improvement has been made in social capital, which prior to the action research was characterized by a lack of trust and reciprocity between community members and their leaders (i.e., *adat* [customary] or formal), certain individuals (i.e., community leaders), and groups of community stakeholders (i.e., original people and settlers). Improvement in social capital is illustrated by a lower incidence of conflicts between *adat* leaders (due to a more open attitude towards each other), mutual contributions by individuals towards voluntary work on agricultural plots, and more informal contact and support between original and settler community members. Relationships that contribute to social capital have also been developed between different communities, which prior to the action research clearly lacked good relations. Improved relationships have laid a basis for the development of collective knowledge *within* given groups of stakeholders (e.g., women or a particular community), which in turn has affected the sharing of knowledge *among* stakeholder groups. Improved relationships have led to the development of social networks or platforms.
>
> —ACM PAR Team Indonesia (2002, 66–67)

Based on the very qualitative assessments of each ACM site reflected above, Table 10-4 shows the links between these levels of social capital and overall ACM success.

Among the sites where social capital was abundant, we found a bimodal distribution: three had high overall success, one had an intermediate level of success, and three had low levels. Among the sites with intermediate levels of social capital, two performed very well, four were intermediate, and four performed comparatively badly. None of these relationships were statistically significant at the 1 percent level. Consistent with our growing conclusion that ACM does best under adverse conditions, those sites characterized as having low levels of social capital did the best, with nine high levels of success, two intermediate and one low. The likelihood that ACM thrives on adversity seems increasingly higher.

Table 10-4. *Level of Social Capital and ACM Impacts*

	Overall success		
Level of social capital	High	Medium	Low
High	3	1	3
Medium	2	4	4
Low	9	2	1

Scale, Diversity, and Social Capital

Although we expected to be able to identify a site as having high or low social capital, we did not envision the complexity we encountered. Within one site, multiple villages may each have a different level of internal social capital. Even within one village, there are subgroups whose level of social capital varies. And subgroups and communities can have differing levels of bridging social capital with other stakeholders. There are additional, broad-scale layers of social capital, including bonding social capital among staff of any given organization, and bridging social capital among members of government departments, industry, NGOs, research organizations, and other stakeholder groups.

Each of those "pots" of social capital represents an opportunity for improving collective action and other development- or conservation-related activities. If social capital is low within one group, it may be possible to build on high levels of social capital in a nearby or related group. If a positive feedback loop is involved, there is always the possibility of catalyzing a virtuous circle—a situation in which existing social capital leads to successful collective action that enhances social capital, which encourages more ambitious collective action, and so on. Kusumanto talks of the snowballing of social capital in Baru Pelepat (Kusumanto et al. 2002a); Senge (1990) describes similar processes in management; Uphoff (1996) provides examples from an irrigation project in Sri Lanka.[8] The idea that social capital is contagious is appealing.

Many of Merrill-Sands et al.'s (2000) observations about how diversity affects relations within an organization (see Box 10-12) can be observed in our research sites. The communication advantages that come from shared perspectives are to some extent balanced by the increasing access to information that can come from interaction between groups with different views and experiences.

Differences in power require careful management and facilitation, and the teams have struggled to perform such roles faithfully. In Nepal, considerable effort was devoted to ensuring meaningful participation of women and marginalized castes and ethnic groups in formal forest management activities (Dangol 2005; Sitaula 2001; Sharma 2002), with a fair amount of success. It appears that their efforts—in the most diverse of our sites—involved a significant increase in social capital within the forest user groups, along with bridging social capital among the hamlets (see McDougall et al. 2002b).

In the Pará sites in Brazil, the ACM team did not want to dominate the research process. They clarified their own views (thus strengthening their within-team social capital) before working with communities toward enhanced community empowerment. Table 10-5 shows the overlap between empowerment as the Pará team saw it and social capital as we have been using the term here (particularly organization, confidence building, and communication).

In Cameroon, the teams effectively used workshops oriented around criteria and indicators as mechanisms whereby diverse stakeholders, with different levels of power, could come together and share perspectives in a safe, carefully facilitated environment. Each forum had a steering committee composed of participants from the various stakeholder groups; the participants could give a yellow card (as in soccer) to the facilitator if they felt the process was going awry. The straightforward

Box 10-12. Diversity in Organizations

The very differences that enrich the potential for teams and partnerships to innovate and do new kinds of work are the same differences that can undermine team cohesion, member satisfaction and overall team functioning …

[A first principle is that since] group cohesion is defined by the attraction of members to others in their group, homogeneous groups will be more cohesive. The second is that demographic differences (e.g., race, nationality, gender, sexual orientation) evoke expectations by other group members that can result in in-group biasing and stereotyping of others. And third, members of different demographic categories come to the group with varying statuses and levels of power. These are based on differential access to resources and influence both within the organization and in the larger society. Members of dominant groups have greater influence in shaping interactions and outcomes. Members of subordinate groups may lose their voice and become marginalized within the group. Steps that have been found to mitigate such "process losses" include explicitly recognizing differences rather than ignoring them; building shared values and norms; defining superordinate goals for the group; establishing process and decisionmaking rules; reducing hierarchy and status differences; sharing power; providing external feedback to the group on team functioning; ensuring group accountability; fostering equal participation and mutual respect; and developing effective communications.

—Deborah Merrill-Sands et al. 2000 (11, 19)

discussions of power, management, conflict, responsibilities, and benefits strengthened the social capital among these widely divergent groups (Diaw and Kusumanto 2005).

Essentially, our experience suggests that diversity complicates the creation and nurturance of social capital but is a blessing in disguise, since it contains the seeds of creative interchange, enhanced networking capacity, and more expansive knowledge. What we have seen in our sites (and within our teams) suggests that the harmony associated with social capital, the tension associated with bounded conflict, and the creative energy associated with agency are important in the pursuit of benign, effective collective action.

Whereas Krishna found that social capital was useful without effective agency, but better with it, our results lead to another question: Is effective agency in fact

Table 10-5. *Potential Empowering Effects of Participatory Methods*

Categories	Empowering effects
Organization	Planning and distributing tasks, learning how to make decisions
Confidence building	Gaining self-confidence, respecting the group, respecting local knowledge, gaining technical training, guaranteeing ownership
Learning processes	Enhancing creativity, getting access to new ideas, learning to learn, finding stimulation for own initiatives
Sensitization	Critical perception of one's own environment, self-positioning, reflection
Communication	Information exchange, articulation of one's own opinions, constructive discussion

Source: Pokorny et al. 2004b.

deleterious without the social capital needed to keep it under control? A balance of power may be needed whereby strong social capital serves to monitor and control such "agents" (with links to more powerful stakeholders). In Cameroon, Indonesia, and Nepal, the lack of functioning social capital may have allowed local leaders to take advantage of the community rather than be accountable to it.

Commentary

This brings to a close my examination of the conditions we identified as likely to be important in the process of catalyzing adaptive collaborative management. For each condition, I have provided a summary assessment of its effects on the level of success in our ACM sites. We do not find convincing links between any of the dimensions and ACM success. Indeed, probably the most important conclusion that can be drawn is the apparent applicability of the ACM approach to conditions of adversity.

I hope this exercise has shown the immense variety in site conditions—and made clear the dangers of relying exclusively on unilinear analysis of standardized, quantitative variables. This analysis strengthens my conviction that the reductionist approach, taken alone, can lead us to the wrong conclusions when we are looking at complex, dynamic systems. I believe that scientific expertise can be put to better uses: we can conduct holistic analyses, oriented toward improving conditions, rather than simply making scientific pronouncements that will be proven invalid tomorrow; we can look at the connections among parts of these systems; and we can improve our ability to facilitate interactions, manage conflicts more effectively, work with different segments of communities, and make our bureaucracies more flexible and responsive.

Reductionist science has made enormous contributions to our understanding of specific aspects of our world and the people within it, and it will continue to do so. But the paradigm proves inadequate for understanding and improving complex and dynamic systems—precisely the kind we must deal with if we are concerned about natural resources management and human well-being.

Appendix 10-1. Measures of Social Capital and Collaboration

Social capital appears to be related to the increased collaboration that we were seeking to catalyze in these forest communities. Our teams struggled with measures of both concepts. In addition to the simple guidelines I provided in Box 10-6, here are several approaches to the measurement of social capital and collaboration, most from ACM sites.

Measuring Social Capital

While our research was under way, Krishna and Uphoff (2001) were devising a handy, quantitative way to measure social capital in the context of Rajasthan, northern India. They used six specific questions, with multiple-choice answers, designed to fit the cultural context. The first three focused on structural social capital: who would come forward to help deal with a community-wide crop disease, who historically looked after the commons, how intravillage disputes were resolved. The last three were designed to measure cognitive social capital: who felt free to correct errant children, what motivated people to plant trees and grow fodder (desired by the community at large), and whether people trusted one another (as demonstrated by their preference for owning 10 units of land alone or owning 25 units with someone else). They tested alternative questions that reflected similar structural and cognitive aspects of social capital (how predisposed people were to act cooperatively in support of one another) and found that such substitutions did not adversely affect the utility of the index.

Four ACM teams developed measurement schemes for social capital or related concepts that seemed to work in their sites: one in Indonesia, two in Nepal, and one in the Philippines. The fact that these are the oldest ACM sites may reflect the evolving understanding that the teams gain as their work progresses.

The Indonesian measurement scheme is the simplest and most elegant. Wollenberg and her team identified three components of what they called village capacity:

- strength of leadership;
- cohesiveness of the community; and
- access to information.

Measures of strength of leadership included economic status, external connections, community support, and education. Cohesiveness of the community was measured by internal loyalties, economic status, human resources, and external connections. And access to information included knowledge of the territory, knowledge of the mapping activities under way, and community discussion. The team used these measures in their evaluation of level of success in a community mapping effort involving 27 villages in Bulungan, East Kalimantan, Indonesia (Anau et al. 2005).

The primary ACM team in Nepal, where each researcher was working in one community, took a somewhat different tack. McDougall and her associates (2001, 10), for instance, looked for changes in five main areas:

1. Relations with external stakeholders.

2. Relations among internal stakeholders (caste, ethnic, and income groups):

- level of trust and connectedness;
- levels of conflict versus collaboration and perception of associated satisfaction;
- level of social mobilization/activeness for joint efforts;
- distribution of power;
- perception of equity within the forest user group;
- awareness and attention to gender, diversity, and equity; and
- culture of decisionmaking (e.g., top-down, chairperson- or elite-led vs. participatory).

3. Sense of ownership of the forest user group.
4. Attitudes toward outside stakeholders (including interdependence versus dependency).
5. Ability to manage conflict internally.

A second ACM team in Nepal, under the auspices of the NGO ForestAction, surveyed eight communities, and the team identified variables that they considered useful in measuring collaboration:

- extent of the forest user group's links and communication with outside stakeholders;
- extent of the group's vertical linkage with macrolevel regulating and enabling institutions, including any feedback to the macro level;
- representation of low-status occupational castes (*dalit*);
- extent of participation of group members in decisionmaking;
- internal communication; and
- links with the district forest officer, networks and federations, and other institutions.

From the standpoint of internal collaboration, they looked at these signs:

- perceptions of legitimacy and autonomy;
- trust;
- power relations;
- communication and knowledge sharing;
- collaborative action, costs, and benefits; and
- other institutional mechanisms.

Lorenzo (2002, 5) also came up with indicators of collaboration in the Palawan site in the Philippines:

- Major stakeholders all have an opportunity to participate in a manner satisfactory to them (including decisionmaking).
- There is effective information flow (communication) among stakeholders.
- New understanding and knowledge are produced among the stakeholders (joint learning).
- Management actions are undertaken cooperatively.

We anticipate that more measures will be forthcoming as the other sites progress.

CHAPTER ELEVEN

Catalyzing Creativity, Learning, and Equity

*I*N ASSESSING OUR APPROACH, I LOOKED first at the results we obtained, focusing specifically on the creativity, learning, and equity that the teams were able to catalyze in many sites. I examined our impacts, first by looking at specific methods or approaches used in the field, and then by examining the livelihoods, environmental, and empowerment-related effects of our work. I next looked at the various conditions under which the adaptive collaborative management (ACM) process was implemented, focusing first on larger scale issues (devolution, forest type, population pressure, and management goals) and then on the microlevel, with its many cross-scale linkages (human diversity, level of conflict, and social capital). In this chapter, I look at aspects of our own team behavior that might influence the ACM process (Box C in Figure 1-4, the unidirectional model of ACM, presented in Chapter 1). All the field teams began by trying to follow the series of steps outlined in Chapter 1—context studies, participatory action research, institutionalization of monitoring or social learning, and strengthening of links with other stakeholders both horizontally and vertically. Here, I portray the variation in just how they did that. In making fudge, when and how you stir the pot makes a big difference; I think it does for ACM as well.

Process is not an easy feature to describe, and this is doubly true if sites and activities are dispersed throughout the globe and one is methodologically committed to addressing different issues in different sites. I must infer a great deal from discussions, e-mails, reports, papers, and field visits. Yet, despite the uncertainties and incompleteness of the information, the issue is too important to ignore. The most fundamental truth in this chapter is the diversity that characterizes a "standardized" process involving complex adaptive systems (including human beings). The discussion is divided into three sections: team characteristics, qualitative approaches, and conclusions and implications for future work.

Team Characteristics

Four process variables—pertaining to the individuals conducting the research and the relationships among us—seem potentially pertinent to the process of conducting ACM:

- qualifications of the field researcher;
- discipline of the field researcher (including research interests);
- research partners; and
- integration into the global team.

Qualifications of the Field Researcher

The qualifications of ACM facilitators range from Ph.D.s to high school diplomas (see Table 11-1). And there have been outstanding ACM facilitators at all levels. The expertise of individuals with master's degrees and doctorates has been important. Such people understand research itself and the importance of being systematic. They have access to a broad range of comparative data, based on their studies, and they can typically draw on fairly extensive personal research experience. They are trained to be self-critical, to analyze results, and to present the findings in forms that are acceptable to other scientists. These have been critical talents in our efforts to monitor and evaluate what we are doing.

On the downside, some may also have difficulty escaping the positivist leanings in their backgrounds. They may have difficulty seeing and dealing with the complex interactions among systems. They may be too wedded to scientific protocol to respond to unusual opportunities that emerge in the field. Having so much higher education and status, they may have difficulty relating to villagers (or vice versa). And perhaps most important (especially in developing countries), other pressing demands on their time may interfere with their ability to spend time in the field.

High school and college graduates have the advantage of being less removed by status and education from villagers. They also may find the idea of working in villages more appropriate or acceptable for their level of education than do the highly educated. Most of the researchers who fall into this category are young, national staff, full of enthusiasm, and excited to try new approaches. Many are willing to withstand the rigors of difficult living conditions and likely to have fewer family responsibilities pulling them away from the field. They bring energy, vitality, and openness to new ideas.

On the downside, they may have a limited repertoire of methods and skills to use with communities. They may have unrealistic perceptions of just how much they really know, sometimes incorrectly and inappropriately answering villagers' questions with unwarranted confidence when they should be listening and exchanging views in a more egalitarian mode.

The potential for ACM-type approaches will be greatly enhanced if it can be applied by people without high educational achievements. Our experience suggests that this is indeed possible but will require devoting more time than we did to training ACM facilitators. We had a few formal training workshops (e.g., a few days for the Bogor-based researchers in 1999, with Bob Fisher and Mohammad Hossein Emadi on participatory action research; two weeks of joint planning for the new Asia-based teams in February 2000 in Bogor; a few days in Nepal with a selection of ACM facilitators at an International Scientific Steering Committee meeting in November 2001). Informal training included some lectures by more experienced team members and ongoing mentoring. A longer, structured course would be beneficial for junior researchers being sent out to communities to catalyze adaptive collaborative management. There are crucial social science skills that can be shared with others and that can enhance our effectiveness at participatory action research and facilitation (e.g., Russell and Harshbarger 2003). Ongoing mentoring is also critical to this approach.

Discipline of the Field Researcher

Our teams represented a wide range of disciplines (Table 11-1). One of the first discipline-related patterns that emerged had to do with the definition of a research site. Social scientists tended to consider a research site to be a village or some other human-defined entity; biophysical scientists were more likely to think in terms of a landscape-based entity. In Indonesia, for instance, Trikurnianti Kusumanto, whose topic for her doctoral dissertation was social learning in the context of ACM, identified the village of Baru Pelepat as her research site. Stepi Hakim, trained in zoology and natural resource management, identified the Lumut Mountain Protection Forest as his site, focusing on a particular watershed.[1] This kind of difference is evident in the graphic representation of important local interactions: the Zimbabwe team (led at that stage by a forester) put the forest reserve at the center of the diagram; the Nepal team (led by a social scientist) put the forest user group at the center of its diagram.

Both approaches work. In the African sites under the direct supervision of Ravi Prabhu (a forester), all sites are identified as formal forest reserves, yet work in villages remains central. Chimère Diaw (an anthropologist) and his team formalized our interest in working at various levels—micro, meso, and macro—in the context of their mesolevel policy work. His group identified important policy domains and selected the research sites in Cameroon accordingly (minority groups, conservation areas, the forest frontier, and council and community forests as policy issues), but none of the teams abandoned research in villages and forests.

A researcher's background affects the kinds of data he or she thinks are important but appears unrelated to the success or failure of our work. In Indonesia, for instance, we had a social learning specialist (Kusumanto), a zoologist and conservation specialist (Hakim), and an anthropologist and natural resource management specialist (Wollenberg) acting as primary facilitators on the three respective sites. Each started out with different interests—Kusumanto began looking at social learning and gender and equity issues; Hakim was interested in strengthening communities' involvement in development and conservation; and Wollenberg was working on future scenarios and mapping as an empowerment strategy—but all became involved in the vibrant discussions about decentralization in Indonesia and began shaping local responses to this policy opportunity.

Table 11-1. *Qualifications and Global Integration of Field Researchers*

Researcher	Site	Qualifications	Discipline	Global integration
Carol J. Pierce Colfer	Global	Ph.D., M.P.H.	Anthropology, public health	H
Ravi Prabhu	Global	Ph.D.	Forestry	H
Cameroon				
Chimère Diaw	All Cameroon	Ph.D.	Anthropology	H
Cyprain Jum	Ottotomo	Bachelor's	Anthropology	H
Anne Marie Tiani	Ottotomo, Campo Ma'an	Ph.D.	Ecology	H
George Akwah	Campo Ma'an	Bachelor's	Anthropology	H
Joachim Nguiébouri	Campo Ma'an	Technician	Agronomy	M
René Oyono	Lomié, Dimako	M.A., Ph.D. candidate	Rural sociology	H
Samuel Assembe	Lomié, Dimako	Bachelor's	Law	L
Charlotte Kouna	Lomié, Dimako	High school, technician	Forestry	L
William Mala	Akok	Master's	Agriculture	L
Patrice Bigombe	Dimako	M.A., Ph.D. candidate	Political science	L
Patrice Etoungou	Lomié	M.A., Ph.D. candidate	Anthropology	L
Samuel Efoua	Lomié	High school	Humanities	L
Yvette Ebene	Ottotomo	M.A.	Social sciences	L
Ghana				
Dominic Blay	All Ghana	Ph.D.	Forest ecology, silviculture	L
Kofi Diaw	All Ghana	Ph.D.	Development planning	L
Janet Boateng	All Ghana	Certificate	Forestry	L
Sophia Anorba Sarpei	All Ghana	Certificate	Forestry	L
Alex Asare	All Ghana	M.Sc.	Forestry	L
Valerie Kwami	All Ghana	B.Sc.	Natural resources management	L
Frank Adomako	All Ghana	B.Sc.	Natural resources management	L

Name	Site	Degree	Field	
Madagascar				
Louise Buck	Ranomafana	Ph.D.	Social sciences	L
Lalaina Rokotoson	Ranomafana	M.A.	Environmental law	L
Bruno Ramamonjisoa	Ranomafana	Ph.D.	Forest economics	L
Jean Rakotoarison	Ranomafana	J.D.D.	Law	L
Maminiaina Razafindrabe	Ranomafana	Ph.D.	Rural sociology	L
Malawi				
Lusayo Mwabumba	Malawi sites	M.Sc.	Environmental forestry	L
James Milner	Ntonya Hill	?		L
Judith Kamoto	Chimaliro	M.Sc.	Social forestry	M
Edward Sambo	Chimaliro	High school ++	Village collaborator	L
Samuel Chirwa	Chimaliro	Primary school ++	Village collaborator	L
Malaitcha Banda	Chimaliro	Primary school ++	Forest guard	L
Linly Ngwenya	Chimaliro	Primary school	Village collaborator	L
Mr Chamanza/ Gondwe	Chimaliro	College certificate	Forestry extension	L
Zimbabwe				
Ravi Prabhu	Africa and Asia	Ph.D.	Forestry	H
Frank Matose	Mafungautsi	Ph.D.	Development studies, rural sociology	H
Wavell Standa-Gunda	Mafungautsi	M.Sc.	Agricultural economics	H
Richard Nyirenda	Mafungautsi	Diploma (B.Sc. equivalent)	Forestry	H
Tendayi Mutimukuru	Mafungautsi	M.Sc.	Agricultural knowledge systems	H
Witness Kozanayi	Chivi	Diploma (B.Sc. equivalent)	Agriculture	L
Hilton Madevu	Mafungautsi	B.Sc.	Agricultural economics	L
Indonesia				
Linda Yuliani	All Indonesia	M.Sc.	Natural resources management	H
Stepi Hakim	Lumut Mountain	M.Sc.	Natural resources management	H
Amin Jafar	Lumut Mountain	High school +		L
Suprihatin	Lumut Mountain	B.Sc.	Forestry	L
Trikumianti Kusumanto	Baru Pelepat	M.Sc.	Agronomy, forest policy	H

Continued on next page.

Table 11-1. *Continued*

Researcher	Site	Qualifications	Discipline	Global integration
Indonesia (Continued)				
Yayan Indriatmoko	Baru Pelepat	B.A.	Anthropology	H
Marzoni	Baru Pelepat	B.Sc.	Agronomy	L
Effi Permata Sari	Baru Pelepat	B.Sc.	Agronomy	L
Fauzi Syam	Baru Pelepat	Master's	Law	L
Helmi	Baru Pelepat	Bachelor's	Law	L
Akhmad Albar	Baru Pelepat	Bachelor's	Law	L
Eva Wollenberg	Bulungan	Ph.D.	Natural resources management, human ecology	M
Godwin Limberg	Bulungan	M.Sc.	Agronomy	L
Ramses Iwan	Bulungan	B.A. candidate	English	L
Made Sudana	Bulungan	B.Sc.	Forestry	L
Kyrgyzstan				
Kaspar Schmidt	All Kyrgyzstan	Ph.D. candidate	Forestry, rural development	M
Nurbek N. Mamatov	All Kyrgyzstan	M.Sc.	Environmental science	L
Mahabat T. Karajeva	Gava[a]	Ph.D. candidate	Biology	L
Nurlan R. Akenshaev	Ortok, Üzgen	M.Sc. candidate	Economics	L
Nepal				
Cynthia McDougall	All Nepal	M.Phil.	Natural resources management, political science	H
Netra Tumbahangphe	Kaski District	M.Sc.	Anthropology, environmental management	L
Sushma Dangol	Bamdibhirkhoria	B.Sc.	Forestry	H
Chimjeewee Khadka	Bamdibhirkhoria	B.Sc.	Forestry	L
Raj Kumar Pandey	Deurali-Baghedanda	B.Sc.	Forestry	L
Laya Prasad Uprety	Sangkhuwasabha Dist.	M.Sc.	Anthropology, social development	H
Narayan Sitaula	Manakamana	M.A.	Sociology	H
Kalpana Sharma	Andheribhajana	M.A.	Home science	H

Name	Location	Degree	Field	
Philippines				
Herlina Hartanto	All Philippines	M.Sc.	Natural resources management	H
Cristina Lorenzo	Palawan	M.A.	Development management	H
Asuzena Estanol	Palawan	B.Sc.	Forestry	L
Erlinda M. Burton	Basac	Ph.D.	Anthropology	M
Cecil Valmores	Basac	M.A. candidate	Anthropology	M
Lani Arda-Minas	Basac	M.A. candidate	Public administration	M
Bolivia				
Peter Cronkleton	Guarayos TCO	Ph.D.	Anthropology	M
Omaira Bolaños	Santa María	M.A. candidate	Anthropology	M
Tracy van Holt	Salvatierra	M.A. candidate	Biology	L
Robert Keating	Guarayos TCO	B.A.	Small enterprise	L
Marco Antonio Albornoz	Guarayos TCO	B.Sc.	Forester	L
Orlando Melgarejo	Guarayos TCO	College certificate	Forester	L
Brazil				
Magna Cunha dos Santos	Porto Dias, Acre	B.Sc.	Forestry, social science	L
Nívea Geórgia Marcondes	Porto Dias, Acre	B.Sc.	Forestry	L
Benno Pokorny	All Pará sites	Ph.D.	Forest engineering	M
Guilhermina Cayres	Tailândia, Muaná	M.Sc.	Small-scale agriculture, sustainable development	L
Westphalen Nunes Lobato	Muaná	M.Sc.	Small-scale agriculture, forestry	L
Dörte Segebarth	Gurupá	M.Sc.	Geography	L
Rozilda Drude	Gurupá	B.Sc.	Linguistics	L

[a]Gava is between Arstanbap–Ata and Achy.

As a social scientist, I see the strengths of social science and in some cases have bemoaned a facilitator's inexperience in my discipline. Intracommunity variation, though widely attended to in our sites, was more likely to be ignored in sites dominated by biophysical scientists. Some of the "soft," attitudinal variables identified by McDougall et al. (2002b) as crucial to success may not have been adequately addressed by those without social science training. Occasionally, facilitators with a background in forestry or ecology were unaware of their own assumptions, insensitive to critical human issues, and unskilled in some community interactions. Yet the reverse was surely equally true. Social scientists undoubtedly missed important connections between human behavior and ecological impacts, or opportunities to build on local customs that had the potential to improve the environment—things that would have been clear to a biophysical scientist.

In this truly interdisciplinary endeavor, facilitators complemented each other's personal strengths—skill in community interactions, understanding of interactions with the local bureaucracy, knowledge of biophysical issues, competence in developing monitoring systems—in a joint effort. To work toward forest sustainability in complex systems, the teams needed expertise in both biophysical and social components—again suggesting that we pay more attention to staff training. This also suggests, given a common tradition of attributing lower prestige and importance to the social sciences, that perhaps having a social scientist as a team leader may be sensible.[2]

Research Partners

Our most central partners in nearly all the sites were communities. Our intent from the start was to conduct our research in cooperation with other organizations and, if possible, to piggyback on ongoing, village-based research, thereby both leveraging our own resources and increasing the long-term impact of our research findings.

Of the 30 sites, activities in 19 were intimately linked with major national environmental or forestry agencies from the start (see Table 11-2). In 1 site (Mafungautsi), the governmental agency was the only major research partner (besides the communities). Eight sites had major interactions with one or more NGOs. Eight more sites linked cooperation with the government with close cooperation with a major internationally funded project; 2 of these cooperated with a local university or research institution. At least one ACM team in all countries except Malawi and Bolivia worked closely with governmental agencies.[3]

In 15 sites, close links were maintained with NGOs. In 3 of the Indonesian sites (Rantau Layung, Rantau Buta, and Baru Pelepat), these were the only dominant research partners. In 8 more, involvement with NGOs was complemented by close involvement with national environmental agencies. Two of these also included complementary involvement on major projects; 2 more included research or academic institutions; and 1 included all categories of partners (Campo Ma'an). Of the remaining 4 sites with close NGO links, 2 worked only with international projects, 1 worked only with a university, and 1 worked with a project and research institutions. Teams in Brazil, Cameroon, Indonesia, Nepal, and the Philippines worked closely with NGOs.

Fourteen sites were intimately linked with major international projects. The 4 Kyrgyz sites linked only with the project and the government. In 2 sites, teams worked with projects, government, and research institutions; in 2 more, they worked with projects, government, and NGOs, and 1 worked with all four types of institutions. Two teams worked with projects and NGOs only, and 3 worked with projects, NGOs and research institutions. These 14 sites were in Bolivia, Brazil, Cameroon, Kyrgyzstan, Madagascar, and Nepal.

Finally, 14 sites (in Bolivia, Brazil, Cameroon, Ghana, Madagascar, Malawi, and the Philippines) had universities or research institutes as major partners. The 2 Malawi sites were closely linked only with a research institute. Six were also linked with international projects; 3 of these also worked with a government agency. Three sites focused on work with a university and a government agency, and 1, on a university and an NGO. Of the final 2 sites, 1 combined a research institution, a project, and an NGO (Akok), and the other included all categories (Campo Ma'an).

Only one site—Tailândia, in Pará, Brazil—partnered with a logging company.

Not surprisingly, there are advantages and disadvantages to all of these partnerships.[4] Our involvement with NGOs has in many cases allowed us to build on an existing knowledge base and NGO rapport with communities. It has given us access to the particular skills of that NGO, as well as their labor and creativity.

However, there have been significant problems on several sites. In Indonesia, for instance, relations were easily strained, because of the NGO personnel's negative stereotypes of CIFOR (as a well-funded, international organization) and some unwarranted assumptions about what we might want to do. The inequitable access to resources between national NGOs and international research centers is another, understandable bone of contention. ACM team members spent considerable time and effort assuaging NGO concerns and keeping collaboration on track.[5] In Acre, the NGO had previously emphasized environmental quality, without paying much attention to working equitably with communities. Conveying the importance of doing this, as well as transferring the skills to do so, took time and persistent effort from our ACM facilitator, backed up by social scientists from the University of Florida. In both Campo Ma'an and Lomié, in Cameroon, cooperation with NGOs started slowly, partly because of different institutional cultures and suspicion directed at the ACM team. Etoungou (2002, 24–25) reflects one side of this periodic tension, in her description of the situation in Lomié, Cameroon:

> The current staff of these NGOs are clear proof of this [interest in income for their own families rather than goals of the organization]. All the secretaries, accountants, deputy directors, development workers, night watchmen, and project officials, are related or very close to the Director or equivalent. Two former university students who have returned to the village see them as "accomplices of the Sustainable Enslavement Program."

One might equally imagine a quote from an NGO collaborator, arguing that CIFOR personnel are interested only in the cushy salaries we are able to obtain as employees of an international center, or at best that we are dupes of globalization, inadvertently or unwillingly playing into the hands of the capitalist market economy.

Table 11-2. *Major Partners for ACM Research*

Site	Government departments	NGOs	Major projects	Research institutions, universities	Timber company
Mafungautsi, Zimbabwe	FC				
Muaná, Brazil	District Government			Museu Paraense Emilio Goeldi	
Adwenaase, Ghana	FORIG/For. Serv. Div.			Forest Research Inst./Univ. Kumasi	
Namtee, Ghana	FORIG/For. Serv. Div.			Forest Research Inst./Univ. Kumasi	
Raromafana, Madagascar	Forest Dept., Park		U.S. AID, Ranomafano NP	U./Antananarivo/Fianarantsoa Law School/ MICET	
Dimako, Cameroon	MINEF		Forêts et Terroir/CIRAD	CIRAD	
Arstanbap-Ata, Kyrgyzstan	SFS, leshoz		KIRFOR		
Achy, Kyrgyzstan	SFS, leshoz		KIRFOR		
Ortok, Kyrgyzstan	SFS, leshoz		KIRFOR		
Uzgen, Kyrgyzstan	SFS, leshoz		KIRFOR		
Manakamana, Nepal	MoFSC	NEW ERA	NUKCFP (LFP)		
Andheribhajana, Nepal	MoFSC	NEW ERA	NUKCFP (LFP)		
San Rafael–Tanabag–Concepción in Palawan, Philippines	DENR, LGUs	Budyong Rural Development Foundation			
Bulungan, Indonesia	MoF	WWF			
Bamdibhirkhoria, Nepal	MoFSC	NORMS			
Deurali-Baghedanda, Nepal	MoFSC	NORMS			
Basac, Lantapan, Philippines	DENR, LGUs	ATD		ICRAF, Heifer Int'l	
Ottotomo, Cameroon	ONADEF			ICRAF	
Campo Ma'an, Cameroon	DGIS, MINEF	SNV, Tropenbos, WWF	Campo Ma'an NP Proj.	Tropenbos, ICRAF	
Rantau Layung, Indonesia		Padi			
Rantau Buta, Indonesia		Padi			
Baru Pelepat, Indonesia		Gita Buana, WWF			
Lomié, Cameroon		SNV	Community Forests		
Canta-Galo, Gurupá, Brazil		FASE	FASE-Gurupá (EU funded)		
Akok, Cameroon		FORCE	ASB	IITA, IRAD	
Porto Dias, Brazil		CTA, PESACRE		U./Florida	

Guarayos, Bolivia
Tailândia, Brazil
Ntonya Hill, Malawi
Chimaliro, Malawi

BOLFOR
ITTO

U./Florida
EMBRAPA
Ctr. For Soc. Res./Zomba
FRIM, Mzuzu Univ.

[anonymous]

In Nepal, one of the NGOs was persona non grata with our principal government partner agency, and only careful diplomacy and creative solutions kept this organization involved in our work. But overall, the process has been worthwhile.

Partnerships with government agencies legitimize one's activities in the country; they provide access to educated, formally qualified personnel; they can provide an institutional home; and they can become the users of one's final results. But government personnel are not always the most highly motivated or most effective in community work. In Ghana, for instance, ongoing responsibilities within the government department have made it difficult for some personnel to perform ACM functions; and the on-site partners from the community forest management unit of the Forest Services Department essentially washed their hands of local problems (Blay 2002a). This was also true, to a much lesser degree, in Zimbabwe, in the initial phase of the research period. Although one of our initial concerns was that government personnel would interfere with our work because of its focus on community empowerment, such interference was generally absent. In fact, in Nepal and Ottotomo (Cameroon), and later in Zimbabwe, governmental input was highly valued by the team members.

Links with major projects had mixed results. In Bolivia and Kyrgyzstan, the ACM teams have been effectively integrated into large projects. These teams have been able to access input from other disciplines and project support of various kinds, as well as contributing to the ongoing activities of those projects without serious problems. But in Cameroon, ACM involvement with the Campo Ma'an project has been fraught with conflict and mutual recrimination (perhaps because six major partners are involved!). This appears to have been the case with the Madagascar project as well.

Similarly, with universities and research institutes there has been great variation. The Brazil work was greatly assisted by both the University of Florida and the *Museu Paraense Emílio Goeldi,* and the Bolivian ACM team also benefited from inputs from Florida. In Malawi, on the other hand, the teams do not seem to have been supported very effectively by their partner research institutions.

Integration into the Global Team

When I speak of integration into the global team, I am thinking of the following issues:

- level of interaction between the local researchers and headquarters;
- involvement in the creation and implementation of the broader research framework;
- cooperation, completeness, and timeliness in providing reports;
- responsiveness to queries from colleagues; and
- voluntary contributions and leadership in ACM affairs.

It is important to remember that I have seen team integration from a particular perspective. Although as program leader, a regular traveler, and editor of *ACM News*, I probably have a better sense of this issue than most team members, there may be regional interactions of which I was unaware (see Table 11-1, for my perceptions of each team's global involvement).

There are several reasons for the differing integration of field teams into the global research endeavor. One important variable was the location of the site coordinator. Site coordinators for Indonesia, Nepal, and the Philippines lived in Bogor, making collaboration among them and between them and CIFOR management easier than for any other sites: problems could be resolved in person, questions could be answered quickly, pressure could be applied and input provided easily.

A second rung of involvement occurred in the Zimbabwe and Cameroon sites, where the leaders were persons who had been located in Bogor, and who had control of significant project finances. These leaders had been centrally involved in the conceptualization of ACM: they understood and were committed wholeheartedly to the goals and were able to pass along their enthusiasm and understanding to their team members, by and large. The French-English language barrier in Cameroon worked against integration, however, slowing the pace of communication among field teams, supervisors, and headquarters.

A medium level of integration occurred in the South American sites. No one who had helped design the ACM approach was directly involved in these sites (though I supervised them from afar). Additionally, the Brazil sites had only minuscule amounts of funding, and our work depended on the inherent enthusiasm and commitment of individuals who were basically working for a pittance. One of these ACM researchers noted with dismay the minimal responses he got from others in the field, in his attempts to increase his integration in the overall ACM work. Fortunately, we also had informal but extremely valuable backup support from Marianne Schmink, Noemi Porro, and Samantha Stone at the University of Florida, who kept the Portuguese-English language barrier from completely disrupting our work. The Bolivian work followed the initiation of the Brazilian work and involved people who had participated in ACM-Brazil. Their experience may have complemented the positive impacts of the University of Florida presence in both countries and strengthened the Bolivia coordinator's commitment and understanding of the approach. The South American work began later than all the Asia sites and some of the African sites, as well.

Similarly, the Kyrgyzstan efforts represent an intermediate level of involvement. These sites were the last to begin, and they were well-funded from a source outside CIFOR. We recognized that external funding and the involvement of the primary researcher in a Ph.D. program might limit his freedom to carry out our approach. However, the enthusiasm of the team and the donors for the ACM approach resulted in initially encouraging integration of their work with ours. Although we have had excellent cooperation in terms of information provision from this team, the ACM approach there has fallen by the wayside.[6]

Our lowest levels of success at integration occurred in Ghana,[7] Malawi, and Madagascar. The Ghana and Malawi sites were supervised from Cameroon and Zimbabwe, respectively. No one who had been involved in developing the ACM ideas was directly involved in the Ghana or Malawi research; and activities there began considerably later than in the Asian and other African sites. Unlike the Cameroon and Zimbabwe sites, there were no CIFOR offices from which to work, few colleagues involved in related research, and presumably little of the financial flexibility that can derive from being near those who hold the purse strings. Since Malawi and Ghana are English-speaking countries, language barriers were not a problem. But in Ghana, the cancellation of a planned sabbatical for the social scientist was reported to have resulted in his unavailability for ongoing work, though he remained on the payroll, thus making it difficult for the rest of the team to get the needed input and effort.

The Madagascar activities predated ACM, and although activities there were never intended to replicate the formal ACM steps, the researcher who served as coordinator had been involved in the development of ACM and fully understood the approach. However, she could not deal effectively from afar—Cornell University—with the difficult problems that emerged (specifically death and dishonesty). Team members also all had language problems, and funding for the Madagascar work was already winding down.

Finally, issues of personal motivation and opportunity were important on all the sites. Some people easily and comfortably fit into a broad umbrella framework; others prefer to strike out on their own. Some people love doing research or community action or have passionate commitments to development or conservation; others do not. Some researchers are enmeshed in bureaucracies that interfere with their performance; others are not. Some people love to analyze and write up their research results (making it available thereby to other team members); others dislike this task. In any group of researchers, there will be variations. It is important to remember that interesting results have come from those who were not so engaged in our original plan.

Qualitative Approaches

There are a variety of ways in which our ACM teams have run with the "squishy" ball we gave them. Our framework was designed to be flexible and loose, an umbrella rather than a straitjacket. The following observations describe eight kinds of differences in the ways that different teams approached their task.

Micro, Meso

Some teams (Jambi, Indonesia; Nepal; Brazil) focused immediately on a village and began establishing rapport with community members, in preparation for the intensive process of participatory action research. The ACM facilitators here wanted to initiate an intensive process of (self-)directed change in the communities.

Other teams (Cameroon, Bolivia, Ghana, Zimbabwe) worried that this approach would be too narrow, resulting only in case studies; they took a wider view. In Cameroon, the team identified important policy issues, selected sites that would provide insights on those policy issues, and then conducted workshops among multiple stakeholder groups (such early workshops were also conducted in Bolivia; East Kalimantan, Indonesia; Ghana; and Zimbabwe), working "down" toward communities. In Nepal, the teams were also interested in the mesolevel, but approached it from below, with a process that involved "spiraling up and out" from the communities. McDougall et al. (2002b, 140) note,

> ... [M]eso level networks already existed, but the considerable distance and difficult terrain between the FUGs [forest user groups] involved acted as a barrier to their activeness (amongst other factors such as weak leadership). This was the rationale for developing the more locally based networks (e.g., in Hansapur, the FUG neighborhood network).

Similar geographical constraints to scaling up exist in Porto Dias, in Acre.

Kamoto (2003c, 10)[8] describes the changes that she witnessed over the course of the ACM project in Malawi:

> There are stakeholders at macro, meso and micro level involved in forest management at Chimaliro. Their interests vary; some are interested in providing inputs for tree nursery establishment, others in providing extension services and others in technical management and still others in the use of the forest resources. These stakeholders had minimal interaction and they were not collaborating as they do now, before the ACM project was introduced at the site.

Process Oriented vs. Goal Oriented

Some team members (Nepal; Jambi and Bulungan, Indonesia; Lantapan, Philippines) were interested in the mechanisms for social learning. They wanted to catalyze a process and see how it unfolded. Others (Pasir, Indonesia; Palawan, Philippines) focused on monitoring systems or identifying priorities collaboratively. They were more goal-directed. Still others (Acre, Brazil; Madagascar; Zimbabwe) did both. A small minority seemed to see ACM as a means to implement existing governmental goals (Ghana, and to a lesser extent, Malawi).

Hartanto and her team summarize the ACM approaches that led to the kinds of outcomes they obtained in the Philippines, in Box 11-1. These observations, though not duplicated on all sites, do reflect common themes.

Attention to Equity Issues

Some teams focused a great deal of effort on intracommunity equity issues. They, and the communities with which they worked, were concerned that forest benefits were distributed inequitably, that some players had virtually no voice in management and other decisions. Working with community members, they addressed these concerns explicitly (Bolivia; Kyrgyzstan; Nepal; Zimbabwe; and Campo Ma'an, Ottotomo, Lomié, and Dimako, Cameroon). The Madagascar team was concerned about giving marginalized groups more control over local resources to reduce their dependence on the forest (e.g., by shifting to alternative livelihood strategies). In Nepal and Bolivia, for instance, special gender workshops took place. In others, equity was less of a concern (Brazil; Akok, Cameroon; Ghana[9]; Malawi).

Box 11-1. Approaches in the Philippines

1. Approaches that enhance information sharing and communication horizontally (includes internally within the members of their institutions and external communication to other organisations and institutions), and vertically. The latter includes information sharing between the lower level institutions and its central office, and between the People's Organizations and policymakers. We found that information flow would be effective if it took place using several different communication platforms (meetings, workshops, training sessions, dialogues, working groups, etc.) and different media (radio programs, newsletters, bulletin boards, etc.) so that it would reach the intended target audience through different channels.

2. Approaches that enhance and improve the skills, knowledge, and attitude of the members of People's Organizations and communities so that they become more forward and future looking, self-reliant, and confident. These include mentoring, formal and informal training sessions, and experimentation. The tools such as problem tree analysis and SWOT (strengths, weaknesses, opportunities, and threats) analysis improved their analytical thinking, while future and anticipatory scenarios encouraged them to anticipate and plan for the future.

3. Monitoring and processes that encourage deliberate and analytical reflections on past experience and other people's experience to extract lessons learned and their implications on the next actions. Tools such as problem tree analysis, monitoring, and venues for learning such as cross-visits, and community-to-community exchanges are effective to activate reflections and learning.

4. Approaches that allow different stakeholders to discuss, negotiate, and subsequently come up with a mutual goal and jointly work together towards it, such as criteria and indicators and future scenarios, have proven to be effective in facilitating discussions and negotiations among different stakeholders.

5. Joint planning and action strengthen cohesiveness and transparency and provide a platform for different people and stakeholders to participate, contribute, and access the benefits. This approach is very applicable in the Philippines, as collective action is embedded within their culture and traditions, i.e. *bayanihan, pahina*, etc.

6. The complexity and interconnectedness of the issues that the People's Organizations have been facing were often overwhelming and made it difficult for them to strategize their actions to solve the issue in a systematic way. Tools such as criteria and indicators, causal diagramming or webbing techniques are useful in unlocking the complexities and revealing the relationships among several issues so that future actions can be targeted to address the heart of the problems.

7. In implementing ACM, we encouraged the People's Organizations to identify several key issues that they would address in a collaborative and adaptive manner. Prioritization proved to be effective as resource limitation does not allow all the issues to be addressed at once. An approach that encourages people to focus on several key issues is essential. Due to the interconnectedness of the systems, other connected issues would be addressed eventually.

—Herlina Hartanto et al. (2002a, 60–61)

Conservation vs. Utilitarianism

Some researchers saw their role as one of providing balance between two competing orientations—conservation and utilitarianism. In Nepal, the ACM researchers considered themselves a balancing force, urging active use and equitable distribution of benefits, while the forestry establishment focused on forest protection. The Madagascar mantra was conservation through sustainable use. In both Philippine sites, the ACM researchers served as a comparatively neutral facilitator among mesolevel stakeholders with strongly polarized views about conservation versus use.

Cunha dos Santos, the ACM facilitator in Acre, Brazil, was an employee of a conservation-oriented NGO (the Group for Research and Extension in Agroforestry Systems of Acre, PESACRE) and worked with an NGO with an even greater conservation orientation (the Center for Amazonian Workers, CTA), in a context where conservation issues had been important politically throughout the state but particularly in the Porto Dias Agro-Extractive Reserve.

Other teams were more clearly development-oriented. The ACM team in Indonesia likens the Pasir team's participatory action research activities to a community development project. A similar focus seemed clear in the Jambi site during the early days, but that has since shifted to a focus on governance.

To some extent, these differences were mandated by the concerns of the communities as well as the policy contexts. Whereas the Acre state government is explicitly environmentally oriented, the interests of the newly empowered districts in Indonesia are focused almost exclusively on income generation.

Policy Involvement

Policy issues were central to the concerns of researchers in Cameroon, Indonesia, and Madagascar. These countries were implementing forest policy changes that elicited wide interest. Although such changes were also under way in Bolivia, Ghana, and Kyrgyzstan, the ACM teams there did not get as directly involved in policy as they did in Cameroon and Indonesia. In Nepal and the Philippines, the forest policy had been in place for some time, and the focus of the research was on working out the kinks, fiddling with unanticipated effects, and adding value at the local level.

Facilitators' Approach

Although facilitation of the ACM process was crucial on all sites, there were differences of approach. In Zimbabwe, Chimaliro, and Malawi and in Pará, Brazil, for instance, members of the local communities became actively involved in the facilitation process (in Malawi and Zimbabwe, they were paid). In Andheribhajana and Manakamana, Nepal; Palawan, Philippines; and Pará, Brazil, formal training in facilitation was given to appropriate local and mesolevel stakeholders. This training also had important equity implications, since one emphasis was on the skill to obtain input from all participants. In most sites, the ACM facilitator retained this function—partially related to the amount of time the research has been under way. In Andheribhajana, Nepal, training on leadership development and empowerment was also provided, in the hopes that this would reduce the level of chaos that characterized typical meetings of the forest user group committee. Similar training took place in Lantapan, Philippines.

Building on Existing Institutions or Externally Created Ones

Facilitators in Kyrgyzstan, Malawi, Nepal, Philippines, and Zimbabwe structured their efforts around external institutions. The advantage of this approach is that one has a ready-made potential user in the governmental institution that created the local-level institution. The down side is that such institutions are likely to be upwardly accountable, and are often marginally relevant for daily life in communities.

ACM teams in Bolivia, Brazil, Ghana, Madagascar, and Indonesia worked through institutions developed by local communities. Such institutions are likely to be more vibrant and meaningful to local communities but of course may provide minimal potential and insights for wider use of research approaches and results.

Levels of Funding and Effort

We initially planned to have one to two sites in each of the countries where we conducted ACM research. However, site selection teams opted to use their budgets differently (and those in Brazil had significantly smaller budgets than any of the other sites). In some locales (Malawi and Acre, Brazil) a single facilitator-researcher managed the research process alone on a single site.

In Cameroon, the site coordinator undertook research on six sites (later reduced to five, and since increased to seven), each covering multiple villages, with cross-cutting miniteams working on each site. Similarly, a four-person team in Kyrgyzstan worked in four multivillage sites, and a four-person team in Madagascar worked in several sites in one region. The Nepal work was undertaken by two separate teams, each of which looked after two sites.

One common approach (Indonesia; Pará, Brazil; Bolivia; Zimbabwe; Philippines) was to have a small team of two to three people working on a given site. Of these sites, Baru Pelepat in Indonesia was the most labor intensive, with one village having the attention of seven researcher-facilitators (not all full time). In the other cases, a site included from 2 to 27 villages.

In Ghana and Malawi, the lead researchers were officially half-time. In reality, however, the Ghana lead researcher reportedly devoted only 10 to 15 percent of his time to ACM; in Chimaliro, Malawi, the lead researcher employed a villager to help her in each of the three "blocks" and had some help from the forestry extensionist.

The Nepal teams compared their own financial investments across sites and found that whereas the teams' financial or in-kind support to the forest user groups ranged from 10,000 Nepali rupees (NR) to 48,000 NR[10] (a difference of almost a factor of five), the outcomes across the four sites were fairly consistent. They concluded that although "financial input was useful, it is not a primary driver in the positive outcomes we have seen associated with the ACM approach ... which was relatively consistently 'applied' across the four sites" (McDougall et al. 2002b, 134–35). Comparing the financial inputs to all the ACM sites, we see that the financial incentives were not a primary driver of ACM success.

Box 11-2. Facilitators' Skills, Attitudes, and Behavior

Skills

- Thinking/analytical skills in analyzing issues, understanding, and synthesizing information in a logical way.
- Communication skills to articulate and translate different paradigms, mental modes for communicating to different stakeholders with mixed levels of literacy and education; skills in listening, encouraging participation, and making observations.
- Interpersonal skills in dealing with different people and institutions
- Management skills for planning, managing time, and resources.

Attitude and behavior

- Open.
- Honest.
- Just.
- Sensitive to culture.
- Attentive to people and process.

—Herlina Hartanto et al. (2002c)

Conclusions and Implications for Future Work

The implementation of our global program in adaptive collaborative management involved a great deal of site-by-site variation. We purposely selected sites with different conditions and ACM facilitators with a wide range of skills and disciplines because of our interest in approaching the human and environmental problems from an interdisciplinary perspective. The differences in approach are thus not particularly surprising. Our team management approach, replicating our approach in villages, also varied because of the autonomy we encouraged in field teams— autonomy granted in recognition of the diversity of contexts, talents, and interests the researchers would bring to and encounter in the field.

The skills, attitudes, and behavior that Hartanto et al. (2002c, 55) identify as impor- tant for ACM facilitators (Box 11-2) reflect the importance in the ACM process of independent thought. The nature of this process, as we have broadly defined it, requires a great deal from the facilitators, in terms of cognitive and emotional input and per- sonal attitudes. The variety of methods and approaches outlined in Chapter 5 demon- strates the creativity that was required of these facilitators: they chose methods and approaches that were compatible with the people and conditions they confronted.

Furthermore, the necessity to deal with change in the field context means that no blueprint from headquarters can provide reliable answers to the questions that will emerge. What White et al. (2001, 227) say about protected areas applies more widely: "Social tastes and preferences change over time, as do ecological systems, and for this reason protected area management is a dynamic and sustained process, and not a movement toward a predetermined, fixed state."

Change is as constant in logging concessions and villages as in protected areas; and the ACM facilitators were very much aware of the continual change they confronted—both change that they helped to catalyze and change that impinged upon them beyond their control. They were enmeshed in panarchies of mutually interacting systems operating at different scales and changing at varying rates, and they had to respond adaptively.

Uphoff saw the importance of looking at possibilities as well as probabilities, viewing events and people in a positive light, and recognizing the role of ideals in benign and effective development processes (Box 11-3). His observations about Gal Oya, in Sri Lanka, are relevant for the ACM process as well.

The diversity of approaches in a global program might be considered a short-coming, but I see it as our greatest strength. Although I have carefully analyzed our results in reductionist terms (consistent with the plans made in a context where reductionist science rules), the findings indicate only that such an approach alone is inadequate for complex systems. The ACM facilitators were, by and large, able to assess conditions in their sites and develop, together with local communities and other stakeholders, approaches that worked. The ACM approach allowed them to build on the interests and strengths of local community members. The best results emerged in sites where the ACM facilitators were most motivated and most committed to working with communities, building on the strengths of all.

Box 11-3. Lessons from Gal Oya

In our Gal Oya work, we needed to keep one eye focused on near-run probabilities while the other eye kept longer-run possibilities always in view. Successful vision merges the respective images, combining realism and imagination ….

To the extent that we treated things according to abstract categories (as if their "essence" was homogeneous), and especially when we judged people, plans, ideas, or motivations to be wholly unworthy, we were reducing the possibilities for getting positive contributions, which were always needed. We were led to begin using categories more cautiously, more tactically. We needed to make judgments, of course, but we learned to keep them tentative, being continually surprised by support from unanticipated sources and by some unexpected disappointments. Regarding ambiguous situations (or persons) in positive terms created expectations of a mutually beneficial outcome and tended to make such outcomes more likely. We were humbled by discovering that we sometimes made the "right" decisions for the "wrong" reasons, just as "right" reasons could produce "wrong" results.

To try to understand all these energized efforts without some reference to ideals would be mistaken, since the most active institutional organizers, farmers, and officials were the most idealistic in each group. Expressions of idealism were infectious, changing the way others defined and carried out their tasks. We were all catalysts in our own ways. The desire to see long-standing disadvantages eliminated was a motive force not to be underestimated or subsumed under reductionist notions that this activism was really only an expression of our egoism. We were not self-sacrificers or saints, but neither were we thinking only of ourselves. The influence of idealism was substantial even if by itself it was neither a necessary nor a sufficient cause. This realization was a useful insight coming from both—and thinking, self-interest, and altruism can and do coexist. One should not ignore ideals because they are not by themselves determinant and their results are not always predictable.

—Norman Uphoff (1996, 197, 301, 364–65)

CHAPTER TWELVE

Commentary and Conclusions

The experience of managing in complex adaptive systems is more similar to catching waves or looking for emergent corridors for action than pulling strings or working levers.

—Westley (2002, 354)

*T*HIS CHAPTER BEGINS WITH A SUMMARY of the conclusions that follow from the reductionist analysis of the conditions under which ACM was implemented and their respective levels of success, including some discussion of the world views of scientists. I then allow myself some commentary on several aspects of our work, including indigenous vs. settler populations, swidden agriculture, cooperation vs. competition, governance, and the pace of change. The next section provides an answer to the question, "does ACM work?" followed by suggestions about the context in and mechanisms by which it could be expanded. I conclude with a summary of the story line of this book.

Conditions, ACM Success, and Scientific World Views

Analyses linking particular conditions to ACM success are problematic: the same phenomena can have different meanings in different contexts, dynamic conditions at various scales can affect outcomes, the quality and quantity of information available for analysis differ, and our ability to predict results in complex adaptive systems is inadequate. In our data set for the 30 ACM sites, "success" rates were in many cases the same for multiple sites within a given country, a fact that seemed to skew some of the results, potentially leading to inappropriate conclusions. Nevertheless, I would like to summarize here the outcomes of these analyses, linking each condition to each level of success (from Appendix 1-1).[1]

For *devolution*, contrary to our initial expectation, ACM processes seemed most effective in sites with no formal devolution, second best in sites with strong, formal devolution, and worst in sites with weak, formal devolution.

For *forest type*, among the humid and dry tropical forest contexts we see a fairly even distribution of levels of success, including high, medium, and low. ACM in the temperate zone forests performed uniformly poorly, and ACM in the subtropical montane forests performed uniformly well—both artifacts of other factors, in my view.

For *population pressure*, ACM was least successful in the densely populated sites (though there were some high performers), and most successful in the sparsely populated sites (with no poor performances). Those sites with intermediate levels of population were distributed fairly evenly among the three categories of success.

For *management type*, timber (both large scale and small) and conservation areas did fairly well, each having only a small number of sites with low levels of success (2 or less). The sites characterized by multipurpose management included almost half high achievers and half low achievers, with only one intermediate site. The prediction that powerful stakeholders in large-scale timber and conservation areas might thwart ACM processes does not seem to hold true (though it is hard to deny that such actors represented difficult barriers in several cases and affected the issues addressed in others).

For *external diversity of external stakeholders*, sites with the least diversity did best, and those with the most diversity did second best. Sites with intermediate diversity had a bimodal distribution, with almost as many high achievers as low achievers and only one in the middle. There are examples of obvious successes and at least one case of dramatic lack of progress in all three categories.

For *internal diversity*, both high and intermediate levels of caste and ethnic diversity did well, with all the high-diversity sites doing well. The success rate declines with diversity, but even the 16 sites with low diversity levels included 5 high achievers and 3 medium achievers. Those sites characterized by low gender differentiation did best. Intermediate levels of gender differentiation resulted in fairly even distribution among the three levels of ACM success; and high gender differentiation sites had a bimodal distribution, with mostly high and low success.

For *national-level conflict*, the results are most dramatic—and the direct causal connections most implausible. Of the 12 sites in high-conflict countries, 10 had high levels of success; of the 8 sites in low-conflict countries, 7 had low levels of success. The sites in countries with intermediate levels of conflict were fairly evenly distributed among the three success categories. This finding suggests that adaptiveness and collaboration may be most easily catalyzed (or most emergent) in chaotic conditions.

For *local-level conflict*, again, good ACM performance is linked to significant conflict. The 3 sites with low levels of conflict were evenly split among the three categories of success. For intermediate levels of conflict, we obtained a bimodal distribution, with 8 quite successful, 8 quite unsuccessful, and 1 in the middle. The 10 high-conflict sites were almost all in the high and intermediate levels of success (5 and 4, respectively).

Finally, for *social capital*, the results are surprising and therefore interesting. Those sites characterized by low levels of social capital did the best, with nine having high overall success, one intermediate, and one low. The sites with intermediate levels of social capital were fairly evenly distributed among the success categories: two high, four intermediate, and five low. Where social capital was reported to be abundant, we found a bimodal distribution of success: three high, three low, and one intermediate.

Those results may give us some ideas for further research, but I would argue that they are less useful for improving forest management than the more systemic understanding we can gain from looking at the interactions among these variables (and others). The most thought-provoking observation is the apparent tendency for ACM to work better in chaotic and difficult settings.

In interpreting what has happened on the 30 ACM sites, it is worthwhile to return to the period of program planning at CIFOR and consider our scientific worldviews. Although the team agreed on our orientation—involving surprise, change occurring at various rates, interacting systems operating at different scales, a focus on opportunities—there was no general agreement within CIFOR that this was a reasonable or "scientific" way to approach research. A sizable number of researchers, both within the institution and in its governing Board of Directors, are committed to reductionist science as the only "real" science. Others interested in applying an ACM approach are likely to encounter similar obstacles.[2]

The result for our team, in my mind at least, has been a bifurcated approach.[3] On the one hand, we assess our ability to catalyze a process involving communities and other stakeholders—making plans together, addressing issues they identified, monitoring their progress, and moving toward some sort of shared vision of their future landscape, both human and "natural." This was an exploratory kind of re-search, designed to develop new methods and approaches while effecting real change in the forests and communities involved. Success would be defined as positive change in the communities along the lines they identified as important, the devel-opment of approaches that catalyzed the process that brought about the change, and improved understanding of that process. This approach was not hypothesis testing, in the conventional sense.[4]

On the other hand, the second approach, basically required by the social con-text of CIFOR science, involved fitting our effort into the reductionist mold. In this view, the project would be successful if we could identify the conditions—the independent variables—that caused the ACM process to work and thus recom-mend it wherever those conditions existed in the world.

I have now thoroughly examined each of the conditions we identified as likely to be important in each of the sites. In my view, none of the conditions we identified (Box A in Figure 1-4), based on extensive literature review and personal experience among the team members, proved to be linked, in a determinant way, with the impacts seen in the communities and forests (Box D in Figure 1-4; see Appendix 1-1; and Colfer and Byron's 2001 collection for similar supportive results in previous research). Factors like length of time on the site and motivation of facilitators proved to be more relevant (Box C in Figure 1-5), though these are also not determinant.

Each place and group of people are unique. It is therefore often impossible to develop standardized solutions to the problems that specific communities or groups within communities will encounter. Holling and Gunderson (2002) describe the changes in uncertainty as a system moves through the four stages of their adaptive cycle (see Figure 1-5). As we move from the exploitation phase (r) to the conserva-tion stage (K), connectivity increases, as does predictability. Perhaps my conviction about the comparative unpredictability of the forest and human systems we have been studying derives from the fact that all of these systems are moving from K toward Ω, or "release" (when things fall apart).

Governments and scientists routinely seek standardized solutions for forested ar-eas despite the overwhelming evidence of the task's impossibility. My perspective— and our results support this perspective—is that although solutions may not be possible to standardize, a process might be. The ACM approach is a process designed to address the problems of communities in managing their natural resources. In fact, I believe it can be used to address many of the problems communities face in other spheres as well (see below).

The ACM process requires that the facilitators be creative, flexible, and responsive, and that the community or group want to solve some problem. We had a good success rate, partly because we chose sites where the community desired to work on shared problems—and we had the flexibility to pass by several communities that were not interested. This is an important condition (pointed out to me by Terry Tucker of Cornell University).

Observations on Forests and People

After spending so much time looking at 30 forests and the people inhabiting them, it is hard not to see patterns. In this section, I raise some issues that have troubled or intrigued me in the course of the research and analysis.

Indigenous-Settler Differences

The first is the ubiquity of the differentiation between local, indigenous (or at least long-resident) groups and more recently inmigrating settlers. A case is often made for the prior rights of those who came first. I have argued this myself for the Kenyah Dayaks of East Kalimantan. The obliteration of longstanding traditional rights in one's home territory is patently unjust. On the other hand, the situation of immigrants is also typically unenviable. In many cases, they are moved as part of government programs, inadequately prepared, put in difficult living conditions, and resented by local residents.

ACM teams have handled this problem differently in different contexts. In Acre, the team initially worked exclusively with the rubber tappers (the longer resident population), helping them control encroachment by more recent settlers. However, the inequities of this approach soon became clear, and the team began planning how to incorporate attention to settler problems. This case made very clear the impossibility of being a truly neutral facilitator.

In Jambi, Sumatra, the ACM researchers (of Javanese descent) worked hard to strengthen social capital and collaboration between the indigenous Minangkabau and the immigrants from Java and other parts of Jambi. The value of such collaboration and cooperation notwithstanding, I cannot help wondering whether our team has in fact aided the longstanding, neocolonialist efforts by the Indonesian government (long dominated by the Javanese) to Javanize Indonesia's Outer Islands[5]—thus weakening indigenous land claims. Similar dilemmas are clear in the work undertaken in Bulungan, where spontaneous immigrants are common.

In Mafungautsi, the Shangwe were the original inhabitants, but the majority of the population is now Shona and Ndebele, tribes with significantly greater political clout (particularly the Shona, who are also well represented within the ACM team). ACM efforts have focused on these latter groups and paid little attention to the Shangwe, despite their valuable forest-related knowledge and their greater forest dependency. Similarly, in Palawan, Philippines, the ACM team focused almost exclusively on the politically dominant recent migrants along the coast, largely ignoring the indigenous groups in the nearby hinterlands.

In both Bolivia and Madagascar, the teams tried to address both groups. In Bolivia, separate researchers worked with settlers and the indigenous groups (*kollas* and *cambas*, respectively) but focused more on the indigenous Guarayos. In Madagascar, separate villages were selected for collaborative work—one composed of the indigenous Tanala, one of the inmigrating Beteseleo, and one mixed village.

Another dilemma in Cameroon and Indonesia is relations between the dominant indigenous groups (swidden agriculturalists) and the smaller, hunter-gatherer groups. The Minangkabau in Jambi had much more prestige and power than the *Orang Rimba*; similarly the Kenyah and other swiddeners vis-à-vis the Punan in East Kalimantan; and the Bantu groups in Cameroon vis-à-vis the various pygmy groups. In all these cases, the teams' efforts to work with the hunter-gatherer groups to the same degree as the settled groups were complicated by the former's mobility.

Those issues have both equity and environmental dimensions. The indigenous or longer resident groups have more knowledge pertaining to forests, closer cultural links with the forest, and greater motivation to protect the forest, both for their way of life and for their descendants. Similarly, the hunter-gatherers are more fundamentally dependent on and knowledgeable about the forest in which they reside than the swidden agriculturalists. Typically, recent settlers are agriculturally oriented and would clear the forest for cropland. To achieve forest sustainability, it makes more sense to work with indigenous or long-resident populations than with more recent migrants. It is possible that our comparative lack of progress in effecting environmental impacts is related to our relative inattention to these more knowledgeable ethnic groups—though insufficient time has elapsed to confirm this idea.

The attention of ACM teams to equity issues is reflected in our recent collection (Colfer 2005). Although the group addressed the issue diligently, the cultural backgrounds of ACM team members made it easier to work with recent settlers and more dominant ethnic groups, which had greater sophistication (literacy, bureaucratic savvy, knowledge of a common language) and existing ties with important actors at other scales.

It is also important to remember that the teams were overworked. Besides the difficult task of trying to catalyze adaptive and collaborative processes among groups of people, they also had to comply with our research requirements that they painstakingly document the process they were trying to catalyze. The ACM facilitators, by and large, gave it their all. Perhaps in future efforts to implement ACM—without the external research component that formed part of our effort—facilitators will have time to focus more attention on the most forest-dependent populations.

The Role of Swidden Agriculture

The primary source of subsistence in the humid tropical rainforest areas where we worked is swidden agriculture (also called shifting cultivation or slash-and-burn agriculture). There are several important commonalities to this approach to making a living that are worth pointing out.

First, swidden agriculture is often blamed for deforestation. Although scientists, managers, and donors are increasingly realizing the roles of governments, timber companies and plantations, and other powerful actors, blame is often directed at smallholders. Although swiddeners are not guiltless, it is important to acknowledge the complexity of the system and its potential to be sustainable. Most of the swidden systems we have seen are small clearings (often 1 ha) planted for a year or two and then left to regrow. The result is a patchwork landscape, with various stages of forest regrowth, that provides useful products and wildlife habitat. Swiddeners depend on the agricultural fields they create, but they also rely on the surrounding forests for game, fibers, medicinal plants, and foods such as fruits, ferns, and palm hearts. The surrounding forests also serve as insurance in a very risky environment, providing environmental services and food when agricultural efforts fail because of droughts, floods, insect invasions, plant disease, and health-related labor constraints at critical points in the agricultural cycle.

The forest also holds meanings of a symbolic nature, contributing to the meaning of life for forest dwellers—be it the link between the female sex role and rice cultivation among Dayak groups in Borneo or a place for secret age grade ceremonies in West Africa. Peoples whose traditions come from the forest are not the unthinking forest destroyers they are often made out to be—though population pressure or external competition may drive them to engage in practices they know are not sustainable.

Another important commonality is the linking of forest clearing with land claims. In many of our sites (Indonesia, Cameroon, Bolivia, Brazil), the person who first claims old-growth or primary forest lays claim to the land he clears. That land usually also becomes available to his children (Diaw 1997; Warner 1991). Recognition of this system of land tenure would help right some of the inequities of access to resources in these tropical forest areas. Until the legitimate claims of comparatively powerless forest peoples are recognized, it will be difficult to manage tropical forests sustainably.

Cooperation and Competition

At a talk on ACM that I gave in September 2002, Steven Wolf, a faculty member in the Department of Natural Resources at Cornell University, commended our emphasis on collaboration but also asked about the role of competition, reminding me of the (very American) view that competition breeds excellence. It was a thought-provoking comment.

Our concept had been born and grown to fruition in Indonesia. Although I am skeptical about pan-Asian cultural traits, I was intrigued by a discussion with Norman Uphoff, as well as his treatment of Asian perspectives in *Learning from Gal Oya* (1996): he described his first exposure to "both-and" thinking in Asia, specifically Sri Lanka (as opposed to the more common "either-or" thinking he's found in the West). Both-and thinking—in which multiple interpretations and methods are understood and acceptable—reduces the centrality of competition; there is room for more than one conclusion, perspective, or individual. It is possible that the Asian context may have encouraged our emphasis on cooperation.

Additionally, we had observed the inability of rural forest peoples to compete with the wealthy and powerful outsiders—timber companies, conservation groups, and governments. An individual Dayak, without title to her land, with minimal education, without any significant source of income beyond subsistence, was unlikely to be able to protect her traditional resources from such onslaughts. Competition was not an attractive option; the outcome of such competition would be assured, and not in her favor. Our initial thinking was that cooperation seemed more likely to succeed: community groups and communities could jointly question the harmful actions of outsiders, or community members and the outsiders might communicate and cooperate directly.

Stall and Stoecker's (1998) article comparing two views of empowerment via community organizing is relevant here. A dominant view (deriving from the well-known community organizer, Saul Alinsky) is competitive and male-centered. It is based on the assumption that people will not get involved in something unless it is explicitly in their own self-interest to do so. Stall and Stoecker identify another, more women-centered thread of community action focused on relationships and networks as building blocks for communities and community action. Such networks are based on an ethic of caring and the kind of shared ideals that Uphoff (1996) identified as important in the Gal Oya case. Stall and Stoecker argue that a combination of the two approaches is the most likely to succeed.

What we have seen in our field sites has actually been an alternation between cooperation and competition in many instances (as predicted by Ramírez 2001). Nepali villagers together chose to monitor the growth or stagnation of their own knowledge of their forest management regulations. This was a cooperative decision, taken as a group. Later, competition came into play, when a woman realized she lagged in learning the regulations. Her desire to compete resulted in her learning the regulations quite thoroughly. A similar example was given relating to toilet construction, where a group that had set themselves a collaborative target outdid their own goal, as they enthusiastically competed with other castes and ethnic groups in their village to improve hygiene (Pandey 2002).

In Pasir, Indonesia, the community and government officials initially perceived the relations between the community and the timber company to be too conflictual—based essentially on competition for local resources—to address their central problems in a joint planning process. Over time, as informal relationships were built and visions shared among the competitors, it became possible to begin to deal with timber-harvesting issues in a joint workshop, and make some plans together, designed to improve overall resource management in the local area (Hakim 2002, 2004).

The strengths that come from cooperating to achieve common goals can indeed be enhanced by the competitive spirit. Both competition and cooperation have a place in ACM—as, I suspect, in all human endeavor.

Governance

Governance issues were addressed at several ACM sites. In Nepal, the system of governing local community forests was improved by more systematically involving all segments of the community, by devolving some management responsibilities to the hamlet level, and by strengthening monitoring and checks and balances within the formal process determined by the government's regulations. In Zimbabwe, community members received training that strengthened their ability to analyze their own social and economic conditions resulting again in the active involvement of a wider segment of the community in forest-related decisionmaking. Community members were then able (and willing) to organize themselves to address their resource-related problems more creatively and had the courage to negotiate their wishes more effectively with government officials. In Bolivia, community meetings were conducted in ways that encouraged women to speak up. The women's input revealed a greater interest in sustaining wildlife and wildlife habitats than had been clear when only the men spoke, opening up new possibilities for collaboration with the BOLFOR project. The various processes to involve more segments of the community resulted in more transparent benefit distribution schemes (with resulting better oversight of community leaders) and better mutual understanding about priorities for community funds. In Indonesia, where governance issues have been a top priority recently, all the ACM communities were able to strengthen both their skills at negotiating and their social capital with outside stakeholders. In Baru Pelepat, the community managed to elect a diverse group of community members to their local governing body (rather than only the traditional elite Minangkabau men).[6] In Bulungan, communities were able to obtain training in legal literacy, an important element if communities are to provide checks and balances in the new, decentralized political system. In Lantapan, Philippines, officers of the village and the upland farmers' association addressed their lack of leadership skills, lax enforcement of village rules and regulations, inactive village officers, and inadequate financial reporting systems.

The interest expressed by local communities in governance issues is telling, in light of the common assumption that forest peoples participate only in activities that directly increase their incomes. This interest in governance issues is also consistent with the idea that a focus on connections or links is important (Stacey et al. 2000). These community members have made good use of the available links with other people—both within the community and outside.

One friendly reviewer questioned my use of the term *social capital* to refer to the relations established in ACM sites between community members and government officials, asking, "What does government have to do with this?" In different ACM sites, the governments had different roles. But one conclusion that I draw from all our ACM experiences is that we must stop seeing "the government" in monolithic terms: it comprises individuals, in varying units, with whom community members can establish and strengthen beneficial links, as opportunities present themselves. The kinds of relationships that will be most beneficial, in my view, are those characterized by strong social capital (including trust, leadership, norms of reciprocity, and so on)—with a propensity for "mutually beneficial collective action" (following Uphoff and Krishna). Uphoff (1996) discusses the value of friendship in the Gal Oya experience, and that is exactly the kind of link that has proved useful when community members work collaboratively with other stakeholders—including government officials—to improve forest management and human well-being. The secret is to retain the initiative with the community, at least until the government officials have been sensitized and liberated from their own bureaucratic prisons.

I also see value in including government officials in the research process, as suggested by Chambers (1997; Chambers et al. 1993), particularly with regard to rural appraisal exercises. The ACM processes that have succeeded to this point appear to have greater potential for expansion and continuation where government officials were successfully drawn into the process—in Zimbabwe, Cameroon, Nepal, Indonesia, and the Philippines. This applies at all levels. In Mafungautsi, Zimbabwe, and in Ottotomo, Cameroon, the local forestry officials have become real advocates; in Pasir, East Kalimantan, local officials have become involved through repeated workshops and other more informal interactions, and the district-level advisory group formed as part of the ACM process seems likely to continue to function. The contacts made through the various national steering committees seem likely to continue to provide ongoing channels through which our results can be more widely disseminated and used. Most experienced researchers have seen their written analyses on important topics ignored by policymakers; direct involvement of policymakers throughout the process offers an antidote.

The Pace of Change

I am convinced that we need to create conditions that allow adaptive collaborative management to emerge. That means we have to get out of the driver's seat and rely on the people, relationships, constraints, ideas, and systems around us to get closer to where we all want to go. I do not mean we can do nothing; but rather that we cannot control everything.

The ACM team knew that the process we were trying to catalyze would take time—precisely because of this lack of external control. As noted in our definition of ACM, "… plans often fail to fulfill their stated objectives." Identifying problems and correcting the course—whether for researchers, managers, or villagers— inevitably take time. We anticipated that we would need at least 5 or 6 years to make a good assessment of the process, and ideally 10 to 15. Instead, of course, we've had 3-year funding cycles, with gaps between phases I and II.

Many things can slow the ACM process. In Madagascar, for instance, an important leader, the *mpanjaka* of Sahavoemba, died, ending the progress being made on partnerships between the community and the national park. The selection of his replacement would have many implications for the community's well-being, and the people were unwilling to hurry the process. There was little the ACM team could do (Rakotoson et al. 1999).

In Chimaliro, Malawi, routine slowdowns occur because of the dramatic number of deaths due to HIV/AIDS and the cultural prescriptions bringing the whole community together for every funeral.

In Ghana, one researcher was reportedly hired with the understanding that he would have a sabbatical from his regular job and be able to devote his time to the ACM work. At the last minute, his sabbatical was canceled. His contract made it impossible to reallocate the funds to someone else, and the ACM team and its research suffered accordingly.

Those kinds of problems and unforeseen eventualities are part of life. I believe, in fact, that the people in those ACM sites that are presented as unsuccessful in this book are quite capable of solving the problems that have prevented their success to date. The problems that stymied them—untimely deaths and divorces, dishonest researchers, bureaucratic requirements, funding shortfalls—are all both common and solvable. But solving such problems takes time and sustained effort. And it takes a funding context in which decisionmakers recognize the long-term nature of such processes. In my view, we must all struggle to create the conditions that grant people the time to solve such problems, if we want to nurture empowerment of local communities and more sustainable management of forests.

Does ACM Work?

Paul Gellert, a rural sociologist from Cornell, asked me what exactly it would mean to say that ACM works. My answer is that ACM works if a process is catalyzed within a community whereby groups work together to accomplish significant common goals, taking into account their progress or lack thereof, and developing and implementing alternatives when their original idea does not go as planned. ACM works even better if mechanisms for collaboration and social learning are institutionalized within the community or group; and better still if such collaboration and social learning can be institutionalized among a wider set of stakeholders. I see the ACM process as starting small and if successful, expanding outward to affect and incorporate actors on a larger scale, as time goes by.

On the 30 ACM sites, the ubiquity of significant, short-term impacts should be clear. The inhabitants of virtually every site made impressive leaps forward—in terms of self-confidence and ability to mobilize, or transparency and equity in community decisionmaking, or ability to deal effectively with other and more powerful stakeholders, or social learning or social capital (see other ACM-based collections by Colfer 2005; Diaw et al. forthcoming; Fisher et al. forthcoming; Matose and Prabhu forthcoming; Hartanto et al. 2003; Kusumanto et al. 2005; see also reports to funding agencies, available from CIFOR).

This diversity of outcomes precludes the typical standardized approach to assessment. We cannot meaningfully say that incomes increased on average by x dollars per year, or that y hectares of forest were saved per site. Instead, we can say, for instance, that

- in Nepal and Zimbabwe, the number of women and marginalized users involved in community forest management decisionmaking soared (Dangol 2005; Mutimukuru et al. 2005);
- in Indonesia and Zimbabwe, the ability of people in study communities to negotiate successfully with district-level policymakers increased dramatically, as did the frequency of such negotiations (ACM PAR-Team Indonesia 2002; CIFOR 2002b; Nyirenda et al. 2001);
- in Cameroon and Bolivia, the level of understanding among stakeholders rose dramatically (Tiani et al. 2002b; Oyono and Efoua 2001), resulting in improved conflict management (Jum et al. 2003; Cronkleton 2005).
- In Brazil, Indonesia, and the Philippines, community members learned to approach management iteratively, building in explicit monitoring and feedback

mechanisms (Cunha dos Santos et al. 2002; Hakim 2004; Hartanto et al. 2002b, 2003a).

• In Nepal, community members were able to detect and sanction dishonest community elites whose positions had previously allowed them a free hand in the management of community forest funds (McDougall et al. 2002b).

Additional results are detailed in Chapter 5.

We are convinced that such measures are much more significant and meaningful than standardized measures, for two main reasons. First, the significance of any given level (whether of income or tree cover or self-confidence) depends in large part on the starting point from which change is measured. The "normal" number of trees per hectare in Mafungautsi, Zimbabwe, is far lower than the "normal" number of trees per hectare in Guarayos, Bolivia, or Bulungan, Indonesia, rendering a standard measure meaningless. Socially acceptable levels of day-to-day conflict are much higher in South America than in Asia. These contextual differences are even more important in spheres where measures themselves are controversial (like determining levels of social capital or amounts of self-confidence).

Second, issues had different importance in different contexts. In Nepal, women and marginalized caste groups have been barred from involvement in formal community forest management, but in Indonesia, women have had comparatively active roles in traditional, informal forms of forest management, and there are no castes (Colfer et al. 1997). Equity issues therefore demanded more attention in Nepal than in Indonesia. Conversely, Indonesia has been undergoing chaotic political change, and local and regional governance issues were on everyone's mind, but in Nepal, the community forest governance structure had been operating comparatively smoothly for more than a decade. Forest governance was therefore more important in Indonesia than in Nepal. A conceptual framework like ACM allows for such a local determination of focus.[7] The freedom (and obligation) to address locally relevant issues is far more important than the constraint of not having standardized measures for comparison.

The answer to the question? Yes, ACM does seem to work under many unpredictable and complex conditions.

Future Directions

The implications of our findings and our interpretations are far-reaching and difficult to implement. We began by testing the ACM approach in 30 forested sites around the world, but we believe that this same ACM process is applicable to the collaborative management of any natural resources, and indeed may be useful in developing solutions to any human problems.

Most fundamentally, if our scientific, reductionist strategy for dealing with social and environmental change is inadequate, we must fashion a more appropriate strategy. The attempt to predict and direct our future course is natural (Kelly 1963). We cannot resist the kind of proactive thinking exemplified in AtKisson (1999), Axelrod and Cohen (1999), or the collection by Gunderson and Holling (2002). As Yorque et al. (2002, 419) observe, "the most pragmatic approach is one based on learning our way to sustainable futures, rather than planning our way" (see also Wollenberg et al. 2004). These authors represent efforts to cope with the uncertainty and surprise that characterize our worlds.

Stacey et al. (2000, 114) argue for a focus on *connections* between agents, noting,

> The ... internal properties of the network [of agents] are the connections between its entities, and these connections create conflicting constraints. The internal dynamic is thus one of enabling cooperation and of conflicting constraints at the same time, a paradoxical dynamic of cooperation and competition. In human terms, connections between agents may be taken as analogous to relationships between people, and relationships immediately constrain those in relationship. Power is constraint; conflicting constraints, therefore, translate in human terms to power relations.

These connections among people have also been identified as central by Uphoff (1996), who attributes much of the dramatic success of the irrigation project at Gal Oya to friendship, along with ideas and ideals. Krishna's results (2002a) suggest the importance of links between rural people and proactive local agents who have further links to the wider society. Holling et al. (2002a, 413) say,

> It is clear, therefore, that to manage adaptively is a question of creating the right links, at the right time, around the right issues to create a responsive system ... it is not a question of identifying best practices or institutional arrangements.

Within the ACM team, there have been some early efforts to analyze connections among people (Haggith et al. 2003; Standa-Gunda et al. 2003a), but most of our attention to connections has taken the form of routine, daily interactions between ACM facilitators and the communities. These may well have been the keys to success.

If we really want to improve conditions both for people and for the environment, we need to work toward a situation in which all people's power to act is recognized and built into change processes. This applies within scientific communities, government bureaucracies, villages, and among all parties—and it implies drastic changes in the attitudes and standard operating procedures in bureaucracies and in science.

The evidence from ACM has been repeatedly replicated in individual case studies. Whereas from a reductionist perspective, the uniqueness of case studies is a flaw, an inconvenience that interferes with determining unidirectional causation, from my perspective, such uniqueness is simply the nature of things. It is time to acknowledge meaningfully the agency—the ability and propensity to act—of the people for whom decades of development and conservation efforts have been planned so unsuccessfully. ACM is one path: it has demonstrated that people without high educational qualifications can catalyze impressive results in and with forest communities. Implementing the approach within developing countries' extension agencies is therefore possible, but recreating several features of the ACM context will require effort.

One simple requirement is that the communities be interested in working with ACM facilitators. Success with one community can easily spread to others: several facilitators have reported that despite our short time in the field, neighboring communities already want to adopt or participate in similar activities.

Another fairly simple requirement is that ACM facilitators receive periodic training in facilitation, conflict management, simple research methods, monitoring, and social learning techniques. They also need training in attitudes: in respecting local knowledge and input, listening to and incorporating diverse views in community planning, identifying opportunities proactively, strengthening equity, and supporting local people's leadership rather than leading themselves (Box 12-1; van den Berg and Biesbrouck 2000; Kayambazinthu 2000; Chambers 1997). Such training is unlikely to be a one-shot deal; refresher courses are important, and mentors are essential.

More difficult requirements stem from the typical work environment of extension agents—a broader scale system in which both communities and agents take part (Pretty and Uphoff 2002). Most extension agents are part of formal bureaucracies. The idea that those higher in the bureaucracy should make the decisions both for those lower in the bureaucracy and for its "client" population is deeply entrenched and constitutes what Senge (1990) calls "a learning disability."

Box 12-1. Importance of Attitudes in ACM in Nepal

Attitudes are a critical driving force for ACM (and are enhanced by it):

1. a "learning attitude" to management, including a willingness to experiment, and the perception of "failures" as opportunities to learn;
2. openness and understanding between gender and caste and wealth groups;
3. confidence (to participate and try new ways of operating);
4. openness to cooperation, sharing of information and communication;
5. honesty and transparency in leadership and fund-management;
6. willingness to share power and benefits; and
7. sense of ownership of the community forest.

Skills are needed for effective ACM:

1. leadership;
2. facilitation and participatory process skills;
3. understanding of the concept and practice of self-monitoring as a learning and planning tool;
4. an understanding of the concept of learning and experimentation in community forest management; and
5. technical forest management, bookkeeping, and recordkeeping.

Other conditions were found to be supportive of ACM approaches in forest user groups:

1. adequate incentives and time (as well as skills) for facilitators to learn together with forest user groups (conditions currently lacking among district forest officer staff);
2. a perception of resource scarcity or potential environmental threat;
3. presence of multiplicity of external stakeholders and accessibility and exposure to them; and
4. supportive interpretation and implementation of forest policies by district forest officers.

—adapted from Cynthia McDougall (2002b, 9)

Three elements of our management strategy are probably central to the success of ACM and will require changes in most bureaucracies if ACM is to be implemented more widely:

1. Make ongoing attempts to prioritize field activities, and be responsive to facilitators' and communities' needs.
2. Strengthen two-way communication between headquarters and the field on substantive issues.
3. Maintain a horizontal management approach, in which each team member's role is respected and valued, whether driver, facilitator, country coordinator, or team leader.

Two points are important to add. First, we were never completely satisfied that we had sufficiently prioritized fieldworkers' needs or developed adequate two-way communication (cf. Senge's (1990) observations about the perpetual processes involved in being part of a "learning organization"). These are difficult goals to realize; continual effort is essential. Second, our efforts to be responsive to community needs did not mean trying to solve the communities' financial needs, except indirectly. ACM facilitators helped write proposals, made links to possible donors, and occasionally funded training. But most community activities had to be self-funded, in the interests of sustainability of effort. The facilitators' linking function fits nicely with the idea that connections are important in making and understanding change.

Another difficulty one can anticipate, fitting ACM into conventional bureaucracies, is the necessity that facilitators respond to the interests of the community. A forestry extension agent might be called on by community members to help with a health-related project, as was the case of building toilets in Nepal, completely outside the forestry department's mandate (Pandey 2002). Similar problems can arise in connection with commitments to donors. The project in Madagascar, funded to develop partnerships between communities and Ranomafana National Park, ran into difficulties when the park management refused to cooperate. In Campo Ma'an, Cameroon, the park management had its own bureaucratic and donor-related constraints, which specified the form and functions of the park a priori, and that hampered cooperation. In ACM, the answer in both cases is straightforward—try another route—but funding constraints or bureaucratic mandates can make this very difficult.

The "community of practice" is an important concept. If extension agents or other bureaucratic staff are to implement ACM effectively, higher level administrators will probably need training as well. The administrators must understand and take on board the three elements of management strategy: to give field staff the freedom to determine priorities, to facilitate two-way discussions among the staff, and to become more horizontal and less hierarchical.

The idea of "freedom to fail" is also alien in most bureaucracies, but it is an integral part of the adaptive element of ACM. "In complex, adaptive systems, disequilibrium and surprise are the rule, and failure is as instructive as success" (Westley 2002, 339). Failures must be seen as common and as opportunities to learn and move forward, rather than as evidence of incompetence or stupidity.

Inserting ACM into a bureaucratic context offers the potential for scaling up. Successes within an institution can spread rapidly, building on existing human links and resources. People with various levels of training and experience who can be mentors and sources of information for field personnel can be valuable. Several ACM teams are now testing whether ACM can loosen the social constraints of a bureaucracy.[8]

Westley (2002, 359) uses one midlevel American bureaucrat, trying to implement adaptive management as a case study and concludes,

> This perhaps is the chief source of surprise for the adaptive manager: the unanticipated consequences of contagion between problem domains that coexist temporally. If this is true, it may be that surprise, from the perspective of the manager, occurs more frequently at higher hierarchical positions, when the manager becomes responsible for more projects simultaneously. This is a paradox of resilience: more hierarchical control, less system control.

Another option, with less scaling-up potential, involves working through NGOs, as we have done in many of our sites. NGOs tend to be less rigid institutions, and creating an institutional context more compatible with the ACM approach may be easier. People drawn to work in NGOs often reject bureaucratic approaches and may be more willing and able to be flexible. Moreover, many NGOs already have a bottom-up philosophy, consistent with ACM. Perhaps the most important potential advantage is that NGO workers can be more neutral facilitators than government officials (remembering that no one can be truly neutral). CIFOR's perceived neutrality gave the ACM facilitators credibility that other, more entrenched actors did not have. Being perceived as a sympathetic and helpful outsider is probably an important feature of the ACM facilitator's role. It is also probably genuinely helpful to have people coming at common problems from different perspectives.[9]

A final possibility is to encourage graduate students to conduct research using the ACM approach. Though it offers less potential for scaling up, this approach leverages young adults' motivation, skills, knowledge, and analytical powers. Perhaps more effective ACM could be conducted in a smaller number of locations. This capacity-building approach also takes the longer view, since these students will eventually work in related fields and continue to contribute over their lifetimes.

A combination of those approaches may be the best overall strategy, given the respective, complementary strengths and weaknesses of each.

In Conclusion

In this book, I have tried to tell the following story: Responding to human and environmental problems in tropical rainforests, the global ACM team designed an approach that we hoped would contribute to human and environmental well-being. Part of our plan was to assess the usefulness of our approach. The approach included a horizontal element, in which we encouraged cooperation among the groups within and around the forest; it involved a vertical element, in which we sought mechanisms whereby forest people's voices could be more clearly heard in policymaking circles; and it involved an iterative element that encouraged social learning. We also assessed our own roles. Field testing began in 1999, and continues at this time.

We have assessed our approach in two ways: via holistic, qualitative cases and documentation, and by identifying specific conditions that we anticipated might help or hinder the process (devolution status, forest type, population pressure, management goals, human diversity, level of conflict, and social capital). This book focuses on the latter assessment effort.

We find, first, that approaches like ours, when combined with effective facilitation, can stimulate significant levels of local creativity and learning. Second, this approach can make significant progress in enhancing intracommunity equity. Third, the type of reductionist analysis we have done is inherently inconclusive for studying complex adaptive systems. Finally, what has happened in these forested sites can be applied in other ecological contexts.

These results imply the need for a shift in the way we study the elements of complex adaptive systems—away from reductionist approaches that dissect the systems, examining each element independently, toward approaches that examine the links among parts, feedback loops, and change. Our results also reinforce calls for change in the way we plan conservation and development programs—away from command-and-control, toward approaches that require bureaucratic flexibility and greater local determination of priorities and problem solving.

Although I am pleased with our results to date, the real test of ACM will take place over time. When we began this effort, we knew it would take time—something that donors and employers cannot always guarantee. We believe we have catalyzed a long-lasting process on some of our sites; we are less sure about others. Only time will tell. But I do believe that Margaret Mead was right when she said, "Never doubt that a small group of thoughtful, committed citizens can change the world. Indeed, it is the only thing that ever has."

Adaptive Collaborative Management in Diverse Contexts

THIS APPENDIX IS DESIGNED to provide the reader with an easily ac
cessible summary of conditions in the 11 countries where adaptive collaborative management (ACM) efforts were initiated. The cases are presented in alphabetical order. Each of the cases is organized according to the dimensions we
anticipated might be important in determining under what conditions the adaptive collaborative management process could be most useful. These dimensions are

- devolution status,
- forest type,
- population pressure,
- management goals,
- human diversity (external, internal, age, gender),
- level of conflict (national and local scales), and
- social capital.

As has been argued throughout this book, none of these dimensions in fact
appeared to play a determining role in ACM success or failure, on its own—
although national-level conflict was statistically positively associated with ACM
success, and population pressure and diversity of external stakeholders were negatively associated with such success. We include the descriptive case material here (a)
to provide real-world substance to the more abstract discussion presented in the
main body of the book, and (b) to portray some of the diversity that characterizes
the world's forests and the people in them. We hope these cases can serve as a useful
introduction to, and reference on, conditions in the world's forests.

For some dimensions, the categorization of sites was fairly obvious. Forest types
included humid tropical, dry tropical, temperate, and subtropical montane. Population density simply required data collection on the site and some judgment in
making cross-site comparisons (see Chapter 7). Determination of the management

goal (beyond community management, which existed on all the sites) was also fairly clear, with large- and small-scale logging, national parks, and multipurpose forest reserves as the operative categories (Chapter 8). However, the other topics required more in the way of agreement among team members about what exactly we were looking for. The ACM teams shared sets of guidelines to make our results more comparable. These are reproduced in the relevant book chapters. Guidelines for devolution status are in Chapter 6; for community heterogeneity, in Chapter 9; for level of conflict and social capital in Chapter 10.

In the cases to follow, the categorization of each country and site is indicated, along with the appropriate heading (e.g., Devolution Status High; or Forest Temperate) at the beginning of the case. For the conflict category, it proved necessary to differentiate national and local conflict; and with diversity, we differentiated between external and internal diversity.

CASE I

Bolivia

Contributing Authors
Peter Cronkleton and Omaira Bolaños

Sites in Bolivia
Guarayos (particularly Cururú and Salvatierra in the municipality of Urubichá)
Santa María (in the municipality of Ascensión de Guarayos)

ACM work in Bolivia was concentrated in three communities in the province of Guarayos. These included two Guarayo indigenous communities in the municipality of Urubichá (Cururú and Salvatierra) and one colonist community in the municipality of Ascención de Guarayos). The municipalities represent very distinct contexts. Urubichá is a municipality characterized by a homogenous indigenous population and by expanses of forests. It is almost entirely within the titled area of the indigenous homeland. Cururú and Salvatierra both have forest management plans that have been approved by the national government's *Superintendencia Forestal* (Forest Superintendency).

Ascención de Guarayos, with a diverse mixture of ethnic groups and migrants, is transected by a major interdepartmental highway, suffers from land conflict and insecure tenure conditions, and is characterized by forest fragmentation and high rates of deforestation. Residents of Santa María attempted to develop a forest management plan but were unable to identify a large, undisputed, contiguous patch of forest that they could claim as a management area.

Devolution Status—Intermediate

There have been several major transitions that created the conditions seen today in the Guarayos region. The first European settlement in Guarayos occurred in

Bolivia
Key ACM Characteristics

Devolution status	Intermediate
Forest type	Humid tropical
Population pressure	Low
Management goals	Large- and small-scale timber
External diversity	High
Internal diversity	Medium
National conflict	Medium
Local conflict	Medium
Social capital	High to low

the late nineteenth century as Franciscan missions were established to settle and Christianize nomadic indigenous peoples (primarily Guarayo and Siriano people of the Tupi-Guarani language group). Revolutionary change in the 1950s secularized the missions, but little changed for local indigenous people as power shifted from the Catholic Church to local landed elites, who continued to dominate the indigenous population. From the 1970s through the early 1990s, Guarayos underwent a transition as first a timber frontier followed by an agrarian frontier passed over the region. This generated local conflict as families were displaced and valuable forest resources were extracted, with most of the income transferred out of the region. In the mid-1990s policy changed dramatically, devolving power to local governments. A land titling process to address the demands of indigenous people was set in motion, and a new forestry law recognized the rights of indigenous people to use forests and commercialize forest resources if residents met government-prescribed conditions.

In 1996, Bolivia's agrarian reform law recognized indigenous rights to manage forest resources for profit, representing a major break in forest policy. The Ley INRA (*Ley del Servicio Nacional de Reforma Agraria*, National Agrarian Reform Service of Law) and a forestry law known as the *Nuevo Ley Forestal No. 1700* (Cronkleton 2001b) were introduced. Ley INRA defined a type of property known as *Tierra Comunitario de Origen* (TCO), or indigenous communal lands, which provide indigenous residents with collective land title. While the residents do not own the forest, they control access and can use forest resources for subsistence. If they wish to commercialize forest products, they must submit a management plan to the Forest Superintendency.

These initially promising policy changes have failed to meet full expectations. Opportunity for land titling exists, but the process has been slow and politicized, with mainly undisputed areas receiving title. Forest access rights were granted, but government has maintained strong oversight. Indigenous people have had difficulty acting on these newly granted rights, because bureaucratic procedures for approval are complex and management plans cannot be approved for forests with contested ownership, which is the case with most untitled TCO lands.

The Guarayos TCO demand, which has been the focus of ACM activities, is for 2,205,370 ha, including land in eight municipalities in the departments of Santa Cruz and Beni. The government divided the claimed area into five polygons and has titled the two northernmost areas, which are the most remote, least populated, and least conflict-ridden (Cronkleton 2001b). The titled area encompasses almost the entire Urubichá municipality. The remainder is still subject to conflicting claims and has an uncertain future. The U.S. AID-funded Bolivia Sustainable Forestry Project (BOLFOR) has been working with various levels of government to provide backup in the implementation of the new laws and to support communities to develop sustainable timber management regimes.

Forest Type—Humid Tropical

The province of Guarayos lies in the northwest corner of the department of Santa Cruz bordering on the Beni department. Guarayos is a transitional zone between the arid forests of Chiquitanía and the humid forests and pampas of the Amazon basin. Guarayos has a mean temperature of 22.6°C and an average annual precipitation of 1,600 mm, mostly concentrated in a clear rainy season from November to April. Humid subtropical forests cover much of the province and have drawn the interest of timber companies for decades. In Urubichá the forests are expansive and contiguous. In Ascensión, where Santa María is located, forests have suffered high rates of clearing and are fragmented.

While TCO demands are considered a single unit, the tendency in Guarayos has been to define individual management units near interested communities. Five communal forest management plans have been developed by indigenous communities in Guarayos and approved by the Forest Superintendency. Of the ACM sites, Cururú's management plan for 26,421 ha was approved in 2001 with support from BOLFOR. The Salvatierra plan for 40,866 ha was only approved in 2003. Santa María's residents were unable to identify an undisputed patch of forest of sufficient size to develop a management plan (Cronkleton 2001b).

Previous logging activity by timber industries had eliminated the stock of most high-value timber species such as mahogany and cedar, forcing the community groups to depend on secondary or alternative species. Dauber et al. (1999) identified the 10 most common timber species in the Guarayos area, with the number of trees/ha; see Table A-1. Table A-1 shows the timber volumes harvested by Cururú and Salvatierra in 2002.

While the timber management projects would introduce a new commercial activity for the communities, local residents traditionally depended on forest re-

Table A-1: *Timber Volumes Harvested by Cururú and Salvatierra Communities in 2002*

Common name	Scientific name	Volume harvested, m³ (Cururú)	Volume harvested, m³ (Salvatierra)
Yesquero Blanco	*Cariniana*	1,957.76	1,832.8
Yesquero Negro	*Cariniana*	369.29	316.3
Ochoó	*Hura crepitans*	897.31	719.8
Bibosi	*Ficus* spp.	558.18	1,563
Serebo	*Schizolbium amazonicum*	171.89	423.5
Copaibo	*Copaifera reticulata*	95.73	—
Cedro	*Cedrela odorata*	—	4.8
Paquio	*Hymenaea courbaril*	—	2.6

Source: Community records.

sources for building materials and subsistence. In particular, hunting supplies most families with protein. However, many note that larger mammals are less prevalent due to hunting pressure. Van Holt (2002) reports that 92 percent of families in Salvatierra identified agoutis (*Dasyprocta variegata*) as the most abundant animal in the forest. Everyone considered white-lipped peccaries (*Tayassu pecari*) to be rare, with 75 percent of respondents including spider monkeys (*Ateles paniscus*) in the rare category. Other animals reported in decline included tapirs, armadillos (*Daypus novemcintus, Eupractus sexcinctus*, and *Priodontes maximus*), black howler monkeys (*Alouatta* sp.), coatis (*Nasua nasua*), jaguars (*Panthera onca*), collared peccaries, pacas (*Agouti paca*), parrots, and curassows. The most common animals hunted were the collared peccary and the agouti, followed by tapir, brocket deer, paca, and turtle.

All respondents indicated declining fish populations. Worrying trends include reported smaller size of fish and abundance of *cimbao*, which is a lungfish that derives oxygen from the surface, suggestive of low oxygen levels in the water. In addition, frequent use of fish poisons, primarily the sap of the timber species Ochoó (*Hura crepitans*) has decimated fish stocks near Urubichá and Salvatierra.

Population Pressure—Low

Two of the three communities where ACM was initiated are in the area with a predominately Guarayo population (Salvatierra and Cururú). The other, Santa María, is a mixture of indigenous peoples (Quechuas, Aymaras, and Mojeños), *campesinos*, and migrants. The communities represent the two extremes of population dynamics changing the region: (1) the consolidation of communal lands by local indigenous people and (2) the influx of newcomers attempting to gain access to land.

The Salvatierra population numbers 333 residents in 59 families, with a sex ratio of 1.28. Forty-eight percent of the population is 12 years old or younger (Cronkleton 2003a). In Cururú, there are 205 residents in 24 families, with a sex ratio of 1.15. Salvatierra has been continuously settled since its founding as a mission site in the 1930s. Cururú was founded in the 1990s by Guarayos families moving closer to their farm fields located in an abandoned mission-era cacao plantation.

A 2001 census in Santa María, taken as part of the research project, revealed a population of 365 in 93 families, with a sex ratio of 1.25. Of these 93 families, only 73 live in the community full time. The remainder live in towns and cities and visit the village during harvest time. Most Santa María residents (76 percent) are in-migrants to the department of Santa Cruz from other departments in Bolivia. Only 6.8 percent are natives of Santa María (Bolaños and Schmink 2004). Santa María has experienced high population movement due to its location along the interdepartmental highway that connects the cities of Santa Cruz and Trinidad. This facilitates the influx and establishment of people who come to the area looking for land. This is the case of a group of colonists (approximately 100 people) who arrived in the village in 1999, attracted by the possibility of acquiring land. This group settled in an area of the community reserve set aside for future land distribution among villagers, creating new conflicts within the community. In August

2001, a new group of approximately 80 colonists came to the village asking for available land; the group left the area after community leaders demonstrated the lack of available land (Bolaños 2003). Bolivia is the least densely populated of all countries where ACM has been applied, with an average density of 7.56 persons/km^2 (INPE 2001). In the Guarayos Province, the population density is only 1.07 persons/km^2 (INPE 2001). Within the province, the Municipality of Urubichá, where Cururú and Salvatierra are located, is the most sparsely populated.

Management Goals—Large- and Small-Scale Timber

The Guarayo area has been subject to large-scale timber production. Many of the problems that typically accompany such large-scale production have beleaguered Bolivia in general, including corruption, internal migration, land tenure insecurity, illegal logging, forest degradation, and violent conflict. However, as in Indonesia, since the decentralization legislation that marked the mid-1990s, there have been serious attempts to strengthen local communities' abilities to benefit from and sustainably manage their forests.

Shortly after the 1996 forestry law was ratified, several development projects and NGOs, including BOLFOR, began working with local communities to prepare management plans so that the Guarayo people could commercialize forest resources, with approval from the Forest Superintendency. The Guarayo people were motivated by two major objectives. The first was to gain official recognition of their land use through approval of a forest management plan, which would strengthen their claim to their indigenous territory. Secondly, commercial timber management would allow for an additional income source for rural families in this impoverished region.

Cronkleton et al. (2001) report that the community of Cururú, with effective internal leadership, has made considerable progress in developing the required management plan. In less than a year they formed a *Unidad Forestal Indígena* (UFI)—a forest management association—and completed an inventory and a census of the chosen annual cutting area. Their UFI follows the organizational structure suggested by BOLFOR. This includes a coordinator and working groups with responsibility for census, administration, harvest, and commercialization. Those in responsible positions understand and take their roles seriously. Cururú residents have been motivated by visions of significant cash income accruing from their forest. Cururú's villagers successfully sold timber in 2002 and 2003 but decided to forgo a sale in 2004 due to low prices. Salvatierra's management organization sold timber in 2002 but have not held a second sale due to problems with the Forest Superintendency in 2003 and low timber prices in 2004.

Before any timber sales had taken place, considerable confusion arose among community members about how benefits would be distributed. Equitable distribution of timber profits was recognized as problematic, but villagers were unsure what to do. The communities had kept records of who had worked and how much, but the records were not being used as a basis for a monitoring system. Cronkleton, working with the Salvatierra UFI, brought community members together and developed a system using the work records to plan how the anticipated benefits would be distributed accordingly. The group also went through a process of discussing how to use the profits that would come to the community at large—a process that also allowed women and young people to express their views.

In Santa María, the process of developing a functioning plan for timber harvesting has stalled, due to differences relating to land tenure rights, access to forest resources, and uncertainty about becoming a part of the indigenous Guarayo territory. Although people in Santa María have a long history of participation in forest operations as wage laborers within forest concessions, they continue to experience limitations in securing access to forest resources and lack the rights to manage them. The potential area for timber harvesting has been considerably reduced by several private owners who have taken over the land. The remaining forestland presents other obstacles, such as distance and difficult access to the area. This situation decreased the community's opportunities to receive benefits from the forestry project (Bolaños 2003).

In addition to this, internal conflicts in Santa María related to the division of labor in the timber management process complicate the potential participation of certain community groups in the forestry project. Due to previous working experience with timber companies, men expect to be the main participants in management. Women, on the other hand, are expected to participate in activities related to their traditional roles, such as preparing food for the forest workers; they are not invited into forestry group decisionmaking. Thus, women's opportunity to participate and receive direct benefits was constrained (Bolaños and Schmink 2004).

External Diversity—High

Cronkleton (2001b, c) identifies the following stakeholders: neighboring communities; *Central de Organizaciones de los Pueblos Nativos Guarayos*, or Confederation of Guarayos Native Peoples (COPNAG), an indigenous political umbrella organization; the *Campesino* Federation; the timber sector, including large, medium, and small logging companies; forest service providers such as heavy machinery owners, sawmillers, truckers, sawyers, and manufacturers; government stakeholders, including the Forest Superintendency, INRA—which titles TCOs or indigenous territories—and municipal governments; and development institutions like BOLFOR. Other actors involved in the recent past have included the Dutch Foundation of Netherlands Volunteers (SNV) and the Bolivian NGO CEADES, as well as *Asociación Indigena Forestal Urubichá-Salvatierra*, or the Indigenous Forestry Association of Urubichá and Salvatierra (AIFUS).

Age and Gender—Intermediate

The population of these sites is very young. Nuclear families are the norm, with young couples moving into their own homes once they are married, but extended family ties are very important for production and support networks. The elderly without access to an extended family network in the community usually suffer the most hardship and poverty.

Cronkleton (2004) notes differences in perceptions between older and younger community members about how the benefits from forest management might be used in the Bolivian site, but serious conflict has not been reported. After the 2002 timber harvest, the residents of Cururú set aside funds to buy school supplies and support a small number of students so that they could attend high school in Urubichá. Salvatierra residents also invested in their school, purchasing musical instruments so the children could form a band.

Most direct benefits from timber management in the form of daily wages have flowed to men rather than women, although more women have become involved in the project and have earned wages. Women have suggested that the village management organization pay wages in their presence so that they can know how much their husbands are bringing home and thus have more say in the use of these funds. Women in Cururú expressed satisfaction that the timber project has meant that men have not had to migrate out of the village in search of wage labor but are now able to stay closer to home.

In a division of labor analysis, people from Santa María expressed differences in the kind and amount of time girls and boys participate in domestic activities. Girls, like their mothers, dedicate more time to household responsibilities, and boys, like their fathers, participate in few domestic activities, and only occasionally. Both boys and fathers were more associated to activities outside the household, such as hunting and fishing. An analysis on the impact of the forest management project on family responsibilities did not consider boys and girls as potential or direct participants. Their role was assumed to be that of facilitating their parents' direct involvement in forest management activities, especially that of their fathers (Bolaños and Schmink 2004).

In our comparative examination of gender differentiation we considered Guarayos to be intermediate, with a score of 4 (fairly high) for division of space and labor, and 3 (intermediate) for male dominance and hostility to women in public arenas.

Internal Diversity—Medium

Guarayos is a large area, but our efforts there have focused on Cururú, Salvatierra, and Urubichá as indigenous communities, and on Santa María as a colonist community. As part of formal involvement in forest management, communities are obliged to create forest management associations (UFIs), recognized by the indigenous umbrella organization, COPNAG. The first such organization, AIFUS, no longer exists (see below). Now Cururú's group is called *Asociación Indígena de Manejo Cururú* (AIMCU). Salvatiera's group is called *Asociación Forestal Indígena Salvatierra* (AFIS), and Urubichà's group is called *Manejo Forestal Urubichá* (MANFU) (Cronkleton 2003b).

Whereas the indigenous communities are fairly homogeneous, Santa María has moderate diversity, with a large variety of in-migrants from other areas in Bolivia. There are a number of "excuses" for differentiation within Santa María: (1) membership in one of three unions, among which there is considerable conflict regarding land distribution and leadership; (2) being a lowlander from Santa Cruz Department (*Camba*) or a highlander (*Kolla*), each of which has negative stereotypes about the other; or (3) being on one or the other side of the controversy about involvement in BOLFOR's forest management activities and the inclusion of the community as part of the Guarayos TCO (Bolaños and Schmink 2004).

Land distribution generates the main internal differentiation within Santa María village. Issues relating to field crops and land tenure among members of unions have been conflict-ridden, due to the overlapping settlement areas and the arrival of additional colonists in the community. This situation has created several groups with different interests regarding their tenure rights. The traditional and stereotypical views of *Cambas* and *Kollas* have contributed to differentiating native peoples. The former are seen as lazy and weak, while the other in-migrating group is seen as hard-working and ambitious. Although these perceptions have not led to a critical division within the community, they have strengthened the belief that colonists would contribute to the improvement of the community's economic conditions. The controversy surrounding the community's involvement in the forest management project and whether it should be included within the Guarayo TCO has contributed to internal differentiation by creating two opposing groups led by the two main local leaders. One group represents the interests of natives and the earlier colonists, while the other represents the interests of the most recent colonists, who claim to advocate *campesino* interests (Bolaños 2003).

Level of Conflict—Medium

The Guarayo sites have seen considerable conflict over the past few years. Indigenous communities have made an ongoing and controversial attempt to establish a TCO. COPNAG is the political umbrella for the Guarayo people and initially requested the TCO designation. Because of confusion about pre-existing land tenure—among ranchers, settlers, loggers, and local communities—the process of defining the TCO has been problematic and conflict-ridden. COPNAG itself has also been weakened by lack of transparency and internal conflicts (Cronkleton 2001c). Recent land invasions by migrant colonists have heightened tensions with large landowners and Guarayo communities.

Logging and its attendant conflicts have been part of the landscape for decades. A network of logging roads was constructed in the past, which provided entry for illegal loggers who made off with the remaining, high-value timber after the big companies had harvested the most valuable species. Communities still do have a stock of commercial-grade timber, although this consists primarily of lower value alternative species. Commercial hunters also used these same roads and decimated the faunal population on which local communities have depended for the protein in their diet (Cronkleton 2003c).

More recently, in 1999, the first community forestry project began in Urubichá municipality, under the auspices of AIFUS and with the assistance of aid agencies. A joint management plan combining Urubichá and Salvatierra for the 40,000-ha area was planned and approved in 2000, but serious conflict erupted as soon as the first timber was sold (Cronkleton 2001c, 16). The Salvatierra residents felt that most of the work fell on their shoulders, as they were closer to the management unit, while Urubichá residents captured the power and the benefits. The distribution plan was ignored and some U.S.$1,500 disappeared, with the leadership unable to account for it.

Another serious conflict related to hunting. According to the AIFUS management plan, no hunting was allowed in the management area. However, the inventory and census trails made access easier, and the people of Salvatierra were dependent on hunting for their subsistence needs. Ultimately, the Salvatierra group pulled out of the project. The Urubichá group reorganized as MANFU and received approval in 2001 for a second operating plan, though the group still suffers from serious internal disputes (Cronkleton 2001c, 21). The group was unable to harvest wood at all in 2001 (Cronkleton 2003b), and the Forest Superintendency suspended approval of the AIFUS management plan pending resolution of the conflict and accounting problems. The next steps remain unclear (Cronkleton 2003c).

Santa María is composed primarily of in-migrants, among whom there is an underlying difference of opinion about whether or not the community should support the village's inclusion in the Guarayos TCO, as a means to strengthen their land tenure rights. (Similar disputes are under way elsewhere—see Cases 2 and 12.) Participation in BOLFOR's forest management project—the goals of which were actually only to improve forest management in the area—was seen by some as an indigenous strategy to strengthen the claims of the former group against the settler (*campesino*) interests. Others participated actively. Behind this internal dispute are the two main regional organizations: COPNAG and the *Campesino* Federation. Each has its own agenda and interests regarding forest management and land tenure issues. Fundamental differences between these organizations have not been directly debated in the context of implementation of the INRA and the Forestry Law, yet they highly influence the decisions of the people in Santa María. Both the *campesinos* and the indigenous regional organizations struggle to maintain autonomous legal systems relating to tenure and management of resources. However, these organizations differ in the kind of tenure advocated for the groups they represent. While COPNAG champions collective titling, the *Campesino* Federation supports individual tenure rights. This situation affects the position each group holds in regard to the advantage of participating in forest management as a means to secure land tenure rights. For the recent colonists, the community involvement in the forestry project may provide more benefits to the indigenous rather than to the *campesino* groups because it facilitates the indigenous organization's control over the land where the colonists have settled (Bolaños 2003).

There have also been conflicts about the appropriate roles of women in the forest management project. In Santa María, the few women who participated in the preliminary forest activities such as forest exploration and inventory told of the conflicts they faced at home, especially with their husbands, who disapproved of their involvement in the project. Because forestry operations are considered a male activity, it was suggested that women should concentrate on tasks more related to their female role, such as cooking and washing for forest workers. They should also avoid going into the forest because it is an inappropriate place for women to work. Thus, local perceptions of the appropriate roles for males and females in forestry contributed to conflicts in dividing tasks and responsibilities (Bolaños 2003; Bolaños and Schmink 2004, 286). Cronkleton (2003d) notes that these efforts to stimulate formal forest management in Santa María had, by 2003, come to naught, due to the inability to locate a sufficiently large expanse of uncontested, good-quality forest land near enough to the community.

Social Capital—High

The people of Cururú, indigenous Guarayos, have shared a difficult past. Living in tough physical conditions in an isolated location, the inhabitants were initially subjected to considerable external censure for leaving "civilization"—meaning, most notably, churches and schools for their children. One priest accused them of "returning to a life as forest savages dependent on bows and arrows." Although this phase has passed—they now have an elementary school—the strength of social control and an emphasis on conforming to social norms remains evident (Cronkleton et al. 2001, 3). Five or six Cururú families have titled plots of land, and the whole area is now within a titled portion of the TCO. The TCO, however, gives communal title but says nothing about internal divisions and access. Land rights in the current context (and traditionally) are primarily based on community consensus.

Salvatierra is similarly remote and characterized by cohesive social bonds (Cronkleton 2003d). Both communities have worked well together in their forest management efforts. In less than a year, the people of both Cururú and Salvatierra formed a strong and well-organized management group with a UFI organizational structure suggested by BOLFOR and conducted an inventory and census of their annual harvest area. The UFI has effective leadership and a clear division of responsibilities, known to the community at large (Cronkleton et al. 2001, 6).

Social Capital—Low

The *campesino* community of Santa María has serious divisions. The three separate associations in charge of distributing land and their leaders are divided on a crucial issue: whether to support the TCO request of which their land is a part. A lower level of social capital would be expected in a recently established community composed of in-migrants from many parts of Bolivia than is found in indigenous groups, and that appears to be the case. Those that identify themselves as natives have less power in the community's decisionmaking. The many reasons to be in opposition, along with the lack of commitment of recent in-migrants in regards to resolving differences, contribute importantly to the lower level of social capital (Bolaños and Schmink 2004).

Urubichá is a larger, primarily indigenous town, marked by quite low social capital (Cronkleton 2003d). Cronkleton speculates that the reasons for the marked factionalism within the town may be related to their history as mission settlements:

> When the power of the missions finally disappeared, the region was already undergoing a frontier transition. Rather than having consolidated leadership and representative institutions, their [Guarayos] organizations grew in an era when land speculators and businesses had a vested interest in keeping communities divided. In an atmosphere of bribes and corruption, it was hard for unified movements to develop (Cronkleton 2003d).

CASE 2

Brazil—Acre, the Western Amazon

Contributing Authors
Magna Cunha dos Santos, Noemi Porro,
Marianne Schmink, and Samantha Stone

Sites in Acre
Porto Dias

Devolution Status—High

Brazil is a huge country, with considerable differences in policy from state to state. The State of Acre has long been considered particularly interesting from a policy perspective because of its early acceptance of what Schmink and Wood (1992) characterize as a "green discourse." The extractivists (rubber tappers) have also had unusual success in making powerful alliances and successfully accomplishing their goals, one of which was the creation of extractive reserves.

In 1992, representatives of the Group of Seven (G-7) nations agreed to support the conservation of Brazil's tropical forests by creating a Pilot Program for the Protection of Brazil's Tropical Forests (Hardner and Rice 1999). In Acre, where the state government has decreed itself to be "the government of the forest" or *Governo da Floresta* (Cunha dos Santos 2002a), there are two extractive reserves—Chico Mendes and Alto Juruá—that are part of this pilot program. Conservation is a central goal of all the projects; the state has a goal of keeping deforestation below 10 percent (Gomes and Sassagawa 2002).

The Federal Institute for Colonization and Agrarian Reform (*Instituto Nacional de Colonização e Reforma Agrária*, or INCRA) was involved in setting up the Porto Dias Agro-Extractive Reserve, with which we are concerned here and which differs from the extractive reserves above. Each family with rights in the agro-extractive reserve must own at least two rubber trails, and the association created to manage the reserve must keep deforestation to less than 10 percent of the total area. Although each family's rights are generally considered secure, families do not have title and cannot sell the land. They can, however, sell their rights to live on the land, with the agreement of the other association members. Resource rights in the region are not technically in dispute, although some uncertainty exists, based on fears that the laws will change, that colonists will successfully invade, and that population increases will result in higher levels of conflict. The Brazilian Institute for the Environment and Renewable Resources (*Instituto Brasileiro de Meio Ambiente e dos Recursos Naturais Renováveis* or IBAMA) is also involved in decisions and actions pertaining to local livelihoods.

The Porto Dias Agro-Extractive Settlement Project (PAE) is located in the municipality of Acrelândia, approximately 140 km from the Acre state capital, Rio Branco. Formerly a rubber estate, it became an agroextractive settlement, established by INCRA, in 1989. Its 22,145 ha of tropical forest are home to about 90

Brazil—Acre, the Western Amazon *Key ACM Characteristics*	
Devolution status	High
Forest type	Humid tropical
Population pressure	Low
Management goals	Small-scale timber
External diversity	Medium
Internal diversity	Low
National conflict	Medium
Local conflict	Medium
Social conflict	High

families. The reserve is located between the Abuña River (the border with Bolivia) to the south, and BR-364 (a road), with its ranches and small farms, to the north. Porto Dias is one of the oldest community forest management projects in Acre. It was the first such initiative to be implemented in an agroextractive reserve with rubber tappers. The nongovernmental organization Center for Amazonian Workers (CTA) worked with a group of 10 families in 2004 to help them manage timber on 100-ha plots—10-ha plots to be harvested each year with a 30-year felling cycle—on each 300- to 600-ha holding (Cronkleton 2000; Cunha dos Santos 2001b).

Forest Type—Humid Tropical

The Porto Dias Agro-Extractive Reserve is in the humid equatorial (Amazonian) zone on the western border with Bolivia. Comparatively speaking, forest condition is very good, with little demographic pressure. Ninety-seven percent of the reserve is covered with vegetation, most of which is dense *Umbrofila* forest, with some open forest, some bamboo, and a rich fauna. Topography is relatively flat. Temperatures range from 23°C in July to 25.8°C in October. Soils are red-yellow latosols, with low fertility (Cunha dos Santos 2002a; Cronkleton 2000).

Gomes conducted a study in the Chico Mendes Extractive Reserve, also in southern Acre, and found cause for concern about future forest conditions. Notably, interest among young rubber tappers in cattle ranching and agriculture has increased, due to low prices and hard work involved in rubber tapping; progress on the paved road going from Rio Branco to Puerto Maldonado in Peru (Gomes and Sassagawa 2002) is another factor. Another major road—the state's first highway—goes from Rio Branco to Porto Velho, passing close to Porto Dias. Deforestation, related to ranching and agriculture, was evident along all roads, more so in the eastern part of the study area. Porto Dias is also east of Chico Mendes.

Three management entities exist within the Porto Dias forest. Area I, managed by the Rubber Tappers Association of Porto Dias, is where we have focused our research to date. Area II has been managed by the Agro-Extractive Association of São José (Mossoró). Area III has been managed by the Association of Rural Producers of Acre (ASPOMACRE). Both areas II and III are inhabited by in-migrants of a more agricultural and more mobile bent, with various legal statuses. Some have legal rights from INCRA to settle in the area; many, however, illegally deforest beyond the 10 percent of their land allocation allowed in order to expand their agricultural production within Porto Dias.

Little control or inspection by either the Rubber Tappers Association or IN-

CRA has led to an active land market. Cunha dos Santos (2002a, 17) reports that the sale value of a landholding with three rubber trails is approximately 2,000 Brazilian Reais (U.S.$1 = 3.24 Reais, May 2003). If the sellers divide their landholdings into lots the value is lower.

Population Pressure—Low

Across the whole province of Acre, density is 1.2 people per km^2 (Kainer et al. 2002, 5). Porto Dias Agro-Extractive Reserve is home to 100 registered families and 20 families residing as "squatters." Children under the age of seven make up a quarter of the population. Less than 4 percent of the population is over age 60. Cunha dos Santos (2002a, 4) considers the remainder to be potentially economically active. The sex ratio is significantly skewed; women make up only 33 percent of the population. In nearby Chico Mendes Reserve, 33 percent of the people are migrants. There, the mean number of sons under 14 years of age in a household is 2.03 and the mean number of daughters is 1.62, suggesting a similar skewing (Gomes and Sassagawa 2002). This also suggests that an average household has nearly six workers. See also Campbell et al. (2004) for a discussion of one pattern nearby involving women's disproportionate migration to cities to educate their children and participate in wage labor.

The two settlement projects adjacent to Porto Dias are inhabited by in-migrants from other regions. These settlers have a more agricultural orientation than the traditionally forest-dependent rubber tappers who inhabit the reserve, whose parents or grandparents were also in-migrants; the settlers are also more mobile. The settlers comprise additional potential pressures on the forest, as does a paved road under construction. The price for rubber has also been falling in recent years, eroding the rubber tappers' commitment to their forest-based livelihoods and strengthening their interest in agriculture and livestock alternatives (Cunha dos Santos 2002a).

Management Goals—Small-Scale Timber

Porto Dias is the only formal project specifically designed to encourage small-scale logging among our Brazilian sites, though such a practice is common in most other sites informally. In 1995, CTA proposed the development of an environmentally friendly, sustainable, profitable forest enterprise to the Rubber Tappers Association of Porto Dias. The ACM team (PESACRE/University of Florida/CIFOR) has been working with CTA and the Rubber Tappers Association on this process since 2000 (see Cunha dos Santos 2002a).

Initially nine rubber tapper families (now ten) were involved in the project, which began with a rather top-down approach. CTA brought in trained foresters who tried to show the rubber tappers how to develop a small-scale logging enterprise, but the group was unable to extract any timber until 2000. The involvement of ACM signaled a successful change in the endeavor, when CTA began to realize both the importance of getting more genuine buy-in from local people and the utility of instituting mechanisms for learning together from the problems encountered.

Typical landholdings in the Porto Dias Agro-Extractive Reserve are 300–600 ha, but each project participant sets aside 300 ha for timber management, in order to guarantee a 30-year cutting cycle. Twelve tree species are harvested yearly, with a low intensity of cutting (one to two trees/ha, selected based on ecological and economic factors), from management areas of 10 ha each. The goal is a yearly total volume of 1,000 m^3 for the association. In 2003, Stone found that the yearly volume had been scaled back to 320 m^3, after rubber tappers and CTA experienced high costs and limited local administrative capacity with larger scales of production. Profits are designed to be divided equally among participants. One goal of the group was to become certified, which they did through Forest Stewardship Council in late 2002 (Cunha dos Santos 2002a).

External Diversity—Medium

Cunha dos Santos (2002a) lists the following stakeholders in Porto Dias: National Rubber Tappers Council (CNS); Rubber Tappers Association of Porto Dias; squatters; neighboring settlers in the Directed Settlement Project; Agro-Extractive Association of São José in Area II and Association of Rural Producers of Acre (ASPOMACRE) in Area III (settler areas); the NGO CTA; government institutions (IBAMA, INCRA, Acre state government); PRODEX (credit for rubber) through the Bank of Amazonia (BASA); the Union of Rural Producers of Acrelândia, which issues land titles to colonists; and *Programa Nacional de Fortalecimento da Agricultura Familiar* (National Program to Support Small-Scale Farmers or PROFNAF).

Age and Gender

In all the Latin American sites, the young are expected to support themselves once they get married. Young couples choose their partners, with varying amounts of advice from the parents. Parents in the western Amazon make significant sacrifices to ensure, for instance, that their children obtain an education (Campbell et al. 2004). Whereas in the African sites age-related issues center on relations between young and old adults, in the Latin American sites the responsibilities between adult parents and their young children are emphasized (Campbell et al. 2004; Bolaños and Schmink 2004).

In a global comparison, gender differentiation is intermediate, as in Bolivia, with division of space and labor being scored 4, and male dominance and hostility to women in public arenas scored 3 (1 = low, 5 = high).

Internal Diversity—Low

In Porto Dias, people define themselves as rubber tappers or as colonists (agriculturalists) by occupation, rather than by ethnicity, as in many other parts of the world. The rubber tappers of Porto Dias, though long-term residents, are almost entirely descendants of people from outside the region. They include Northeasterners relocated from the municipalities of Tarauacá, Placido de Castro, Rio Branco, and Assis Brazil—all within the State of Acre—as well as from other regions, such as Porto Velho, the capital of the neighboring State of Rondônia, and Bolivia (Cunha dos Santos 2002a). The rubber tappers differ in ethnicity but consider themselves a comparatively homogeneous community. Similarly, colonists define themselves as having common interests and characteristics, even though they too are a mélange of peoples from various places in Brazil and various races (Cunha dos Santos 2001b).

Level of Conflict—Medium

The violence in Acre's history (Hall 1997) does not characterize present-day conflicts. Today the most striking conflicts are between the Porto Dias rubber tappers and agricultural colonists. Since the creation of the Porto Dias Agro-Extractive Reserve, disagreements about the best way forward have prevailed. Some support the idea of a PAE, with its large plots, 300 ha and larger; others support a more conventional agricultural settlement, with individual titles to smaller areas of 50 to 100 ha.

The PAE's three areas form the shape of a Y. The lower area (I) is dominated by rubber tappers (the official, intended use of the entire PAE), the upper areas (II and III) by would-be agriculturalists. The eastern side of the PAE (III) borders on colonization projects (Santo Antônio do Peixoto and Soão João de Balanceio) (Cunha dos Santos 2002a, 3). Land conflicts are particularly serious in area III, where land has been divided and sold to new settlers without much attention to boundaries with neighbors. These newcomers are able to obtain credit, if they obtain *mansa,* or declarations of possession, from the Union of Rural Producers of Acrelândia. The *mansa* is considered equivalent to a land title from INCRA (Cunha dos Santos 2002a, 12).

The three managing associations—the Rubber Tappers Association, the Mossoró Association, and ASPOMACRE—have had very little dialogue, nor has there been oversight by any governmental institutions to ensure compliance with the PAE's use plan, which is increasingly ignored in areas II and III. The result has been increased misunderstandings and conflicts among these groups (Cunha dos Santos 2002a, 19).

A series of conflicts has emerged relating to a logging road that was built in connection with the Rubber Tappers timber management project. The residents of Mossoró would like the road extended to serve their community better, but the rubber tappers are worried about the increasing number of agriculturally oriented settlers and the potential for changing the PAE into an agricultural settlement project. Cunha dos Santos (2002a, 20) sees the road as "a symbol of power and resistance for the Rubber Tappers Association."

Conflicts among the rubber tappers themselves are again geographically based. Those rubber tappers living in the southernmost part of the PAE have had better access to marketing and to outside support. Those living in the northerly part of the PAE have received fewer benefits from their joint project. These conflicts have been partially resolved through a self-conscious process of collaborative problem solving, in cooperation with the ACM team and others (Cunha dos Santos 2001b).

Social Capital—High

In our initial estimate, the people of Porto Dias were considered to have high levels of social capital, with strong traditional, cultural, and economic ties to the local forest. Local residents' associations represent the community to governmental, NGO, and commercial entities. Although these initial estimates are not wrong, they mask significant issues. First,

> the creation of the PAE was not marked by a process of social mobilization that involved all the families in a fight for land rights [as were other extractive reserves]. In addition, the form in which families are grouped and the geographic distance between landholdings within the PAE has made the work of social organization of the Rubber Tappers Association of Porto Dias more difficult … (Cunha dos Santos 2002a, 13)

In the timber management project discussed above, inequitable distribution of benefits occurred after the first timber harvest. A clear distinction emerged between southern and northeastern rubber tappers, with each group having stronger social capital within than between groups (Cunha dos Santos 2002a, 49). Construction of a logging road also adversely affected social capital within the PAE, when the rubber tappers involved in the timber management project unilaterally chose the road's location, to the general detriment of the settlers.

Additionally, a strong tradition of patron–client relationships persists. Rubber barons were once very powerful, and there remains a former middleman, who has turned to ranching, but who maintains important influence in the area by providing employment opportunities and medical assistance for a small group of isolated rubber tappers. He is also responsible for more than his allotted 10 percent of deforestation within the PAE (Cunha dos Santos 2002a, 33–34).

Strong distinctions exist between the PAE's south-central rubber tappers and the settlers concentrated in its northwest and northeast. The two settler associations—ASPOMACRE and Mossoró—differ in their memberships. ASPOMACRE belongs to a neighboring colonization project, and the majority of its members (34 of 45) reside outside PAE Porto Dias. A small number of isolated families inside the reserve have joined ASPOMACRE, even though their PAE residency prohibits them from sharing in all of ASPOMACRE's benefits, such as credits. Mossoró has 30 members. Both ASPOMACRE and Mossoró have been pushing for easier access to services like education, health, electrification, and income-generating projects. They have obtained equipment like rice threshers and seeds and have gotten a feeder road and a school built. They have become political forces, pushing for a change in status for the area from a PAE to an agricultural settlement, and they have obtained a variety of services and inputs for their members (Cunha dos Santos 2002a).

CASE 3

Pará—Eastern Brazil

Contributing Authors
Benno Pokorny, Guilhermina Cayres, Westphalen Nuñes,
Dörte Segebart, and Rozilda Drude

Sites in Pará
Muaná (including Jaratuba, Recreio, Nova Jerusalém), Canta-Galo,
Gurupáv and Tailândia (including São João Batista, Nova Jericó)

Devolution Status—Low

Authority over forests is rather different in Amazonian Brazil than in the other forest-rich environments like Indonesia or Cameroon. In the Amazon, legal forest management is linked to legal ownership and an official management plan approved by the Brazilian Institute for the Environment and Renewable Resources (*Instituto Brasileiro de Meio Ambiente e dos Recursos Naturais Renováveis* or IBAMA). Only private forests are legally managed, and most timber enterprises do not own forestlands, instead using temporary exploitation contracts. Currently, Brazil is debating whether to declare huge areas of the Amazon as national forests or agroextractive reserves as in Acre (see Case 2), subject to the kinds of large-scale concessions common in Indonesia and Cameroon (Pokorny 2003a; ITTO 2003; Pokorny et al. 2004a).

In the meantime, local communities have great difficulty legally managing "their" forests without legal titles. Some have settled illegally, as in Tailândia. Some are living on public land, as in Gurupá before 2000. Some, like Muaná, are using forest resources belonging to others. Tenure conflicts abound—partly due to governmental inefficiency, partly to population movement and illegal land occupation. Both national government institutions—such as the National Institute for Settlement and Agrarian Reform (*Instituto Nacional de Colonização e Reforma Agrária* or INCRA)—and state government institutions—such as the Tenure Institute of Pará (*Instituto de Terras do Pará* or ITERPA)—make legal forest management complicated. Even those who do have tenure documentation meet challenges, because their management plan must be approved by IBAMA. Created for large-scale forest enterprises, IBAMA regulations are complex. The title acquisition process is slow and bureaucratic, officials have little experience dealing with communities, and

Pará—Eastern Brazil
Key ACM Characteristics

Devolution status	Low
Forest type	Humid tropical
Population pressure	Medium
Management goals	Large-scale timber: Tailândia
	Small-scale timber: Gurupá
	Multipurpose reserve: Muaná
External diversity	Medium: Gurupá, Tailândia
	Low: Muaná
Internal diversity	Low
National conflict	Low
Local conflict	Low: Gurupá and Muaná
	Medium: Tailândia
Social capital	High: Gurupá
	Low: Muaná, Tailândia

local people lack the financial wherewithal to pay off corrupt officials. Even the significantly simplified proceedings for small forest management units recently implemented by IBAMA still suffer from a lack of practicability and local acceptance.

Pokorny (2003a) stresses the local variability in Pará. He outlines the four institutional levels involved in forest management: federal, state, governmental and nongovernmental district (*município*), and local. At the federal level, IBAMA and INCRA are responsible for forest management and tenure questions, respectively; each has offices in all Amazonian states. Both are hierarchical, centralized institutions with inadequate financial and human resources. Both are staffed with political appointments, making for little institutional continuity, and are marred by variable amounts of corruption.

Since Rio 1992, the Pilot Program to Conserve the Brazilian Rain Forest (PPG7) has set the goal of decentralization, resulting in a slight increase in the role of state governments in addressing environmental issues. However, this process has been characterized by strong conflicts, depending on the varying power of actors in the national capital, Brasilia, and in the states. Smaller states like Acre have managed to influence the forestry sector somewhat more, though results in large states are quite variable.

The third governmental level is the *município*, which is even more heterogeneous than the state governments. Some *municípios* in Pará are nearly the size of Acre. Due to that size and also because Pará is distant from Brasilia, many national and international initiatives, including PPG7, have sought to strengthen the *municípios*. One attempt to improve natural resource management at this level has focused on enhancing effective participation of the population. Various *municípios* have established planning committees composed of representatives of the population. These committees support and advise the district government, again with varying success. As at the state level, continuity is a problem, and institutional successes do not always outlast incumbents. But Pokorny (2003a) notes "a clear overall tendency for more democracy and more active involvement of local populations."

The lowest level, the local, also differs from most other ACM sites. In addition to turning to their informal village leaders, the people in the Brazilian Amazon often form associations as legal mechanisms for communication and negotiation with outside organizations. Generally, the president of the association is the village leader. The associations usually address production issues, especially agriculture, and their existence is a precondition for access to credit lines. Where forests traditionally contribute significantly to income—as in Acre or the flood plains of Pará—in Amazonian Pará the associations may address forestry issues. Some associations are effectively linked to regional groups such as the Federation of Agricultural Workers in Pará (*Federação dos Trabalhadores na Agricultura do Estado do Pará* or FETAGRI) and contribute significantly to public policies, even at the federal level. But often such associations exist only on paper, as in Muaná.

In Tailândia's district, dominated by timber enterprises, the district government has shown interest only recently in establishing democratic mechanisms for involving the population in decisionmaking. The communities of Nova Jericó and São João Batista have no formal organization, nor have they been taking part in any formal or informal representative mechanism. One of the ACM activities involved organizing a meeting for the community to elect leaders and identify improved communication channels. However, this effort was unsuccessful, due to the extreme mobility of the community: fewer than 25 percent of the families involved remained the following year. Most sold their homes and left. Forestlands belonging to absentee landlords were illegally invaded by individual families.

In the Gurupá *município,* since 1992 a number of communities have received a collective, legal land title as part of an integrated rural development project. Community efforts succeeded with the cooperation of the nongovernmental organization Federation of Organizations for Social Assistance and Education (*Federação de Orgãos para Assistência Social e Educacional* or FASE) and the Rural Workers Union (*Sindicato dos Trabalhadores Rurais* or STR). Such an accomplishment was perhaps achieved partly because the communities are acknowledged as *Quilombos* or descendants of slaves, who have rights similar to those of indigenous tribes. This joint project paid special attention to the development of democratic structures at the district level, and a regional development plan involved intense participation by the local population. The community and the NGO have even succeeded in gaining certification of community-based management of nontimber forest products under an EU-funded project that ended in 2002. The leaders of the community of Canta-Galo, where ACM focused, took part actively in these activities, including participating in courses on forest management. They have established an informal forest reserve designed to protect their flora and fauna.

The district government of Muaná recently established a committee for rural development. However, the regional representatives were selected by the government, and the communities have said they do not feel properly represented. The committee's main tasks are to allocate public credits from the National Program to Strengthen Family Farmers (*Programa Nacional de Fortalecimento da Agricultura Familiar* or PRONAF). Nova Jerusalém, Recreio, and Jaratuba formed an association designed to access the PRONAF public credit funds, but it has not represented the communities' interests well. The president was selected by the district government, and fewer than 25 percent of the members actually participate, with many lacking motivation to change the situation. Of the three communities, Jaratuba is the most organized, due to a politically ambitious local leader. In 2002, Muaná also received PPG7 funds to establish a local Agenda 21. So far, this has not been implemented effectively.

The communities in Pará have to pay part of their production to absentee land-lords for permission to use the surrounding forest. Each community has a resident representative who looks after the landlord's concerns. No mechanisms—formal or informal—were identified for dealing with the forest, though as part of ACM project activities, the communities established small groups for the management of the palm trees *Açaí* and *Patauá*.

Forest Type—Humid Tropical

Pokorny (2003a) describes the Pará sites thus:

> All sites are characterized by an equatorial humid climate with two seasons, a rainy winter from December/January until June and a drier summer from July to December. In general, soils belong to 4 categories: *terra amarela, terra roxa, terra preta* and *terra branca*. Landscape is strongly influenced by rivers (*Igarapés*). Although similar in biophysical characteristics, selected communities reflect very different realities of the Eastern Amazon region.

Nearly all forests belonging to Nova Jericó and São João Batista communities in Tailândia are located on dry land and have been selectively logged by timber companies, up to three times. Average initial harvest was about 25 m³/ha focusing on *Maçaranduba* and *Angelim vermelho* species, resulting in seriously degraded forests with major gaps in the canopy and few remaining trees of commercial interest with diameters over 60 cm. The communities distinguish virgin forests, high forests, low forests, older secondary forests, and young secondary forests. Although many community members believe the remaining forests hold little commercial value, a forest inventory in a half-hectare plot conducted as part of the ACM activities revealed 42 trees over 30 cm diameter at breast (dbh), and 25 species with a commercial volume of more than 200 m³/ha (Pokorny et al. 2001b, 2003b).

Forest conditions in the village of Canta-Galo, Gurupá, are excellent, with large areas virtually untouched. Common tree species include *Breu Branco* (*Protium sagotianum* March.) and *Breu Vermelho* (*Protium puncticulatum* J.F. Macbr.), the seeds of which are eaten by wildlife. Other species with high commercial timber value include *Sucuruba* (*Trattinnickia burseraefolia* [Mart.] Willd.), *Angelim vermelho* (*Andira coriacea* Pulle), and *Louro branco* (*Nectandra angustifolia* [Schrad.] Nees & Mart.). Vegetation on poor soils is dominated by *Cajú* (*Anacardium excelsum* Skeels.), *Jurú* (*Solanum variabile* Mart.), and *Murucí* (*Byrsonima crispa* A. Juss.). Community members estimate about 50 tree species with commercial value in their forests and recognize the important role of forests in erosion prevention.

Soils in the region vary in fertility and physical characteristics. Most areas of better soil near community homes have been converted from forest to agricultural land. The community forest is at some distance, near the source of the Rio Camutá.

The forest is rich in wildlife, including the rare Amazonian jaguar. Hunting is particularly important during the rainy season (January to July), when fish are scarce. The communities use an impressive array of forest products, including timber, medicinal oils, fruits, leaves, and bark. Illegal logging occurs both by community members and timber companies—some 70 trees were taken in 2000—but the harvested volume remains low (Segebart et al. 2000; Pokorny 2003b).

Açaí (*Euterpe oleracea* Mart.), a palm growing in natural and planted stands along the river, is essential to the economy of the Muaná communities of Jaratuba, Recreio, and Nova Jerusalém. The natural *Açai* stands produce fruit, whereas planted *Açai* stands are used mainly for palm hearts. *Açai* stands co-occur with inundated forests (*várzea*), which flood temporarily in concert with the tides. Trees are only flooded during the rainy season.

Dry land (*terra firme*) forests occur on higher ground and form three stories. Although the most valuable species tend to have already been harvested, many good trees remain, including *Maçaranduba* (*Manilkara huberi* [*Ducke*] *Standl.*), *Acapu* (*Vouacapoua Americana*), and *Jatobá* (*Hymenaea courbaril* L.). Although reaching up to 40 m, these trees do not form a closed canopy. The second story is more closed, with a greater variety of species, including *Sucupira* (*Diplotropis purpurea* [Rich.] *Amsh*), *Acariquara* (*Minquartia guianensis* Aubl.), *Cupiuba* (*Goupia glabra* Aubl.), *Cedro* (*Cedrela odorata* L.), and *Jacareúba* (*Callophyllum brasiliensis* Camb.), which provide timber. Other species are useful for food and medicine: *Bacaba* (*Oenocarpus bacaba* Mart.), *Cupuí* (*Theobroma subincanum* Mart.), *Bacuripari* (*Rheedia macrophylla* [Mart.] Pl. et Jr.), *Muruci* (*Byrsonima crassifolia* [L.] H.B.K.), and *Verônica* (*Dalbergia monetaria* L.F.). The bottom story is sparsely populated in undisturbed areas where little light penetrates; but an enormous variety of trees, bamboos, lianas, and high shrubs quickly fill any gaps. Another poor forest type called *Igapó* lies in the transitional area between *várzea* and *terra firme*. Table A-2 shows the distribution of forest types in the three ACM communities in Muaná.

Although *Açai* is the only commercialized forest product in the area, local people use a vast array of timber and nontimber forest products. They estimate that the forest provides them with 60-70 percent of their food, 70-85 percent of their construction materials, 80-100 percent of their art, and 2-30 percent of their medicines. The forest also offers them 60-80 percent of their income, most of it from *Açai* (Nuñes et al. 2000; Pokorny 2003b).

Population Pressure—Medium

In all these Pará communities, the population is growing quickly, though at an unknown rate. Women marry as young as 15, with all communities having large

Table A-2: *Land Cover by Environmental Type in Three Communities in Muaná*

	Land Cover by Environmental Types (%)				
Community	Inundation forests (várzea)	Dry land old forests (Capoeirão)	Young secondary forests (<5 years)	Poor forests (Igapó)	Open land (natural and transformed)
Recreio	15	55	—	—	30
Nova Jerusalem	15	60	24	—	1
Jaratuba	—	25	30	15	30

Source: Pokorny (2003b).

numbers of children and adolescents. Natural increase is the source of population growth in Muaná and Gurupá; in-migration may be more significant in Tailândia. Out-migration of women is more likely than for men, as men are able to find gainful employment. Young women who do not marry young often go to the cities in search of work or further education.

Citing INPE (2001) data from 2000 that measure Tailândia population density at five persons/km^2, or 3/km^2 in rural areas, Pokorny (2003c) suspects this is an underestimate. São João Batista has a population of 100 or less; Nova Jericó has about 150. Both communities are unusually mobile, with a constant stream of illegal settlers coming and going. Average family size is about five persons (Pokorny et al. 2001b). For Gurupá, INPE data show a population of 2.43/km^2, falling to 1.82 for the rural population. The community of Canta-Galo has about 250 people. For Muaná, INPE found 6/km^2, with the figure falling to 5/km^2 based on the rural population alone. The Jaratuba population is estimated at about 250 people; Recreio has about 500 and Nova Jerusalém, 150.

Management Goals—Small-Scale Timber

As noted, Pará is a huge state, characterized by dramatic differences in policy and its implementation. In Gurupá, people's main economic activity has been the production of manioc, but as with many swidden systems, the people remain dependent on game and wood from their forests as subsistence supplements. More remote than Tailândia, the people have only been involved irregularly in small-scale logging—logging that is technically illegal, because the people lack land titles and management plans. Timber extraction is limited to single trees, using only a chainsaw. Logs are sawn into boards in the forest, and taken out manually.

Canta-Galo has had regular conflicts with timber companies, due both to unfulfilled contractual promises and to unpermitted extraction. Lacking titles to their land, people in the community had little in the way of legal recourse. Working through FASE and the STR, the community recently obtained a collective, legal land title, as did some other communities in Gurupá. As part of these EU-funded activities, the community developed a forest management plan, including timber management, during the ACM involvement there (Pokorny 2002b; Segebart et al. 2000).

Management Goals—Multipurpose Reserve

Muaná differs from the other cases with multipurpose reserves in that there is no formally recognized forest reserve in which the government plays a significant role. In fact, no interaction takes place at all between the community and IBAMA or with the few small sawmills that occasionally buy local timber. Land belongs to a number of absentee landlords, who extract payments in return for the people's use of the forest. Since the 1970s, the local economy has been based primarily on extraction of palm heart and the fruit of the Açai palm (Pokorny 2002b; Nuñes et al. 2000).

External Diversity—Medium

In Tailândia, Pokorny considers the most important external stakeholders to be timber enterprises and small local timber extractors, who are interested in illegal timber extraction. Small-scale traders and governmental institutions also play a role. These include IBAMA, which is responsible for environmental issues; INCRA and ITERPA, relevant for tenure questions; the governmental Institution for Extension and Technical Assistance in Pará (*Empresa Paraense de Assistência Técnica e Extensão Rura* or EMATER), which gives technical assistance; and the governmental Brazilian Organization for Agricultural Research (*Empresa Brasileira de Pesquisa Agropecuária* or EMBRAPA). Some local families specialize in illegal timber extraction (Pokorny 2002b; Pokorny et al. 2001b).

External Diversity—Low

Little information is available on external stakeholders in Muaná, but the original site selection matrix lists local families, landowners, and the municipal government.

Internal Diversity—Low

The two communities in Tailândia, São João Batista and Nova Jericó, experience a high level of mobility. Since the ACM work began, only about 30 percent of the original occupants of these two villages remain, replaced by similar, poor immigrant colonists. Although coming from various places, these colonists perceive themselves to be, and are perceived by outsiders, as very similar.

Gurupá, Canta-Galo—a community descendant from a single family of ex-slaves—is perhaps the least internally diverse of any of our sites.

The inhabitants of Muaná are characterized as typical *ribeirinhos,* or people who live along the riverside (Pokorny 2003a). They form a well-established occupational grouping in Brazil, like rubber tappers or colonists. Their primary economic activity in the Muaná sites is the extraction of palm hearts. Although not indigenous, they have been living in the area for decades and, in some cases, centuries.

Age and Gender—Intermediate

As in Acre and the other Latin American sites, gender differentiation is intermediate, on a global scale. The division of space and labor between the sexes is scored as 4 (on a 5-point scale), and male dominance and hostility to women in public arenas is scored 3.

Level of Conflict—Medium

Although some settlers have received permission to stay on the land where they settled in Tailândia, all settlers have invaded the land illegally. Yet so far no important conflicts have arisen. In Nova Jericó, some dissatisfaction with community leaders is evident. Continual in- and out-migration, with the resultant mixing of populations, combined with the uncertainty of tenure and the desperate living conditions of the families, all suggest a fertile breeding ground for conflict, despite its relative scarcity to date.

Relations between the communities and the neighboring timber harvesting projects have also been reasonably friendly. The timber companies provide regular transport during the harvest season, which moderates some of the potential antagonism. Prostitution, including with children, is seen as a very negative impact of the timber companies' presence. However, a disagreement with the ferryboat owner over high prices is the most notable problem expressed by the communities (Pokorny et al. 2001b *14*).

Level of Conflict—Low

Conflicts within Canta-Galo itself were minimal, with all inhabitants closely tied by kinship and apparently congenial relations. The fact that Canta-Galo was separate from the neighboring community of Terra Preta (20 minutes away by boat) suggests a previous conflict—probably over the decision to affiliate with the municipality of Gurupá, in the case of Canta-Galo, and the municipality of Porto de Moz in the case of Terra Preta. The two communities share a common background, and two brothers are among the oldest inhabitants of the two communities, respectively. Families in each community visit the other regularly, important festivals and other cultural events are organized together, and community members still work together on some daily activities, such as construction of small boats.

The leader of Canta-Galo has political connections in Gurupá, which tie the community to a number of wider conflicts and opportunities. These links include direct contacts with the STR, one of the municipal councilors, and FASE. Strong competition and conflict exist between the municipalities to claim these communities as part of their respective realms.

At the nexus of Gurupá, Melgaço, and Porto de Moz lies a huge forest area with high-quality timber. Nine times in the past 15 years the community has suffered illegal attempts by timber enterprises, primarily from Porto de Moz, to log the community's forest. The immediate problem has been resolved with the help of the municipal government and IBAMA.

Further conflicts are anticipated with the landlord whose lands surround the community. The community is involved in a patron–client relationship with this landlord, who is the principal buyer of community produce, normally paying in products rather than cash. Some residents express concern about his reaction as the community becomes more independent (Segebart et al. 2000).

The low conflict levels in Muaná are only slightly marred, particularly in Jaratuba, by some disagreement between those families directly involved in palm heart extraction and those who are not. The disagreement revolves around the intensity of use of *Açai* stands. Conflict is controlled by regular meetings and informal efforts at conservation. Another source of recurring conflict is uncontrolled grazing of pigs and buffalo, which damages crops.

External conflicts are also rare in Muaná, where the most significant recurring conflicts also relate to palm hearts—in this case, their illegal extraction. Some conflict is attributable to resentment of absentee ownership of much of the agricultural land (Pokorny et al. 2001b). Landlords receive a certain percentage of the yearly production and make land use decisions, restricting the possibilities to increase efficiency (Nuñes et al. 2000).

Social Capital—High

One of the most striking features of Canta-Galo is the charisma of its unofficial leader. He has been active and successful in organizing community life. He communicates effectively both within the community and in the role of liaison with external actors (public institutions, researchers, NGOs). The community itself is remarkably open to and interested in new views, knowledge, and partnerships. They worked hard to improve their educational situation, seeing it as an investment in the future of their children. They also formed a number of local organizations, including a women's group, youth group, and soccer teams. The researchers noted high levels of participation, organization, and confidence, as well as a spirit of collaboration that characterized the whole community. Families were strongly motivated to improve human well-being for themselves and for the entire community.

Despite this apparently high level of social capital, Segebart et al. (2000) noted that even articulate individuals were sometimes reluctant to express their views. The strength of the dominant leader to some extent precluded a wider participation, as some lacked self-confidence and others did not understand the leader's explanations of his interactions with external actors. It was not uncommon for community members to fail to develop their own opinions, relying instead on the views of their charismatic leader.

In contrast to some Cameroonian cases (see Case 4), the people of Canta-Galo showed an obvious consciousness of the future and of preservation of their forest. They started to establish regulations to ensure the sustainable use of natural resources, including restrictions on forest use. They established relations with a whole string of external groups, including the Association of Quilombos, a farmers' association, STR, the Women's Movement, a church group, the district government, FASE, and CIFOR (Segebart et al. 2000).

Social Capital—Low

In the short history of the Tailândia communities, considerable in- and out-migration has resulted in a permanently fluctuating human context. The ACM researchers found no organizations of any kind. The families, isolated in their respective areas, did not gather together for meetings to discuss any themes of common interest. Although an informal, very basic communication system exists, it has not yet evolved into meaningful social capital. Relations with external actors are also minimal. The most important are relations with the timber companies and small-scale, local timber extractors who are involved in illegal timber extraction. Social capital overall is extremely low (Cayres and Pokorny 2000).

Family and longstanding neighborly relations are close and enduring in Muaná, but few formal organizations or collaboration can be found. The church—Catholic in Recreio and Jaratuba; Evangelical in Nova Jerusalém—has been responsible for virtually all groups in Muaná communities, whether for men, women, old or young. Jaratuba had an association for agricultural production. In Menino Jesus, Recreio, and Nova Jerusalém no such association existed, although some families participated in a municipality-inspired and financed association. The few community meetings do not appear well-planned, and public reports or documentation of decisions are lacking. Local organization seems marked by external leadership and little attention to community interests. Indeed, some powerful individuals actively discourage collaborative action when their own interests are not central (Nuñes et al. 2000).

CASE 4

East Cameroon

Contributing Authors
Phil René Oyono, Samuel Assembe, Charlotte Kouna, and Samuel Efoua

Sites in East Cameroon
Lomié and Dimako

Devolution Status—Intermediate

The forests of Cameroon have legally belonged to the State since colonial times, though local populations, due to tradition and their physical proximity to the forest, have always claimed that they are the "owners" of resources and have always been among the main users/managers (van den Berg and Biesbrouck 2000; Oyono 2003c). In many cases, what we call informal or de facto devolution occurs: communities manage their resources with little interference from government or other actors, usually because they are remote and inaccessible.

In 1993, the government prepared a zoning plan for the southern half of Cameroon—where most of the country's forests are located—attributing 80 percent of lands to the state (Nguinguiri 1999, 10). But in 1994, the government passed a new National Forestry Law, *Loi* No. 94/01 of 20/01/94 pertaining to forests, wildlife, and fisheries, and an implementation decree, No. 95/531/PM of 23/8/95 (Kouna 2001a; see also Djeumo 2001; Ekoko 1997; Vabi et al. 2000).

Van den Berg and Biesbrouck (2000, 46) note that "lawmakers declared the State the sole guardian and chief manager of all forests in the country." However, the law also opened up the possibility for communities to manage community forests. Yet the term "communities," the authors note, is used loosely:

Bantu villages are basically administrative units created by German colonial officials with the purpose to get control over Bantu groups who then lived in small settlements dispersed over the forest. The social cohesion of Bantu villages is mainly outwardly directed and most often expressed in boundary disputes with neighboring villages or conflicts with powerful outsiders such as logging operators. Villages consist of a variable number of politically and economically independent lineages (Houses) ... [the villages] generally lack the internal cohesion to jointly design and implement new local forest management regimes. (van den Berg and Biesbrouck 2000, 83)

Such community management of forests under this law, therefore, was a first experiment for the Central African region (Nguinguiri 1999). Djeumo (2001) and Etoungou (2002) document the important roles of conservation projects, nongovernmental organizations (NGOs), and timber companies in the cases where communities were successful in obtaining community forests.

Translations of Cameroon's forest designations deserve explanation: Oyono (2002a) calls a "council forest" a forest derived from the administrative unit called a *commune*—an intermediate, administrative level sometimes called a municipality, a district, or a subdistrict in other countries. "Council forests" differ in structure from "community forests."

Several potential effects of the National Forestry Law and other related legislation were expected (Oyono 2002a). First, powers to manage "council forests" would be transferred to communes. Second, powers to manage "community forests" would be transferred to local communities. The final expected result of the law was "fiscal decentralization in forestry"—the allocation of 40 percent of annual forest taxes to local governments and 10 percent to local communities living near forest concessions. Management powers and payment of a tax of U.S.$1.50/m³ of timber extracted were to be allocated to each village community living in a forest concession. These three developments represent a formal or official devolution.

Before 1994, these forest financial benefits—especially the U.S.$1.50/m³ tax—had been given directly to communities. Some communities used the money well, but when others did not (Oyono 2004) the government reacted in 1998 with a new ministerial order that recentralized these revenues to the *commune* level (MINEFI/MINAT 1998).

East Cameroon
Key ACM Characteristics

Devolution status	Intermediate
Forest type	Humid tropical
Population pressure	Low
Management goals	Large- and small-scale timber
External diversity	High
Internal diversity	Medium
National conflict	Medium
Local conflict	High
Social conflict	Medium

In numerous ways, the new legal instruments are not working as planned (Oyono et al. 2002). At the local levels, problems include non-accountable management committees, free-riding and rent-seeking behavior, and lack of transparency. Lack of information and public dialogue with would-be beneficiaries are problems at higher levels (see also Fomété 2001).

Considerable interest in the idea of community forests has grown, but the process of legally obtaining such a forest has been plagued by bureaucratic complexities and slow implementation (Brown and Schreckenberg 2001). Diaw et al. (2002) report

> By September 2001, seven years into the reform [on Community Forests], only 19 requests (15 percent) for community forests had been granted. This may be related to the fact that the total area concerned by the policy (only 2 percent of the national domain estate) is far too small to accommodate popular expectations.

The reality is that most communities continue to practice swidden agriculture, hunting, and collecting in lands that do not legally belong to them, despite widespread informal recognition of traditional ownership (cf. Diaw 1997).

Lomié, Eastern Province. One of our primary interests in Lomié was the fact that Cameroon's first "pilot" community forests were established there (Klein et al. 2001). As Etoungou (2002, 11–12) writes:

> Community Forests, which are part of the non-permanent forest, are taken from the forests of the national estate and allocated first to the nearest neighboring communities on the basis of an agreement. In contrast to a land ownership title, the management agreement signed by the State and a village community does not confer any title of ownership of the forest on the latter. The community simply enjoys its use, under supervision; it may log it under State authorization, by sale of standing volume (*vente de coupe*), or by logging permit or personal felling permit.

Attempts to implement the new forestry legislation occurred in Lomié, though the area is best known for having the first successfully established community forests in Cameroon. Various authors (Kouna 2001a, b; Oyono et al. n.d.) detail the problems of implementing the fiscal decentralization policies. These include persistent corruption, manipulation of the populace by government and industry officials, trickery by local leaders, and individual squandering of resources.

In August 2000, the country's first management agreements for community forests were signed between the Ministry of Environment and Forests and five village communities in the Lomié region, including Koungoulou (3,100 ha), Kongo (3,000 ha), Moangé-Le-Bosquet (1,662 ha), Echiambor/Malen (4,490 ha) and Ngola/Achip (4,200 ha). The ACM team has worked most with the villages of Kongo and Echiambor/Malen. These forests have been subjected to intense forest exploitation (Oyono 2002a; Assembe 2004). By July 2001, 20 more communities had applied for community forests (Klein et al. 2001).

Dimako, Eastern Province. Fiscal recentralization has also been examined in detail in Dimako. The 1998 regulations require that a forest royalties management committee be formed at the regional level, under the supervision of the *sous-préfet* and presided over by the mayor. In addition, in each village or group of villages adjacent to a forest concession, village management committees, composed of a chair and four or five members, were required by the ministerial order to deal with forest royalties. In fact, the regional committee has now become the de jure manager of the royalties.

Village management committees may request schools, health centers, and other needed infrastructure. But in practice, only the regional committee can now fund village-level projects. Kouna (2001) found that council authorities very often create, on paper, fictitious projects to be funded, thus misappropriating the money.

Oyono et al. (forthcoming) refer to "a 'strategic axis' made up of members of village management committees, municipal authorities, administrative officials and timber companies" whose goal is to retain control over forest royalties at the regional level, in order to share or mismanage them. Shackleton et al. (2001) find a similar pattern in southern Africa. The second category of forest royalties (U.S.$1.50/m³) are to be calculated by both logging companies and the villagers, but because villagers typically do not have skills in estimating cubic meters themselves, they must rely on company personnel who may remit whatever amount they like.

Dimako has also been of interest because it was the only council forest in the country. Local people were suspicious of this initiative by their own council, feeling that the council was taking their land (Mvondo and Sangkwa 2002). Conflicts erupted between ethnic groups, between the population and the council, and between adherents of council forests and those who favored the fiscal decentralization mechanisms (discussed further below).

Forest Type—Humid Tropical

Lomié. The forests in Lomié are called "evergreen Cameroon-Congolese forest," sometimes characterized as "Dja Forest." The Dja Reserve is a well-known neighboring protected area. Oyono (1998, 557) summarizes the situation thus:

> *Gilber tiodendron dewevrei* is the dominant species in an area marked by striking biological diversity, with over 1600 species of trees including *Baillonella toxisperma* and *Irvingia gabonensis*, the kernels of which are important sources of food for local peoples. There are roughly 165 species of mammals, including a number of primate species, 120 species of fishes and 320 species of birds. A number of local mammals are listed as threatened and endangered: white-collared Mangabey (*Cercocebus torquatus*); forest elephant (*Loxodonta Africana cyclotis*); leopard (*Panthera pardus*); gorilla (*Gorilla gorilla*); and chimpanzee (*Pan troglodytes*).

Dimako. The Dimako Council Forest, consisting of 16,240 ha, is located in the Dimako Subdivision of the Upper Nyong Division, in the Eastern Province (between 3°50′ and 4°4′N, and 13° and 14°50′E). The forest is semidegraded, dense, humid tropical rainforest, dominated by *Ayous* and *Fraké* species. To the north the forest is bounded by the Dimako-Mbang highway, and to the west and south by the Dimako-Kagnol highway. The local economy was very dependent on the logging company, *Société Forestière et Industrielle de la Doumé* (SFID), in the early 1990s, but after its withdrawal to richer pastures, the people have returned to farming, hunting, fruit picking, gathering of forest products, and small-scale timber exploitation (Mvondo and Sangkwa 2002).

Population Pressure—Low

Oyono (1998, 25) sees the extensive forest destruction in the Lomié area as related to increasing population; he estimates the rate of natural increase as two percent. A major component of the population growth is the return to villages of people displaced by the ongoing economic crisis in Cameroon (related to low prices for commodities and structural adjustment policies; see also Oyono 1997 and Sunderlin and Pokam 2002).

Kouna (2001a) describes two study villages, also in Lomié. Kongo, situated 23 km from the city of Lomié, has a population of 378, 71 percent of whom are under age 30. Ngola is 38 km from Lomié and 15 km from Kongo. Ngola has 715 people, living in eight quarters. Eighty percent are under 30 years of age. Etoungou (2002, 37) reports 900 people in 10 "families" in the Baka (pygmy) community of Moangé-Le-Bosquet.

The area served by the Dimako Rural Council includes about 15,000 inhabitants in 17 villages, located along the roads from Dimako town to Bertoua, Mbang, Kagnol, and Doumé. The Bakum ethnic group forms the majority, with minority populations of Pol and Baka pygmies (Mvondo and Sangkwa 2002). Based on his participatory action research in three villages, Assembe (2003) estimates that Nkoumadja/Nguinda village has about 100 people, with 51 percent women and 49 percent men; Petit Pol has about 650 inhabitants, with 55 percent women and 45 percent men; and Mayos has 450 inhabitants, with nearly 54 percent women, due to male exodus from rural areas.

Management Goals—Large-Scale Timber

As in many other humid tropical forest areas where large-scale logging is under way, the region is afflicted with corruption, unsustainable logging practices, adverse impacts on biodiversity, lack of governmental oversight, and inattention to existing uses and claims to resources by local peoples.

Lomié. Five timber companies are active in the Lomié region: *Compagnie Assene Nkou*, Pallisco, Kieffer, *Ingénierie Forestière*, and the Forestry Society of Dja and Boumba (*Société Forestière Dja et Boumba* or SFDB). The Forestry Society of Hazim (*Société Forestière Hazim* or SFH) also exploited forest in the region in the mid-1980s. Only the *Ingénierie Forestière* has a processing unit in its production site; the rest of the timber companies cut timber and send it to Europe.

Dimako. Dimako Council Forest is for timber production. The management plan specifies that the Dimako forest be divided into three components: production (14,950 ha); agroforestry (466 ha); and research (466 ha). The area has hosted the long-established timber company, SFID, since the 1950s. But since the mid-1990s, SFID has stopped logging in the region, due to the scarcity of commercial species. Today logging activities in Dimako are carried out by the Cameroonian Society of Timber Transformation (*Société Camerounaise de Transformation des Bois* or SCTB).

Management Goals—Small-Scale Timber

In Cameroon, small-scale logging recognized by the government occurs only in Lomié and Dimako. The formalization of small-scale logging is connected to the new legislation on community forests and council forests. This legal process is complicated by the concurrent legislation requiring those involved in timber exploitation (typically large companies) to make payments to local communities based on volume extracted.

Lomié. The formal, legal community forests of Lomié were established with the help of Netherlands Development Organization (*Stichting Nederlandse Vrijwilligers* or SNV), a Dutch NGO. SNV was involved in long discussions with communities about how to conduct sustainable logging. Supported by the Ministry of Forestry, SNV strongly encouraged the use of an environmentally friendly saw, the *gruminette*, to harvest timber.

After considerable discussion and debate within and among the five communities with community forests, the village of Moangé-Le-Bosquet opted to use the *gruminette* in their logging; the other four opted for a semiindustrial approach using the faster Lucasmill, a machine that can fell three to five cubic meters per day (see Oyono forthcoming).

Some villagers pressured these four villages to make contracts with timber companies quickly, in order to harvest faster for more immediate profits. This has led to the transformation of community forests into *ventes de coupe* (sales of standing volume, a type of forest concession for areas up to 2,500 ha), for which communities receive U.S.$1.50/m³ cut (Djeumo 2001).

In Ngola, Oyono (forthcoming, 50) reports the results of a multiresource inventory in their community forest: "38,693 trees, 27,563 having an exploitable diameter, for a volume of 220,500 m³." The community has 15 projects planned, based on the sale of this wood. From December 2001 until October 2002, Oyono goes on to report that wood production from the Kongo Community Forest was as follows: *Sapelli*, 31,683 m³; *Iroko*, 54,729 m³; *Doussié*, 2,500 m³; and *Sipo*, 9 m³, including numerous trees felled and abandoned in the forest. The total wood production of this community forest from December 2001 to January 2004 was 500 m³. In none of these communities did Oyono find serious concerns about sustainability. Instead, villagers had seen the government and private companies make a great deal of money from forests in traditional community areas, and felt it was now their turn.

Dimako. Dimako's claim to fame is having the first council forest in the country. This forest has an area of 16,240 ha, subdivided into five five-year logging plans. The forest, dominated by *Ayous* and *Fraké,* was granted to the council under the stipulation that the forest be used for timber production (Mvondo and Sangkwa 2002). One of the first steps in securing this council forest involved a classification of lands into the state's permanent forest estate and areas reserved for farming and logging by the local population. The France-Cameroon Cooperation Project *Forêts et Terroirs* helped in the difficult and expensive process, which involved collecting socioeconomic data from each of 17 villages, drawing use maps, developing a geographic information system (GIS) map, and holding consultative meetings to determine exact boundaries (Mvondo and Sangkwa 2002, 4).

The classification process began about the same time that the government required companies to pay communities holding a license to sell standing volume the U.S.$1.50/m³ fee for the timber harvested in their forest. Within this context, local people did not see the value of the council forest, since they saw an opportunity to generate cash quickly by obtaining a license to carry out logging in their areas. Local people also suspect the council forest to be a plot by the municipal authorities to appropriate the forest for their personal gain. The fact that the mayor of the area is also a manager of a timber company lent credence to this perception. While logging in council forests is supposed to be small-scale, the decisionmakers could opt for large-scale operations (Mvondo and Sangkwa 2002, 4–5).

External Diversity—High

The members of the five ACM villages in Lomié are involved in a number of committees set up primarily by external actors, such as community forest management committees, village development committees, village political party committees, and youth associations. For more than six years, the process of establishing community forests was supported by SNV and assisted by *Centre International d'Appui au Développement*, a local NGO based in the town of Lomié. Other external actors include the Ministry of Environment and Forests (*Ministère de l'Environnement et des Forêts* or MINEF); timber companies harvesting community forests through small-scale logging contracts; municipal and administrative authorities; and more NGOs. Among the NGOs are the Center of Help for Feminine Development (*Centre d'Appui au Développement Féminin* or CADEF); Association for the Promotion of Pygmies in their Environment in Cameroon (*Association pour la Promotion des Pygmées dans leur Environnement au Cameroun* or APPEC); and Center for Forest Studies of Dja (*Centre d'Etudes Forestières du Dja* or CEFDJA) (Efoua 2001). A new stakeholder came on the scene in 2002—a U.S.-Cameroon joint cobalt mining venture called GEOVIC, which started preliminary activities in the village of Kongo. Finally, the region overlaps with the eastern portion of the Dja Biosphere Reserve, linking local communities with conservation concerns such as Forested Ecosystems of Central Africa (*Ecosystèmes Forestiers d'Afrique Centrale* or ECOFAC) (Oyono 2003a, c).

In Dimako, in addition to the Baka pygmy, Bakum, and Pol groups, and in-migrants who settled in the area a couple of decades ago, other stakeholders include the municipal council; *Forêts et Terroirs;* MINEF; Dschang University, which is conducting biophysical research in the council forest; SFID; and now the *Unité Technique Opérationnelle,* or Technical Operations Unit, created by MINEF when *Forêts et Terroirs* closed in 2001. The mayor, also a major actor, has a history of logging involvement (Diaw 2002; Oyono 2003a, c; Mvondo and Sangkwa 2002).

Age and Gender

In all of the Cameroon sites, relative to sites in other countries, marked hostility between young adults and the old is present. In the past, the young depended on the old for the parental financial contributions critical to young people's ability to marry. While parental blessings remain important in ensuring a successful marriage (Akwah 2002), "times are changing," and the young may embrace changes that many of the old fear (see Russell and Tchamou 2001; Tiani 2001). One unusual feature is the fear among the young in some areas that the old will curse them. In our studies on intergenerational access to resources, Cameroon stood out as unusual in the degree to which the current generation was felt to be using up resources that would no longer be available to later generations (Porro et al. 2001; Mala et al. 2002).

Gender differentiation is stronger than in the South American countries (see Cases 1–3), but still intermediate. Division of space and labor was scored 5, male dominance 4, and hostility to women being active in the public sphere 3 (with 1 = low and 5 = high).

Internal Diversity—Medium

Lomié and Dimako each represent a fairly large area in eastern Cameroon. Our Lomié site includes the five villages mentioned above (Ngola, Kongo, Moangé-Le-Bosquet, Koungoulou, and Echiembor/Malen), populated by the five clans of the Nzimé ethnic group and pygmies, or Baka (Oyono 2002b).

In Dimako, ACM works primarily in Nkoumadjap/Nguinda, Mayos and Petit-Pol villages. Dimako is populated by the Bakum and Pol ethnic groups, Baka pygmies, and a number of in-migrant groups who settled in the area a couple of decades ago. These groups, particularly the Pol and Baka, and to a lesser degree the settler communities at the northern edge, all have stakes in the council forest but have different power bases. The mayor is Bakum. The Baka were forcibly resettled to this area, and are typically in client relationships with Bantu patrons (Diaw 2002; Mvondo and Sangkwa 2002; Oyono 2002c).

Level of Conflict—High

Etoungou (2002, 15) provides some flavor of the kinds of conflicts that characterize communities in Cameroon:

> As "egalitarian" as they may at first seem to an outside observer, they are a hive of statutory contradictions (men/women, elders/youth, free men/former slaves, indigenes/incomers), all related to the factors of production (control of labor, land ownership, farmer/herder tensions), or power play (chieftancies, cooperatives, important members of society, etc.), or even rivalries, mysterious jealousies, interpersonal or involving formal or informal networks (neighbors, family, friends, or companions, *protegés*, factionalism, etc.).

Lomié and Dimako. Both areas are marked by high levels of conflict among local communities, logging companies, and district level officials. A whole host of ongoing local-level conflicts that are common to community logging company relations have arisen. Sources of conflict include overhunting, damage to animal habitats, cutting of valued local trees, hiring outsiders over locals, illegal logging by community members, and unfulfilled promises to pay damages. See Kouna (2001a) for an exhaustive list of conflicts in Dimako and Etoungou (2002) for Lomié. Van den Berg and Biesbrouck (2000) note that district officials are reluctant to call in the police to resolve conflicts, and farmers' roadblocks in eastern Cameroon are a common avenue to greater power in dealings with logging companies.

Many recent conflicts revolve around Cameroon's forest legislation, described above. Oyono et al. (n.d.) discuss the minimal benefits that communities receive from logging companies, despite the legislation. Fomété and Vermaat (2001) recount how one of the first communities to be granted a formal community forest immediately contracted with a commercial logging firm to fell the trees under a *vente de coupe* arrangement, whereby permission is granted to log an area. Exhaustive high-speed felling resulted, with minimal reimbursement to the community—only 1,000–2,000 CFA per m^3 logged (U.S.\$1 = 708 CFA in 2002). None of these proceeds actually reached the village at large, which led to violent and disruptive conflict within the village, with two opposing leaders eventually being incarcerated.

Mvondo and Sangkwa (2002) outline the complex interethnic conflicts that affect the Dimako Council Forest. The Bakum, a Bantu group and the first settlers in the area, dominate the management council, followed by another Bantu group, the Pol. The Baka pygmies were forcibly resettled from southern Cameroon by the government more recently and are clearly marginalized, receiving few financial benefits from the forest. The Pol are struggling to strengthen their rights to forest resources; ongoing conflict between the council and the communities they are designed to represent persists.

An example of conflict is the classification of lands to be included in the Dimako Council Forest. Village plantation areas near Nkoumadjap/Nguinda village and the farms of some people living in Dimako town were included in the council forest without input from the traditional landowners. The unilateral demarcation resulted in serious conflict between the inhabitants and council authorities (Assembe and Oyono, 2005). Villagers were basically given the choice between having their farms appropriated after harvest with no form of compensation, or before the harvest with a promise of some financial compensation (Mvondo and Sangkwa 2002). Given the council's routine shady dealings (Kouna 2001a), neither were attractive options.

With the exception of the Baka community of Moangé-Le-Bosquet, all the Lomié communities have serious conflicts under way, from disputes between two communities sharing one community forest in Eschiembor/Malen to internal familial disputes over rights in Kougoulou to external elites running roughshod over their neighbors in Ngola. Kongo suffers an even greater variety of ills: suspension of the community forest status due to illegal industrial exploitation by the elites; conflicts between old and young; and confusion among local institutions about management (Efoua 2001; Etoungou 2002). Oyono reports that in Ngola some groups called for the immediate cutting of their forest with equally immediate access to the resulting benefits, while others opposed this plan (forthcoming, 53).

Social Capital—Medium

Oyono (forthcoming), while noting the importance of extra-local problems relating to forest management in southern Cameroon, also attributes some forest management problems to "low organizational instruments, low institutional arrangements for the community management of resources, lack of strong social agreements, and 'free-riding behaviors.'" The perception that the groups living in southern Cameroon—where the forests are—have low levels of community organization is common (Etoungou 2002).

In Cameroon, the teams began work at the middle organizational level, with many never quite reaching the village-level participatory action research processes that characterized our work in other countries. Their information on social capital is, therefore, more oriented toward social capital at this middle level. Many of their findings emphasize the down side of social capital, where timber companies, the state, and local elites collude to create conditions wherein they share the bulk of cash and timber from Cameroon's forests, leaving the less valuable wildlife, medicinal plants, and other nontimber forest products for use by local communities (Porro et al. 2001; Diaw and Kusumanto 2004; Oyono et al. n.d., 9; Tiani 2000).

Lomié. Oyono (forthcoming, 46) summarizes conditions in Lomié, Dimako and Akok as

> diffuse societies in which the power of traditional authorities was emasculated by contact with colonization. The systems of social control in force, weak as they are, cannot, without a truly collective effort, produce a reliable native governance of common pool resources.

Within such a context, other indicators show fairly low levels of social capital, including the ability of the community forest committees to function without effective community representation, to ignore community interests, to abscond with funds routinely, and to avoid sanctions. When they do gain access to some of the funds set aside for community use, the communities themselves often do not act in the communities' or the environment's best interests. Oyono (forthcoming, 50) reports that forestry fees climbed to U.S.$113,000 in Lomié in 1999 and U.S.$28,500 in 2002. None of these fees was invested in a revenue-producing project such as a community credit union, village stewardship project, or agricultural development project.

In late 2000, small-scale development of community forests began in Lomié district and accelerated in 2001. The timber, cut with portable sawmills, was put on the market and elicited a serious conflict, with community management committee members charged with diverting funds. On the brighter side, in the pygmy village of Moangé-Le-Bosquet, SNV has had some success in controlling these problems.

Etoungou (2002, 43–44) quotes a villager from Ngola who is essentially complaining about a loss of social capital:

It's a complete mess since the Lomié Community Forests management agreements were signed. Each family tries to negotiate things to their advantage … And so you end up with three contracts for the same forest. Some farmer leaders spend nights in the inns of Abong/Mbang, and come back with new clothes, foodstuffs, red wine, sometimes even whisky. There is a lot wrong with this. And the worst is that they are accountable to *absolutely* no one.

Turning to the income from community forests, the village of Ngola, for instance, had received U.S.$22,580 by late 2002, and Echiembor, U.S.$17,500. The community forest committee of Echiambor (and Koungoulou as well) was "captured" by an urban-based elite who wields control from the city and negotiates with would-be forest developers. The community filed a complaint in court against the president of the committee. Etoungou (2002, 28) writes "In Echiembor … the establishment of Community Forests has broken down the little cohesion that existed in the village." Kongo, on the other hand—perhaps demonstrating some level of social capital—managed to roof 30 houses with sheet metal, based on a consensus decision regarding the disposition of community forest income. The village also set aside U.S.$10,300 in a savings account (Oyono forthcoming).

Dimako. The propensity of some villages to "take revenge on history" and "eat and drink" the money from their forests (Oyono et al. n.d., 11) suggests a lack of cognitive social capital. Other villages, however, used their forest funds for health centers, classrooms, and other durable investments for community welfare.

In response to some villages' misuse of funds, the government changed the rules to remit forest royalties to regional- and district-level management committees, presided over by mayors and supervised by council heads. Villages in areas where timber was being extracted also formed management committees that could make requests, but the actual funds were released via the district-level committee.

That these district-level committees often create fictitious projects on paper and appropriate the funds (Kouna 2001a) suggests a shortage of cognitive social capital at the middle level as well. Agents along Oyono's "strategic axis"—members of village management committees, government officials, and timber companies—mismanage these funds for their own personal profit. A new "forestry elite" has evolved, composed of retired civil servants and young men excluded from urban employment by Cameroon's longstanding economic crisis. Disconnected from the rest of the people, with little or no accountability to them, this group functions to marginalize locals from access to forest royalties (Oyono et al. n.d.)—much as organized crime entities such as the Mafia function elsewhere (Sciarrone 2002).

Krishna and Uphoff (2002) do not consider this type of cooperation to be "social capital," but it is worth noting. The control exercised by a corrupt forestry elite is particularly interesting in light of Krishna's (2002a) conclusion that social capital is most effective when there are knowledgeable, educated, more worldly "agents" to serve as links between communities and outside resources such as knowledge, funds, and so on. It may be that in the absence of effective local social capital, such agents "turn bad" (see similar examples in Case 7).

CASE 5

Central and South Cameroon

Contributing Authors
Chimère Diaw, Anne Marie Tiani, Joachim Nguiébouri, Cyprain Jum,
George Akwah, William Mala, and Valentina Robiglio

Sites in Central and South Cameroon
Campo Ma'an (South Cameroon), Ottotomo (Central Cameroon),
and Akok (Central Cameroon)

Devolution Status—Intermediate

For a complete discussion of Cameroon's forest legislation changes and devolution context, see Case 4.

Campo Ma'an, Southern Province. This site, part of which has been a wildlife reserve since 1932 (Jum et al. 2001b), launched the management entity called *l'Unité Technique Opérationnelle de Campo Ma'an*, or Operational Technical Unit (UTO) in 1999. The 771,000 ha of forest are to be managed for three main objectives: biodiversity conservation, rational and sustainable use of the natural resources, and economic development of the region. The first, biodiversity conservation, is primary in the minds of the managers of the Campo Ma'an National Park (Nguiébouri et al. 2001), which forms part of the UTO. The international community's interest in this area is manifest in significant financial contributions from the World Bank (discussed below).

These external visions for the area are in conflict with community perceptions of the territory, which Akwah (2002, 2) describes as

a space for social interactions that had (and continue to have) specific ties between forest communities, which are physically linked to each other by a myriad of forest paths. Community ownership and mastery of this milieu is indicated by the way local people trace the history of any single path (recent or left by forefathers [sic]) that crosses the forest; by the way they retrace peoples' settlement movements that have been shaped by battles and alliances, etc. There is a local geo-strategic and geopolitical reading of the space, which is not yet understood by [those who implement] outside interventions.

Such concerns are in addition to the more mundane but vital subsistence needs relating to swidden agriculture, hunting, fishing, and collecting of nontimber forest products (NTFPs) (Tiani et al. 2002b).

The area is inhabited by indigenous people, including both various Bantu groups and Bakola pygmies (Nguiébouri et al. 2001; Jum et al. 2001a), whose legal rights to be there remain in question. The entire Campo Ma'an management area is estimated to include around 60,000 inhabitants (Nguiébouri et al. 2002), including in-migrants working in agroindustrial complexes like HEVECAM (rubber), SOCAPALM (oil palm), and timber companies.

Ottotomo, Mefou Ako-
no Department, Central
Province. Ottotomo
Forest Reserve—about
two hours' drive west of
the nation's capital,
Yaoundé—is a production
forest management unit
under the control of the
National Office of Forest
Development (*Office Na-
tional de Développement des
Forêts* or ONADEF), the
implementing agency of
Cameroon's Ministry of
Environment and Forests.
Seven villages border the
reserve, and one lies within, with about 1,300 total
inhabitants, primarily ethnic Ewondo (Tiani et al.
2002a).

Central and South Cameroon
Key ACM Characteristics

Devolution status	Intermediate
Forest type	Humid tropical
Population pressure	Low: Ottotomo
	Medium: Campo Ma'an and Akok
Management goals	Large-scale timber: Akok
	Large-scale timber, large-scale conservation: Campo Ma'an
	Multipurpose reserve: Ottotomo
External diversity	High: Campo Ma'an
	Medium: Ottotomo
	Low: Akok
Internal diversity	Medium: Campo Ma'an
	Low: Ottotomo, Akok
National conflict	Medium
Local conflict	High: Campo Ma'an, Akok
	Medium: Ottotomo
Social capital	Medium: Ottotomo, Akok
	Low: Campo Ma'an

After decades of "command-and-control" approaches to the management of Ottotomo, in 2000, based on the 1994 National Forestry Law, ONADEF request-ed CIFOR's services in facilitating an adaptive, collaborative management agree-ment among the local stakeholders. This decision was taken based on the poor record of repressive management, which was felt to have generated resentment and conflict, even provoking increasing encroachment on the reserve.

Akok, Central Province. Although Akok has been the subject of research by the International Institute for Tropical Agriculture (IITA) for some years, no special government program of relevance for devolution is under way. The traditional system (described by Diaw 1997) holds, wherein ownership goes to the person who first cleared the land (and his descendants). In Akok, which consists of six villages (Mala et al. 2001), a patchwork of forest cover from agricultural fields to young fallows to secondary forest covers perhaps 30 percent of the land, with the rest in dense forest. Alternate land uses include cocoa, food crops, and logging, with low population density and low pressure on natural resources. Extension of area used for food crops in Akok is a critical issue for forest sustainability (Mala et al. 2001).

Forest Type—Humid Tropical

Campo Ma'an. Campo Ma'an nestles in the southwestern corner of Cameroon, bordering the Atlantic Ocean to the west and Equatorial Guinea to the south. The climate is coastal equatorial, divided into four seasons: a long dry season (Novem-ber to March), and two rainy seasons from April to June and August to October, separated by a short dry season in July. The main large rivers are the Ntem and the Lobé, each with numerous tributaries.

The area is marked by outstanding biological diversity, with Atlantic Biafran, Atlantic littoral, mixed Atlantic, and semicaducifoliated, subtropical montane, degraded, and swamp forests. Campo–Ma'an is home to 80 species of large and medium-sized mammals, including elephant (*Loxodonta african cyclotis*), buffalo (*Syncerus caffer-nanus*), gorilla (*Gorilla gorilla*), giant pangolin (*Manis gigantea*), chimpanzee (*Pan troglodytes*), and panther (*Panthera pardus*). Of the 29 species of primates that occur in Cameroon, 19 are found in Campo Ma'an. There are an estimated 300 species of birds. Campo Ma'an has 46 percent of the fish species known to Cameroon, as well as 23 species of amphibian, 122 species of reptiles, and 28 of chiropters (Nguiébouri et al. 2002, 2–3).

Ottotomo. The forest reserve, consisting of 2,950 ha (subject to a management plan), has been classified as a forest reserve since colonial days. It is located between 3°37′ and 3°43′N and 11°16′ and 11°20′E. The vegetation is representative of the central southern plateau of Cameroon: a dense, semideciduous equatorial forest. The altitude varies from 300 to 950 m, with red soils on crystallized rock; annual rainfall ranges between 1,450 and 1,750 mm, with less in the month of October (288–306 mm). On average, rain falls 100 days per year (Tiani 2000; Efoua n.d.; Jum 2003).

The dominant timber species in the reserve are *Azobe* (*Lophira alata*), *Ayous* (*Triplochiton scleroxylon*), *Iroko* (*Chlorophora excelsa*), *Movingui* (*Distemonanthus benthamianus*), and *Eyong* (*Eribroma oblongum*). Regeneration activities have been implemented during many periods. Between 1930 and 1940, 13 ha of *sapelli* (*Entandrophragma cylindricum*), *bibolo* (*Lovoa trichilioides*), and *Ngollon* (*Khaya ivorensis*) were planted. Between 1972 and 1978, the National Office of Forest Regeneration (ONAREF) planted 499 ha of forest in Ottotomo Reserve. ONADEF replaced ONAREF and continued planting.

Jum et al. list important products derived from the forest:

Dried endocarps of the seeds of *Ricinodendron heudelotii* (Euphorbiaceae), those of the *Panda oleosa* (Pandaceae) known as "shell nut" and of *Irvingia gabonensis* (Irvingiaceae) … are collected seasonally and may be sold. There are a number of other plants, which provide fruits and nuts, including African plum (*Dacroydes edulis*) and bitter cola (*Cola pachycarpa*). (In addition there are several herb and shrub species providing green leaf vegetables: *Achornea* spp., *Piper guineensis, Amaranthus* spp., *Celosia* spp., *Hibiscus esculentus*.) Plants yielding medicinal products used by the local people include *Hibiscus rosa-sinensis, Terminalia ivorensis, Staudtia stipitata*. Forest spices and stimulants are produced by *Piper guineensis, Afrostyrax lepidophyllus, Garcinia cola, Cola acuminata* and many others. Rattan cane (*Ancistrophyllum secundiflorum*), chewing-sticks of *Garcinia* sp., and saplings of various forest trees for use as building poles are collected for village use and may be traded. The sap of the raffia palm and to some extent that of the oil palm, *Elaeis guineensis,* is collected and used as a beverage (Jum et al. 2001b, 3).

Akok. Seriously degraded by shifting cultivation and logging, Akok was deforested by 12 percent between 1970 and 2000 (Mala et al. 2001). Mala noted the disappearance of many animals (e.g., elephants, gorillas). Water is readily available from the many rivers like Ngo'o and Nsame, which have their source in neighboring hills (*Nkol mekam, Nkol megom, Nkol abang*). Despite abundant forest products, only a few are used commercially, such as *Irvingia gabonensis* (Mala et al. 2001).

Legg (2003) describes three distinct agricultural systems in the area: (1) mixed food–fallow cycles of up to eight years in fallow, with rare new cutting from forest; (2) fairly stable cocoa agroforests, occasionally developed from newly cleared forest; and (3) forest melon plantain fields that form part of a long fallow cycle in secondary forest. Villagers obtain meat, fish, timber, medicinal plants, and other NTFPs from large areas of logged-over and secondary forest. Based on preliminary comparisons of aerial photographs from the 1970s with recent satellite imagery, Legg concludes that this land-use mosaic has been relatively stable for the past 25 years.

Mala and Legg differ in their conclusions about forest sustainability, perhaps because Mala compares Akok with other ACM sites, all of which are comparatively well-forested, while Legg is working from within a more agricultural context.

Population Pressure—Medium

Nguiébouri et al. (2002) report a doubling of Campo Ma'an's population since 1987, with population growth rate high at 5.3 percent (versus 2.9 percent nationally), due primarily to plantation agriculture and timber exploitation. Nasi et al. (2001), on the other hand, report a growth rate of 1.7 percent for the rural population and 3.7 percent for the population directly linked to the industrial plantations. The sex ratio for the population at large is remarkably even: 1.001. People under age 25 constitute 60 percent of the population, with 20 percent over 50, leaving only 20 percent in the main productive years. Population density is low (7.3 people/km^2) compared to the national average, listed as 33.9/km^2 (CIA 2003). Local density falls to 4.4 people/km^2 when people in the agroindustrial zones are excluded (Nguiébouri et al. 2002). However, Diaw et al. (2002) show that unrealistic land use planning has led population density to jump nearly fourfold, from 8 to 29 people per km^2, in areas available to the local population.

Akok residents are ethnically Bulu, a Bantu group, and settled there between 1905 and 1910 (Mala et al. 2001). Census data suggest stability of population, with out-migration balancing the effects of natural increase (Legg 2003). Akok's population totals 1,126 (according to Robiglio, quoted by Mala 2003a) in seven hamlets, ranging from 61 to 245 people. Four clans predominate: Yemvan, Yemekak, Esalane, and Yemisem, in decreasing order of dominance. Yemvan accounts for 521 individuals, and Yemekak for 426. Table A-3 gives a more detailed analysis of three hamlets.

Robiglio et al. (2003, 11) portray the complexity of the local context:

> Existing borders mirror the settlement history of the groups; clans have territories of different sizes not related to their actual population size. (Hence, the pattern and density of cultivated patches relate strictly to each clan's internal population density rather than to the average population density in the area.) For this reason, patterns of cultivated patches are not uniform, and

Table A-3. *Spatial Extent, Inhabitants, and Households by Hamlet in Akok*

Hamlet	Land area (ha)	Arable area (ha)	Cultivation (ha)	Total inhabitants	Total households	No. of people/km^2	Population per ha cultivation
Nkolemvon	7	5	4	142	29	28	41
Eyek	9	7	2	122	14	17	64
Adjap	20	16	4	194	38	12	45

Source: Adapted from Robiglio et al. 2003.

may vary from one clan to another. As shown in Table [A-3], this results in a discontinuous density of agricultural land use. In some hamlets, complex mixtures (*bikotok y mefub bidi*) comprising mixed food crops of various seasons and young (4-5 year old) fallows within a matrix of older fallows (10-15 years) are concentrated along the road. In other hamlets, *mbiam* (mature forest) is part of the pattern, while crops and young fallows are dispersed in the matrix. These representations confirm the varying population pressure at a micro-scale and correspond to the two opposite ways people perceive their land in the area: abundant or scarce.

Population Pressure—Low

The population in and around the Ottotomo Reserve numbers 1,239 inhabitants with an average density of two to three persons per km^2 (ONADEF 1999). No details are available relating to sex ratio and other population variables (Jum 2003b). Three villages were selected for the participatory action research: Koli and Nkolbibadan in the *groupement de Nkong-abok*; and Ottotomo village in the *groupement de Nkongmeyos*. These villages range in size from 60 to 100 inhabitants (Jum and Abega 2003). The proximity to Yaoundé metropolis (60 km), with over one million inhabitants, is an indicator of the difficulties faced by forest managers in conserving this forest island.

Management Goals—Large-Scale Timber

As with the other Cameroonian sites—as well as other forest-rich, humid tropical settings—the effects of large-scale logging on the environment and on local communities have been controversial. While providing some benefits (improved roads in the short term, some employment, usually small and unpredictable contributions to village infrastructure), the presence of large-scale logging in Cameroon has often meant increased corruption, unsustainable logging practices, collusion with elites at various levels, financial losses to local communities and the central government, and a lack of attention to existing forest and land uses by local communities.

Akok. Mala et al. (2002) note that the state's plans for sustainable protection and use of forests have not been followed in Akok either in companies' logging practices or in the system of awarding forest concessions. The only forest management is traditional management undertaken by local communities, while logging companies exploit the forest. In 2000, large quantities of felled logs were abandoned by loggers. Villagers would have liked to recover the logs, but the work was too heavy and costly without external assistance. Similarly, the communities have shown their interest in participating in the formal governmental Community Forestry program: The Akok village of Adjap I applied for such a forest in 1999 but has received no reply to its application. The relations between the Akok communities and logging companies have included various negotiations and conflicts (see below).

Recently, another important management goal surfaced. The PEPIPALM project, designed to support cocoa cultivation, offered financial and technical support for a community nursery of 5,000 seedlings.

Management Goals—Large-Scale Conservation

Campo Ma'an National Park. The Campo Ma'an Management Area, which includes the national park among other land classifications mentioned below, covers 7,710 km^2 on the southern border with Equatorial Guinea. The UTO was created on August 6, 1999, by Decision No. 054/CAB/PM (Nguiébouri et al. 2002). The park has been developed as a sort of environmental quid pro quo for permission to run a pipeline through the center of Cameroon. Conservation of the region's impressive biological diversity is a central concern. ACM work has been under way in a number of the communities and with all major stakeholders since late 2000. The management area includes four main types of land use: protected areas (the national park itself); production forests; industrial plantations (oil palm and rubber); and "agroforestry zones," the area remaining for local communities to use (Nasi et al. 2001).

The Campo Ma'an area has also been subjected to large-scale logging. Communities have benefited, for instance, from a road built for use by the logging company *La Société Forestiere de Campo* (HFC). HFC and the communities agreed on the importance of keeping a major logging road open through the park and prevailed against the wishes of the park managers (see below). Logging's negative consequences for the park itself have been far worse than logging's impact on local people.

Five logging companies close to the national park work in forest management units covering 235.5 km^2 of the Campo-Ma'an UTO. They are *Wijma* in the northwest, *La Compagnie Forestiere Assam* (COFA) near Ma'an, HFC in the south-central area, *Compagnie Forestiere Kritikos* (CFK), and Bubinga in the north around Akom II. The annual logging potential of this forest is estimated at 150,000 m^3.

Management Goals—Multipurpose Reserve

Ottotomo Reserve. A central governmental interest has been to maintain a rich forest reserve close to the railway, from which timber would be available for use in public works and railway construction (ONADEF 1999, 14). This management goal was to ensure medium-term planning and the sustainable use of the Ottotomo forest zone. Ottotomo Reserve had both conservation and development goals, with two specific objectives: to ensure sustainable management of resources and to develop ecotourism. The reserve's management plan encompasses a production zone (1,575 ha), a regeneration zone made of forest plantations and areas of encroachment of slash and burn agriculture in the reserve (516 ha), and a protection zone (796 ha).

The logging company COGINEX EXPLOFOR logged 2,500 ha of Ottotomo Reserve in 1983. Conflict occurred between the state and the company, which left 4,000 m³ of timber behind. One more round of logging was planned to take place in 2000 by *Tranformation Industrielle du Bois* (TIB), but ONADEF and TIB never reached an agreement, although the local population had already planned on using the forest fees.

Ottotomo Reserve harbors a wide range of forest products used by local people. Agricultural encroachment and exploitation of forest products such as fuelwood, medicinal plants, and fruits has been high, partially due to population growth, despite official prohibition of hunting, agriculture, and timber harvesting. Contrary to law, local people claim ownership of the forest adjacent to their villages, regardless of any formal state ownership (Jum et al. 2001b).

Jum and Abega (2003) document the cooperation between the ACM team, an NGO, and the formal managers of the Reserve. Tiani (2003b) provides evidence of the increasing involvement of women in formal management. Government officials initially invited the ACM team to help them, and relations have been cordial throughout the process. The work has proceeded comparatively smoothly and productively.

External Diversity—High

Many ethnic groups live in the Campo Ma'an area, including Bulu, Mvaé, Mabi, Bagyeli, Yassa, Batanga, and a heterogeneous block formed by the workers for various companies and their families (Oyono 2002b). Poachers—both from within the country and from neighboring Equatorial Guinea—cause significant ecological and social repercussions.

In addition to the logging companies already mentioned, the multinational agroindustrial companies HEVECAM and SOCAPALM operate in the northwest of the area. Other external players are Cameroon's Ministry of Environment and Forests (MINEF) and the international community, notably the Dutch Cooperation Agency (DGIS), the Netherlands Development Organization (SNV), Tropenbos, the World Bank, and *Avenir des Peuples des Forêts Tropicales* (APFT) (Jum et al. 2001a).

In Akok, involved parties include governmental entities, logging companies, poachers, traditional healers, neighboring villages, in-migrants, and research institutions such as International Institute for Tropical Agriculture (IITA) and the Institute for Agronomic Research for Development (*Institut de Recherche Agronomique pour le Développement* or IRAD). Rural organizations such as Common Initiative Groups (*Groupe d'Initiative Commune* or GIC) aim to unify socioeconomic efforts at the village level and have growing influence. The PEPIPALM cocoa cultivation project is also a factor (Mala et al. 2002, 5–7).

External Diversity—Medium

Tiani (2000) lists the following important Ottotomo stakeholders. ONADEF manages the "Integrated and Participatory Management of the Ottotomo Forest Area" project and has longstanding, exclusive legal rights to manage the forest. A nonfunctioning *Comité de Suivi du Projet* (Project Management Committee), originally designed to contribute to the above project's management, is composed of local notables and government officials. The Ministry of Agriculture (*Ministère de l'Agriculture* or MINAGRI) and the Ministry of Animal Husbandry and Industries (*Ministère de l'Elevage et d'Industries Animales* or MINEPIA) oversee agriculture and animal husbandry as alternatives to conservation in the area. Local government is the *commune*. The prefect, subprefect, district chief, and other administrative support services represent the national government. Additionally, NGOs, particularly *l'Association Terre et Développement* (ATD), undertake projects like the domestication of timber trees, the processing and storing of NTFPs, plant health, and the promotion of palm oil. Finally, the role of powerful, external elites cannot be ignored.

Age and Gender

For details about age and gender in southern and western Cameroon, see Case 4.

Internal Diversity—Medium

Most communities in Campo Ma'an consist of one of the region's five ethnic groups, with clan structures providing further internal divisions, and patron–client relations with affiliated pygmy groups (Nasi et al. 2001). In some communities outside the ACM group focus area, numerous in-migrants from all over Cameroon and even from neighboring countries have participated in government and private development programs such as plantations, logging, and, now, conservation.

Internal Diversity—Low

Ottotomo is made up of people from two clans of the Ewondo ethnic group. Akok has one ethnic group (Bulu), composed of five clans (Oyono 2002).

Level of Conflict—High

The many contradictions in Cameroon's "egalitarian" society as outlined in Case 4 and by Etoungou (2002, 15) apply to southern and western Cameroon as well.

Campo Ma'an. Campo Ma'an has many ethnic groups, many conflicting traditional and government-mandated land uses, many external stakeholders, and an equally great number of desires for the area. A few examples demonstrate the flavor.

Bagyeli (pygmies) live in Campo Ma'an, typically in patron–client relations with their Bantu neighbors. The Bantu have a range of potential conflict resolution mechanisms: "the family council; the 'Customary' Village Court; the 'Customary' Court of the *Groupement*; religious authorities; the *Gendarmes*; the District Officer; and such 'modern' courts of justice as the *Tribunal de première instance* in the district's capital, as well as the higher judiciary levels" (van den Berg and Biesbrouck 2000, 40). The first three authorities deal with matters concerning land and natural resources. Yet the Bagyeli rarely make use of such conflict resolution mechanisms, relying instead on good relations, mobility, and threats of witchcraft both to prevent and resolve conflicts. However,

> there is a strong plea among increasing numbers of Bantu people to exclude neighbouring villagers and Bagyeli from all types of utilization of forest resources found on their village domains. This situation, together with diminishing tolerance against strangers and in some cases organized control to exclude and sanction intruders, harbours a substantial potential for social conflict. Young men of the village of Yem for instance, have established a *comité de vigilance* to control poaching in the forest that belongs to their villages. This informally organized committee, which is actively supported by the village chief, primarily aims to exclude and sanction non-authorised strangers. Similar initiatives were recorded in other villages, not only with regard to hunting activities, but also with regard to commercial [sic] valuable NTFPs. (van den Berg and Biesbrouck 2000, 55)

The fiscal decentralization law described in Case 4 also has an effect. Van den Berg and Biesbrouck predict more potential for social conflict, now that the new forest law requires agreements between villages and logging companies.

> [V]illage boundaries in the high forest become even more important … [T]he effect of compensation of individuals for logging commercial tree species by [the logging company] Wijma was that individuals increasingly claimed private ownership of trees which were previously held under common property arrangements and created conflicts between different right holders. (van den Berg and Biesbrouck 2000, 55–56).

Bifa, one of the 116 villages and hamlets surrounding Campo Ma'an National Park, had part of its village territory usurped in 1975 by the rubber plantation company HEVECAM. Some of the adverse and long-lasting effects recounted by Tiani et al. (2004) include exacerbation of inter- and intracommunity conflict for the remaining land; competition between neighboring villages for the remaining resources; destruction of large areas of forest and reduction in resources and incomes; influx of strangers into the area in search of jobs; and increased poaching and illegal occupation of land by plantation laborers and their families.

Positive effects of the plantation are employment, a market for local produce, and income for women, who sell what their husbands hunt. However, since the park became a reality in 2000, ecoguards have patrolled the area, confiscating all meat, no matter its source. Park boundaries are unmarked, and local people are uninformed about any hunting regulations. The necessity to hunt clandestinely has resulted in buyers going to the forest to get the meat directly from the men, effectively removing the women from the process and from their source of revenue.

Another recurring conflict near the village of Ebianemeyong is a logging road that has served off and on as a vital link between this enclave and the outside world. The Campo Ma'an project and the World Bank wanted the road closed; the HFC logging company sided with the community in its wish to keep it open. Local women were particularly adversely affected by the road closure because they used the road to market low-value agricultural produce in Campo, the nearest town. By March 2004, the road was again open, with two new barriers to entry, controlled by HFC agents. Each logging vehicle is now supposed to be accompanied by two MINEF guards, in convoy. Local people maintain that the friendliness of HFC toward community members evaporated once the road was opened. HFC no longer respects the old barrier maintained and guarded by the villagers of Nko'elon. The villagers complained in a meeting on September 3, 2003, that HFC vehicles had destroyed the road through the Nko'elon barrier.

Akok. In Akok, local conflicts are submitted first for resolution within the family. A common topic involving conflict is the harvest of *ndo'o* (*Irvingia gabonensis*) in July and August of each year. Harvest of the fruits—used both for subsistence and commercially—occurs in fallows or in cocoa farms, and many disagreements about user rights coincide with such harvests.

The usual problems with timber companies also are evident. For example, deterioration of essential bridges on the road to the town of Ebolowa caused by SOBIEL's logging trucks (Mala et al. 2002, 6) led to conflict. The villagers had asked foresters to ensure that the bridges were repaired without success. In the end, the villagers set up barricades blocking the roads and preventing the trucks from passing. The crisis was resolved with the involvement of the divisional and subdivisional officers (*prefet* and *sous-prefet*) who facilitated the prompt execution of the community request so that the logging companies repaired the bridge.

Forest ownership, NTFP use, and access to forest financial benefits are all subject to conflict (Oyono et al. n.d., 5). One bone of contention revolved around the second installment of forestry royalties, which were never paid to the community (as in Lomié and Dimako, Case 4). Mala quotes a villager as saying

> That money is shuttling around from the Divisional office, from the Council office to the Council Revenue Collector's office. None of these three officers, who are well aware of the whereabouts of this money, has ever said anything to us in that respect. We have several times delegated some village representatives from the Management Committee of forest revenues but they were always asked to go back and keep on waiting. Maybe we will never get this money.

Other conflicts characterized local relations with the employees of the National Agricultural Research and Extension Programme (PNVRA). Upon request, villagers set up common initiative groups in order to gain access to assistance and support. But in four years, no help came from PNVRA employees, nor did IITA (a sister center to CIFOR) follow through on pig and fish farming initiatives. Those who began the work "vanished" (Mala et al. 2002, 6). In 2000, a three-year program was begun by the Alternatives to Slash and Burn initiative (like CIFOR and IITA, a part of the Consultative Group on International Agricultural Research).

Makak. Research on another ACM site, Makak, where oil palm plantations were an important feature, was discontinued for some time due to high levels of conflict, though some research was in fact carried out in the end (Mala et al. 2002).

Level of Conflict—Medium

The eight communities in and around Ottotomo Forest Reserve routinely use the forest for agriculture, hunting, firewood, building materials, medicinal plants, and other NTFPs. Community members have been legally banned from use of the forest since the 1930s, but they have consistently contested their banishment. Encroachment was increasing prior to ACM involvement. As a result of this conflict, ONADEF began experimenting with comanagement in cooperation with the ACM team and an NGO in 2000 (Jum et al. 2001b; Jum n.d.)

According to Jum (n.d.) and ATD (2001), existing conflicts revolved around the following issues: discrepancies between modern and customary land laws; differing definitions of appropriate roles and responsibilities regarding reserve and forest management; and sharing of forest benefits. Local communities have routine conflicts with ONADEF, loggers, the municipal council, and among themselves. Additional conflicts have occurred between ONADEF and MINEF and between loggers and the municipal council.

Tiani (2000, 11) lists more specific sources of conflict. Agricultural land is in short supply due to population pressure. Some clans have more land than others and withhold permission to plant perennial crops. Community members and ONADEF blame each other for fraudulent exploitation of timber. Community members charge that urban sawmillers work in the forest reserve with the complicity of ONADEF staff, while ONADEF argues the reverse. Likewise, hunting and fishing within the reserve are contested, with each side accusing the other of complicity in illegal operations. Finally, of course, logging creates concerns about the future distribution of forestry benefits.

Social Capital—Medium

Oyono et al. (n.d., 14–15), while noting the importance of extralocal problems relating to forest management in southern Cameroon, also attribute some forest management problems to low social capital. This perception is common (Etoungou 2002). As in eastern Cameroon, the ACM teams focused on the middle administration level (see Case 4).

Ottotomo. Ottotomo represents the most village-based ACM team in Cameroon. Efoua (2000) describes the leadership roles in the village of Nkong-Abok, where a committee of the wise, including a village chief and elders, regulates village affairs.

Several groups of youth and women undertake joint action (see also Jum 2003). Ottotomo's villages fall under two formal governmental administrative structures (*groupements*), where each village's development association meets monthly to plan activities (Jum 2003). Indigenous rotating credit and rotating work groups related to shifting cultivation also exist. Such groups represent a framework within which social capital could flourish, though it is not yet particularly apparent (Jum 2003).

The NGO *Association Terre et Développement* (ATD) reports on the dynamism of the community of Nkolbibanda, which they consider to be well organized, committed to dialogue, and having the general interests of their village in mind (ATD 2002). In this area EBAMAN, a federation of women belonging to several local initiative groups in Nkong-Abok, serves a *groupement* made up of six villages adjoining the Ottotomo Forest Reserve. EBAMAN brings together over 300 women from 20 different working groups or self-help associations to develop income-generating activities. Recently, EBAMAN bought a cassava-grinding mill with funds they have accumulated, and they expect to earn more income and put it toward additional development activities. The World Agroforestry Center (ICRAF) is also jointly undertaking development activities with local communities (Jum 2003).

Tiani considers social and institutional capital in Ottotomo to have been weak when ACM work began. She found some structured groups in Nkong-Abok, for instance, which were organized by the political elite, partly to obtain funds unavailable to individuals, and partly to further the political ambitions of some. She does note, however, that their skills in finding partners, negotiating with the state, and managing their own resources have improved in the intervening two years (Tiani 2003a). A number of parameters appear to have relevance, including some related to bridging social capital. In Table 10-3, she analyzes the social capital of some crucial stakeholders.

Oyono et al. (n.d., 14) report a desire to abolish the forest, especially among the radical young. These youth believe that when the forest is gone, the state and the companies will be the real losers. As in eastern Cameroon (see Case 4), such a belief suggests a shortage of "cognitive social capital." Jum (2003), however, has a more optimistic perspective, seeing villagers' propensity to seek aid and advice from outside actors as a means to strengthen their social capital.

Akok. Initially estimated as having a medium level of social capital, the community was able to set up several microcredit organizations (*tontines*), displaying a desire and ability to locally reinvest money saved through mutual assistance and solidarity. The village was involved in the creation of a management committee in 1999, assigned to manage forest fees from industrial logging activity. The membership of this committee was controlled by outsiders who managed its finances: the president was the subdivisional officer, and the treasurer was the mayor of the Ebolowa rural council. Some of the funds obtained through forest exploitation are used for community activities, like building wells, a meeting hall, and a classroom—contrasting with some neighboring villages that used such money to buy food (Mala et al. 2002, 16–17; Diaw et al. 2002; see Case 4).

Despite the absence of a direct study, a fair amount of social capital can be inferred. Many issues pertaining to natural resources are organized within the village. Where land is sharply divided between families or lineages, women are still free to fish using a "dam and drain" method in communal streams passing through others' land, with these fishing rights being transferable from a woman to her daughter-in-law. Similarly, rights to hunt are regulated at the village level. Hunting is one of men's roles, and provides protein, cash, and protection for community crops. National laws require permits, but no Akok hunters have them, suggesting that regulation is truly local. Official land titles do not exist, and ownership is based on actual work done on the land or evidence that it was done in the past. Such land transactions as do exist among the populace involve the loan of land to cultivate food crops (Robiglio et al. 2003). These observations suggest a functioning indigenous system in which people are able to cooperate reasonably effectively. The young people who coalesced to block trucks carrying wood in protest against the inconsistency of receipts of timber royalties locally are another example of social capital (Oyono forthcoming).

Mala (2003a) has identified several rotating agricultural work groups (*eka'a*) and over 11 *tontines* and other groups. All are based on strong, mutually enforcing social and historical relationships among the members. Three church groups (one Catholic and two Protestant) are also active and able to mobilize more people for meetings than other groups. These churches are closely linked to the leadership and power positions of families (*nda bot*) of the same lineage (*mvog*) and sometimes between lineages of the same clan. There are also active burial groups and youth organizations, the latter being particularly active during holidays.

Social Capital—Low

Campo Ma'an. A complex site surely including a wide range of levels of social capital, Campo Ma'an falls into the lowest category because extreme levels of distrust have permeated the area since the national park and its accompanying management unit were formed. Longstanding local communities have their own histories of conflict and cooperation, such as with in-migrants of other ethnic groups who work in the plantations and logging companies that surround the park. Distrust of park officials is high. And similar distrust exists among other stakeholders—the national park, the plantations, the logging companies, and the governmental agencies responsible for regulating these actors.

Van den Berg and Biesbrouck (2000, 36) write about the role of what we would call social capital in resource management among the Bagyeli, the pygmy group living within the park:

> Bagyeli right-holders can allow others to exploit and utilize those [forest] resources. Here, good relationships play an important role. These 'others' can be distant relatives, or friends who live in another village. The privileges may range from the possibility to temporarily utilize one particular resource in a limited area, to a more general possibility to exploit resources in the entire area. Such a person announcing himself and expressing the wish to hunt with a rifle in the area, has the "moral" obligation "to satisfy the village," or

at least his host or guide, for example by sharing the yield or distributing small gifts. The approval of those having the right to control the exploitation is generally considered a precondition for having good luck during the hunt, and a protection against "accidents."

Akwah (2003b) lists four common kinds of groupings throughout the forest regions of Cameroon. Two common indigenous grouping mechanisms that undoubtedly vary from community to community in their levels of social capital are rotating work groups (*tontine de travail*) for sharing of agricultural labor and rotating credit groups (*tontines*). Colfer's [admittedly long-distance] perception is that the potential of such groupings may have been underestimated by the ACM team in Cameroon, though Nguiébouri argues that the weight of the debate on forest issues was sufficient to bring these diverse groups within the communities together and enable them to speak with one voice. A third kind of group, *Groupe d'Initiative Commune,* is a "modern" grouping, subject to legal control and stimulated by the government (Oyono forthcoming, 9). Fourth, village development committees designed to take charge of local development typically work on natural resource management, health, politics, education, and so on. These all occur in nearly all the villages with which we work in Campo Ma'an. Some of the villages have joint village development committees. For example, Messama I, Messama II, Messama III, and Bindem form one such group; Bifa, Zingui, and Akok form another grouping (Akwah 2003b).

CASE 6

Ghana

Contributing Authors
Dominic Blay and Kofi Diaw

Sites in Ghana
Adwenaase and Namtee

Devolution Status—High

Ghana is considered to have a comparatively high level of formal devolution because of its unusual, longstanding recognition of community ownership of land. Indeed, Ribot (2001), in a survey of decentralization in Africa, paints a comparatively rosy picture for Ghana. Kotey et al. (1998, iv–vii) document the evolution of policy in Ghana from a consultative phase (1874–1939) through a "timberization" phase (1940–1953) to a "diktat" phase (1954–early 1990s) to the current collaborative phase (since 1994). Traditional landowners were able to use the courts to successfully stop the nationalization of forestlands and negotiate their own concession agreements with loggers, including determining and collecting royalties during the consultative phase, indicating strong community involvement in natural resource management at that time. Throughout the middle of the twentieth centu-

Ghana Key ACM Characteristics	
Devolution status	High
Forest type	Humid tropical
Population pressure	High
Management goals	Multipurpose reserve
External diversity	Low
Internal diversity	Low
National conflict	Low
Local conflict	Medium
Social capital	Medium

ry, however, the power of local communities waned, and environmental destruction reigned.

Although Ghanaian land has technically been owned by chiefs—symbolized by a "stool"—who have represented their communities since colonial times, forest management rights have been vested in the Forestry Commission. Diaw (2001, 11) summarizes

All trees and timber in Ghana are owned by the appropriate stool but these are vested in the President of the Republic on behalf of the stools and people … [T]he naturally occurring trees are vested in the President in trust for the people … All extraction of trees/timber either naturally occurring or planted for commercial purposes must be authorised by the District Forestry Office following an application.

The government provided for partially elected district assemblies in 1987 and devolved some authority to district assemblies—if not the funds to implement programs (Ribot 2001). Then in 1994, Ghana promulgated a New Forest Policy (replacing the previous policy of 1948). Blay et al. (2002, 1) note a significant improvement in the new law, which included a requirement for consultation among landowning communities, other stakeholders, and foresters. They also describe the Timber Resources Management Act 547, which prescribes consultation with landowners and allows landowners a stake in selecting concessionaires who are to operate on their lands via "Social Responsibility Agreements." Blay (2002a, 22) adds that "The legal, organisational and administrative framework for decentralised development and governance in Ghana is provided by the Local Government Act 1993 (Act 462)."

Still, there remain some significant growing pains not too dissimilar to those encountered among countries with lower devolution status. Specifically, Blay et al. lament a continuing top-down approach by the Forestry Commission; ignorance among the general populace about their rights; unwillingness on the part of the National Forest Directorate to devolve powers to the district level; and lack of incentives and motivation for communities to cooperate. Illegal logging, occurring in almost all communities, provides a useful example of the lack of incentives. Official royalties deriving from legal timber are distributed as follows: Forest Services Division, 60 percent; Administrator of Stool Lands, 10 percent; district assemblies, 25 percent; paramount chiefs and stool landowners, 5 percent (Blay et al. 2002). Such a distribution provides an obvious incentive to collaborate with illegal loggers, who can provide more direct and substantial benefits to communities—or at least to local elites. This distribution may also account for the unwillingness of the National Forest Directorate to devolve its power.

ACM sites—Adwenaase and Namtee—are both in Assin Fosu District in the Central Region of Ghana. They were "created" formally in 1995, as community forests at the initiative of the local community triggered by an external researcher to the communities (Diaw 2001, 8). They cover 190.5 ha for Namtee forest and 171 ha for the Adwenaase forest (Blay 2002a). Diaw (2001, 9) describes the legal status of the two community forest reserves:

> The community reserve is the property of the community people them-selves. The forest is under no concession and at this stage cannot be given out as a concession. It is not part of any forest reserve gazetted under the forest ordinance, and therefore outside of any government managed reserve.

The Adwenaase site includes the Adwenaase forest itself, the Assin Akropong township, and communities fringing or situated inside the forest. The community forest is located in the northwestern part of the Assin district, about 2.5 km off the main road between Assin Brofoyedur and Akim Oda. The forest is located on Assin Akropong stool land and, together with the Worakese and Akenkausu stool lands, constitutes the Akoti stool lands. The stools owe allegiance to the paramount chief of Assin Apimanim. The forest is bounded on the east by the river Fum, by the river Kanta Kuma on the south, and to the northwest by the river Subin (Diaw 2001).

We define the Namtee community forest to include Assin Worakese, Assin Akenkausu, and the communities fringing or situated inside the forest (Nkukuasa, Mensa Dovi, Kwasi Kuma, Amete, Bosumpra, and Amanor villages, with the last two located within the forest). The Namtee forest is drained by four streams, which are tributaries of the Pra, a major river. The forest came into existence when the ancestors of Worakese and Akenkausu communities vacated the land on which the forest is currently situated as a result of war and invasion and selected a new site for resettlement. The forest is about 17 km from Worakese/Akenkausu. It is troubled by nearby farmers who illegally enter and farm. Timber contractors have also entered illegally and felled some of the trees without the communities' knowledge (Diaw 2001).

Forest Type—Humid Tropical

In the two forests that constitute ACM sites in Ghana, considered "off-reserve," or community owned, forests, about 90 percent of the area is in natural forest, with cocoa farms constituting nearly four percent. Fallows dominated by *Chromolaena odorata* cover just over five percent. Food crop farms cover less than two percent of the land area (Blay 2003a). In contrast with the Cameroon sites, these forests are largely (and intended to be completely) separate from the communities' farmlands.

Adwenaase's moist, semideciduous forest covers 215 ha and is considered slightly degraded: obviously disturbed, but patchy. Mean annual rainfall ranges between 1,250 and 1,500 mm, with mean daily temperatures, between 25°C in the wet season—March to October—and 27°C in the dry season—November to February. The forest has a moderate, three-story structure in areas not used by farmers, and those used by farmers (illegally) are now regenerating naturally. Soils are naturally acidic, with Adwenaase's soils higher in nutrient content than Namtee's. Forty-one nontimber forest product (NTFP) species were recorded in a recent survey (Blay 2002b), with a total of 2,380 stems/clump, averaging 238/ha. Blay reports that "39 percent of the species are used for herbal medicines, 14.6 percent are for construction, 12 percent are used for wrapping of food and other items, 4.8 percent are used as pestles in the preparation of *fufu*, a staple food in Ghana, and 9.6 percent are used as sponges." Dominant timber species were *Triplochiton screloxylon* and *Celtis mildbraedii*. Conspicuously absent were the redwoods, *Entandrophragmas* and *Khayas*. Other important timber species included *Piptadeniastrum africanum*, *Petersianthus macrocarpus, Blighia sapida* (Blay 2002b).

The Namtee Community Forest covers 190 ha (Blay et al. 2000) and is classified as mostly degraded, with slightly acidic soils. Its structure is open, with few areas of closed canopy. The forest includes some illegal cocoa farms that have not expanded. Patches of naturally degraded areas have been planted with *Cedrella odorata*. In the closed forest areas, an invasive climber believed to be in the family Convolvulaceae has been a problem, covering large areas and reaching the tops of some trees, eventually killing the host. Rainfall ranges between 1,250 and 1,500 mm, and the mean daily temperature ranges from 25°C in the wet season to 27°C in the dry (Blay 2002b). In the same study by Blay, 35 NTFP species were recorded, with 1,454 individual stems/clump, and an average of 208/ha. Forty-eight percent of the species were herbal medicines, 17 percent were for construction, and 17 percent were used as food. As in the Adwenaase forest, *Triplochiton screloxylon* and *Celtis mildbraedii* were dominant species, and *Entandrophragmas* and *Khayas* were absent. *Macaranga spinosa, Ceiba pentandra,* and *Ricinodendron heudelotii* were also important.

Population Pressure—High

The Assin District has grown in population from 46,473 in 1960 to 193,888 in 2000 (Blay 2002a). Adwenaase Forest is used by about 9,200 people, only 83 of whom live inside the forest in the settlements of Bakaapa, with 65 people, and Agyalo, with 18. Subinso I and Dawomanso are additional settlements close to the forest boundary, accounting for 200 and 20 people, respectively. Subinso II is one km from the forest and houses 280 people. Sixty inhabitants of Nsutam live two km from the forest, and 45 inhabitants of Fawomayo live four km from the forest. In addition, 8,540 people live in two nearby towns: Asin Akropong, with 6,540 people at six km distance from the forest, and Brofoyedur, with 2,000 people, three km from the forest (Blay 2003b).

The total population with links to Namtee Community Forest Reserve includes roughly 11,700 people. The forest itself has only 110 people living within it, in the communities of Bosompra (50 people) and Amano (60). Within a kilometer's distance, four more communities account for a total of 1,580 more people: Nkukuasa (1,500), Amete (15), Mensa Dovi (40), and Kwasi Kuma Dovi (30). The towns of Worakase (18 km away) and Akenkausu (24 km away), each with about 5,000 people, also have links to this forest (Blay 2003b).

Management Goals—Multipurpose Reserve

Management of these two community forests is fairly comparable. Three communities are involved: Worakese and Akin Akenkausu manage the Namtee Community Forest Reserve, and Assin Akropong manages the Adwenaase Community Forest. These communities have historical and cultural ties to their respective forests, though located some distance (4–25 km) away (Frost 2002). Most day-to-day management activities are in the hands of a small group of community volunteers, whose enthusiasm for their tasks has waned in response to ongoing responsibilities and minimal personal benefit (Blay 2003b).

Asin Worakese and Asin Akropong are part of Assin North District, with its District Assembly in Assin Fosu. Akenkausu is in South Birim District, with the district headquarters being in Akim Oda. Still, the Adwenaase and Namtee Forests are under the jurisdiction of the Assin Fosu District Forestry Office, which governs one of the largest and, in terms of both legal and illegally harvested timber, one of the most productive in Ghana (Frost 2002, 2).

Frost (2002, 6) sums up the management setting thus:

> The chiefs and elders of Worakese/Akenkausu and Akropong are responsible for managing and administering their forests in collaboration with the Forestry Commission, specifically with the Collaborative Forest Management Unit (CFMU) and the Range Supervisor of the Forest Services Division, both of whom provide technical advice. Following the establishment of the reserves and the formulation of the management plans, the CFMU has reduced its profile to some extent ...

Some community members sense that the governmental forestry establishment has abandoned them. Each community, in cooperation with the Forestry Department, has drawn up management plans that have only been partially implemented and that both parties believe to need revision. Activities have included boundary demarcation, inventory, and establishment of small plantations (Frost 2002, 8).

External Diversity—Low

Blay et al. (2000) list the following significant external stakeholders: timber exploiters, the Ministry of Lands and Forestry, the Forestry Commission, the Forest Services Division, the Ministry of Food and Agriculture Agro-Forestry Unit, the Environmental Protection Agency, the Department for International Development (UK), and the World Bank.

Age and Gender

In Ghana, as in Cameroon, recurrent conflict between the generations exists. A central issue is the sale of community timber resources by the elders, who retain the proceeds. With regard to community forests, "While the elders claim to be in charge because of their social positions, the youth also claim to be in charge because of their commitment to the establishment of institutional structures for the management of the community forests" (Blay 2002c).

Gender differentiation also follows Cameroon, with division of space and labor scoring high differentiation (5), male dominance somewhat high (4), and hostility to women in public arenas intermediate (3).

Internal Diversity—Low

The people of Adwenaase and Namtee in Ghana are primarily from the Assin ethnic group, part of a larger Akan-speaking group (Blay 2002c). Several small hamlets in or near the Namtee Community Forest are occupied by Akuapem, Fante, and Ewe. In and near the Adwenaase Community Forest one can find Fante and Ewe inhabitants, as well (Blay 2003c).

Level of Conflict—Medium

The elders of Adwenaase and the forest volunteers responsible for maintaining the community forest dispute the sharing of forest resources. The volunteers felt they were doing all the work without any recompense. The formation of the community forest committees, on which both elders and volunteers were represented, has been helpful in defusing this internal conflict.

Adwenaase and Namtee also disagreed over planned joint activities. Each community competed to hold planned workshops in their own communities, so joint activities had to be held in Fosu, the district capital, to avoid serious problems.

The most serious conflict pertains to a group of 15 families from Brofoyeduru who have encroached on the community forest. They are said to come from Akropong and be royalty of the Akropong stool. They are also said to think they have sufficient money to deal with any legal issues and to have planted cocoa and other crops within the community forest since the community forest project began. The encroachers are accused of having destroyed the seedlings planted by the community. The community has tried to discuss this issue with the families to no avail and has considered having a letter of eviction drawn up by the elders and delivered to them. The community would like to forcibly evict the encroachers and to destroy their crops either by themselves or with the help of the Forest Service or the Ghanaian military. Allowing the volunteers to harvest the encroachers' cocoa was also suggested (Blay 2003b).

In Namtee, elders and volunteers contested the distribution of cutlasses provided by a previous community forest project. This conflict seems to have been resolved, or at least minimized, with the formation of community forest committees in which both the elders and the volunteers are represented. Officials are also suspected of turning a blind eye to illegal logging (Blay 2002a, 50).

Social Capital—Medium

Evidence on social capital is minimal for the Ghana sites. Adwenaase has mobilized volunteers who patrol the community forest with little to no remuneration, suggesting a fair amount of social capital. In one community meeting, the community vowed never to sell the community forest for a mining concession. Blay (2002a, 65) considers the community groups and the volunteers to be well-organized and led by "selfless" leaders and coordinators.

Social capital appears less evident in Namtee, although volunteers were also available to patrol the community forest. Considerable disagreement erupted in the cutlass distribution dispute (Blay 2002a, 42), but the community is able to organize for communal labor and other forms of joint action (Blay 2002a, 65).

CASE 7

Indonesia

Contributing Authors
Stepi Hakim, Adnan Hasantoha, Yayan Indriatmoko, Ramses Iwan, Trikurnianti Kusumanto, Godwin Limberg, Moira Moeliono, Made Sudana, Eva Wollenberg, and Linda Yuliani

Sites in Indonesia
Rantau Layung and Rantau Buta (Lumut Mountain, East Kalimantan), Baru Pelepat (Jambi, Sumatra), and Bulungan (East Kalimantan)

Devolution Status—Low

Indonesia's policy context changes hourly. For decades, under President Soeharto, the nation's forest estate—comprising 75 to 80 percent of the nation's land area—was centrally controlled and formally managed primarily by the president's cronies for their own benefit. Day-to-day management tended to be in the hands of local communities with strong, traditional, claims to these lands. State power was asserted sporadically and unpredictably. After Soeharto's fall, two critical laws were passed in 1999: one on regional governance (UU 22/1999) and the other on fiscal balance between the central and regional governments (UU 25/1999). Although the former devolved significant authority to the *kabupaten*, or district level, implementing regulations issued in May 2000 retained significant central authority, stimulating uncertainties and disagreements among various levels of government. The law on fiscal balance increases the share of revenue generated by the forestry sector that is allocated to the regions. Several important fees are now to be split 20 percent to the center and 80 percent to the regions (Dermawan and Resosudarmo 2002). On January 1, 2001, significant authority moved from Jakarta to regional governments. Another important law, No. 5/1979 on village administration, was revoked, relaxing the uniformity of village governments throughout the nation and opening the door for considerable regional variation in administrative structures (see e.g., Barr et al. 2001; Casson 2001; Potter and Badcock 2001; Resosudarmo and Dermawan 2002; Thorburn 2002).

However, these decentralizing laws do not grant any legal rights to the smaller units such as communities, which in many cases have resided in the forests for decades or centuries but have not obtained legal tenure since independence (Wrangham 2002). Interest in strengthening forest peoples' rights is increasing (see Colfer and Resosudarmo 2002; Lynch and Harwell 2002), but forest communities still have no legal basis to gain access to the lands and forests they inhabit.

Rantau Layung and Rantau Buta, Pasir, East Kalimantan. The two remote communities of Rantau Layung and Rantau Buta share the watershed of the Kesunge river in the Lumut Mountain area. Rantau Layung covers an area of 18,913 ha (its traditional territories, recognized by neighbors, but not officially recognized by the government). This area includes protection forest, forest gardens, swidden agriculture, and human settlement. Rantau Buta's territory covers 16,546 ha, including 500 ha for shifting cultivation (Hakim 2001c). This area is also representative of the chaos that characterizes the policy context in Indonesia. On July 31, 2000, the local government issued a regulation allowing "communities" rights to harvest timber on a small scale (100 ha). Lacking funds and institutional support, the communities were unable to take advantage of the policy, and private companies offered support to communities. About 100 private logging companies thus obtained permits on behalf of communities for such timber exploitation from the head of Pasir District in 2000 and 2001. Although the district government stopped granting such permission in April 2001, companies that already had obtained permits were not affected. Meanwhile, in the same month, a private logging company paid Rp1,000,000 (U.S.$91, a substantial sum in this economy) to each household head in Rantau Buta to obtain the community's permission to extract timber, in the hopes of thereby influencing the district government to grant it legal permission (Hakim 2001c).

Indonesia
Key ACM Characteristics

Devolution status	Low
Forest type	Humid tropical
Population pressure	Low
Management goals	Large- and small-scale timber: Rantau Layung and Rantau Buta
	Large- and small-scale timber and large-scale conservation: Bulungan (Malinau) and Baru Pelepat
External diversity	High: Baru Pelepat
	Medium: Bulungan (Malinau), Rantau Layung, Rantau Buta
Internal diversity	Medium: Baru Pelepat, Bulungan (Malinau)
	Low: Rantau Layung, Rantau Buta
National conflict	High
Local conflict	High
Social capital	Low

Baru Pelepat, near Kerinci Seblat National Park, Jambi, Sumatra. Baru Pelepat is a village with a territory of 7,265 ha, part of which is classified as state production forest, part agricultural, and part settlement areas (Kusumanto 2001a). Since the mid-1970s, part of the production forest has been logged as a concession area by several large-scale logging companies (Indriatmoko 2002). Production forest within the official village area covers around 4,750 ha. Local people, particularly the original Minangkabau, see their customary area as overlapping with the state forest and extending beyond the officially delineated village area. Baru Pelepat village is also situated in what is officially called the buffer zone of the Kerinci Seblat National Park, located about 40 km to the west.

The people of Baru Pelepat include the original, matrilineal Minangkabau, settlers from Jambi and from Java (both Javanese and Sundanese), and the nomadic *Orang Rimba,* who inhabit the Baru Pelepat village forest. The dominant Minangkabau community practices swidden agriculture based on long fallows with rubber as a main component, creating a patchwork of secondary forest in much of their territory. Over the last couple of decades, government-supported resettlement schemes have increased the local population and pressure on local resources. Another important nearby development has been governmentally instituted large-scale plantation activities, notably oil palm plantations. With policy in flux and clear rules absent, confusion reigns and illegal logging is rife (Kusumanto et al. 2002b).

Bulungan Research Forest, East Kalimantan. This 320,000-ha area, near Kayan Mentarang National Park on the border with Malaysia, was identified as a CIFOR research site shortly after its 1993 founding. Bulungan contains the best remaining dipterocarp forest in Southeast Asia (Bryant et al. 1997; Sayer et al. 2000). Conditions range from the remote, pristine, sparsely populated, upland areas in the north and west to the degraded, increasingly populated timber, fisheries, and mining "developments" in the east and south. These disparate poles illustrate the continuum of trends under way throughout Indonesia's forests: local people with longstanding but unrecognized land claims, newcomers drawn by economic opportunities, large-scale timber and mining concessions, and talk of oil palm plantations and transmigration settlements. It has a frontier atmosphere. Recently, CIFOR work has focused on the Malinau District.

Confusion about which policies are operative is rampant. Different government bodies are vying for access to and control of Bulungan's wealth both through official and illicit channels. In local communities, as longstanding historical differences surface, a similar free-for-all has resulted (see Barr et al. 2001; CIFOR Bulungan Research Forest ACM Team 2002).

Forest Type—Humid Tropical

Rantau Buta and Rantau Layung, Lumut Mountain. Lumut Mountain Forest was designated by the government as a protection forest in 1993, with neighboring areas including official production forest as well. A protected area of 35,350 ha provides conservation functions, including flood, erosion, and landslide control, and habitat protection. Logging companies are active in the area, as are individual entrepreneurs, so considerable concern has arisen that this humid, tropical, lowland dipterocarp forest is being harvested unsustainably, though it has been in fairly good condition over the course of this research. In 1982–1983 and 1997–1998, El Niño weather led to drought and fire, and in 1984, serious flooding occurred.

Rattan collection, timber extraction, and shifting cultivation are important for local people's forest-based livelihoods. Rattan availability has decreased. Community representatives noted declines in fruit trees, honey, and fish between 1950 and 2000. Coffee cultivation has increased, though still on a small scale; and wild boar populations have been comparatively unaffected by the changes underway over the same time period (Hakim et al. 2001b).

Rich in biodiversity, the area's dominant species (>10 cm in diameter) include *Buni* (*Aglaia cp.*), *wayan* (*Aglaia tomentosa*), *terap* (*Artocarpus elasticus*), *nato* (*Madhuca sericea*), and red *meranti* (*Shorea leprosula*). Potential wood production in Rantau Buta, for instance, is estimated at 312 m³/ha, and potential rattan production at 1.2 tons/ha. *Gaharu* (*Aquilasia* sp.)*,* the aromatic wood, is also present (Hakim 2004).

Baru Pelepat. Baru Pelepat, although quite remote until recently, now witnesses timber companies, plantation companies, conservation agencies, and transmigration projects operating nearby (see Chapter 8).

Annual rainfall is around 3,000 mm. Relative humidity ranges between 56 and 85 percent. Located on the equator, its relatively high temperatures range from 25.8°C to 26.4°C. Elevations around the village run from 70 to 1,300 meters above sea level (m asl) (Center for Statistics Bureau of Bungo District 2000).

In local primary forests, the most abundant species were *kelat* (*Eugenia* sp.), *balun ijuk* (*Diospyros curanii*), *medang/medang burung* (*Litsea* sp.), *antui* (*Polyalthia lateriflora*), and *balam berah* (*Palaquium gutta*). The most dominant species were *tenam* (*Anipsotera costata*), *Eugenia* sp., and *tembalun* (*Parashorea aptera*). In secondary forests, *Eugenia* sp. and *Palaquium gutta* were both most abundant and most dominant, with *Kabau* (*Archidendron bubalinum*) and *petaling* (*Ochanostachys amentacea*) quite common as well. Other dominant species in secondary forests included *keruin/rambai* (*Dipterocarpus palembanicus*) and *pauh* (*Swintonia sweinckii*).

In old fallows, *setepung* (*Callicarpa tomentosa*), *terap* (*Artocarpus elasticus*), and *aro kain* (*Ficus variegata*) were the most commonly found species. The first two species plus *kelampaian* (*Anthocephalus chinensis*) were the most dominant species. *Sekubung* (*Macaranga gigantea*) was also found to be quite abundant and dominant. In young fallows, the most abundant species were *narung* (*Parasponia parviflorag*) and *ubah payau* (*Glochidion* sp.), while the most dominant species were *Parasponia parviflora*, *medang kempa* (*Vitex pubsecens*), and *gedebu* (*Ficus ribes*).

Pressure from logging on commercial tree species, such as *meranti* (*Shorea* sp.), *Parashorea aptera*, *kelampaian* (*Anthocephallus stipularis*), *balam* (*Palaquium* sp.), and *kayu kacang* (*Strombosia javanica*), was quite high. In plots located in secondary forests, the only abundant commercial tree species were *Palaquium gutta* with an average of two individuals/1,000 m², and *meranti batu* (*Shorea hopeifolia*) with an average of one individual/1,000 m². Water quality remains good.

Endangered species found in the area included the tiger (*Panthera tigris sumatrae*), endemic to Sumatra. Other protected species included the honey bear (*Helarctos malayanus*), siamang (*Simphalangus syndactilus*), and *kua-kua/wau-wau* (*Hylobates agilis*). Wild pigs are considered pests in the area.

Hunted animals include *rusa* (*Cervus unicolor*), *kijang* (*Muntiacus muntjak*), *napu* (*Tragulus napu*), and *kancil* (*Tragulus javanica*). Bird species captured include *murai batu* (*Copsycus malabaricus*), *balam* (*Streptopelia chinensis*), *ayam hutan* (*Galus galus*), and *punai* (*Treton* sp.). Honey is popular, but recent harvests have not been good. Bees are drawn to *kayu kawan* (*Acitodaphne glomerata*), *jelmu, kayu kundua* (*Mastixia trichotoma), sekabu (Goosampinus valentonil)*, and *kedondong hutan*, all of which are protected locally. Rattan, bamboo, and *damar* are other important nontimber forest products (NTFPs) in the area (Hartanto et al. 2001).

Bulungan Research Forest. In December 1995, Indonesia's Ministry of Forestry designated 321,000 ha of forest in the Bulungan district of East Kalimantan (now divided into three districts) to be developed as a model of exemplary management for sustainability. Since then, CIFOR has carried out interdisciplinary research there to develop effective strategies for multiple uses of tropical forests. The Bulungan Research Forest abuts the Kayan Mentarang National Park. Together, the two areas, which form a natural unit for integrated management, constitute more than 1.7 million ha in the heart of one of Asia's largest remaining tracts of tropical rainforest (Sayer et al. 2000). Lying in the watersheds of the Tubu, Malinau, and Bahau rivers, the area has production, protected, and traditional community forests. Competition for various uses—selective logging, shifting cultivation, collection of NTFPs, coal mining, and oil palm plantations—is increasing, complicating forest management.

The climate is equatorial, with an annual rainfall of around 4,000 mm, varying monthly by 200–400 mm. Relative humidity and temperatures are high; the lowest temperature recorded by the state-owned logging company, Inhutani II, was in unlogged forest: 23.5°C.

Topography is rugged; about 40 percent of the area has slopes of 25 to 40 percent. The most strongly dissected terrain lies to the southwest and west, bordering the national park. Elevations range from 100 to 2,000 m asl. All soils have a low to very low intrinsic fertility, with relatively high acidity and low nutrient content, cation exchange capacity, and base saturation (Shiel 2002). High rainfall leads to intensive weathering, leaching, and biological activity, as is characteristic of most Kalimantan soils.

Lowland dipterocarp forest covers 98 percent of the total area. Agathis (*Agathis borneensis*) trees are considered the most valuable commercially, because of their high basal area. Trees may reach 35–40 m in height and have a diameter at breast height (dbh) of over 50 cm (27 percent). The largest tree recorded in a recent study (Sist et al. 2002) was *Shorea venulosa* with a dbh of 199.6 cm. Secondary forest occurs in small patches along rivers and near villages, resulting primarily from shifting cultivation.

Also richly biodiverse, Bulungan has been noted for its species richness of trees with dbh over 10 cm, a mean of 151 tree species/ha, and a mean density of 407 trees/ha, with dipterocarps dominating. About 60 percent of the mammals and 70 percent of the lowland birds of Borneo are found in the region, many of which are endangered or threatened (Machfudh 2002).

Significant effort has been devoted to the eastern district of Malinau (roughly 2°52′–3°14′N by 116°–116°40′E), with particular focus in and around the 50,000-ha forest concession managed by Inhutani II since 1997, which covers about 40 percent of the Bulungan Research Forest. The Malinau watershed covers 42,000 km^2 and has a population of about 37,000, composed of over 20 ethnic groups (Wollenberg 2003a). In this area, topography is deeply eroded, with elevation ranging from 100 to 300 m asl. Average number of harvestable trees per hectare can exceed

10, and timber harvests often reach 100 m³/ha (Sist et al. 2003, 3). In a recent CIFOR study, the three species with the highest density and basal area were *Agathis borneensis, Shorea elliptica, S. maxwelliana,* and *S. parvifolia.* The highest density for a single species in a plot was recorded for *Shorea maxwelliana,* followed by *Agathis borneensis* and *D. stellatus* (Sist et al. 2002). The indigenous groups inhabiting the area, collectively known as Dayaks, are mainly rice farmers who practice extensive agroforestry and harvest a wide range of NTFPs (CIFOR Bulungan Research Forest ACM Team 2002).

Population Pressure—Low

Rantau Layung and Rantau Buta are small, isolated communities (210 and 85 people, respectively, according to Indonesia's 2000 Village Potential Census; Dewi 2003). There are 55 and 22 households, respectively; average household size is four people. If we compute the population density based on their community territories recognized by the local government, we get densities of 1.1 people/km² for Rantau Layung and 0.5/km² for Rantau Buta. The subdistrict (Batu Sopang) population density is 5.4 persons/km². The subdistrict sex ratio is 1.12, reflecting a general imbalance in East Kalimantan, home to many in-migrant men seeking their fortunes. If we compute the village-level sex ratio using the numbers of men and women who can vote (anyone older than 16 or married), we get still more skewed sex ratios: Rantau Buta yields 1.18 and Rantau Layung 1.49. Eighty percent of the couples in the reproductive age range practice family planning in Rantau Buta, and 79 percent in Rantau Layung (Dewi 2003).

According to the Jambi ACM team, the population of Baru Pelepat, in Jambi, Sumatra, is 557 people, with 270 men and 287 women, yielding a sex ratio of 0.94. Of these 557, 324 are children under 17 (58 percent), with a sex ratio in this group of 0.91. Population density is 0.077 persons/km² (based on the village territory of 7,200 km² as recorded by the National Land Agency, *Badan Pertanahan Nasional* [BPN] 2000). Average family size is five people (Indriatmoko 2003a).

Kalimantan's Bulungan Research Forest, covering a considerably larger area than either of the other Indonesian sites, is more complex. About 30 percent of the villages (and 17 percent of the total population of Malinau) are hunter-gatherers, the Punan. Other important groups include four Dayak groups. The core research area has between 5,000 and 6,000 people living in it, distributed over 300,000 ha (about two people per km²; Wollenberg 2003b). The most populated part of our research area is Long Loreh, a "village" that includes four communities (Loreh, Pelancau, Sengayan, and Bila Bekayuk), each inhabited by one ethnic group. In 1999, the total population of the Loreh area was approximately 1,171, comprising 543 Kenyah, 128 Punan Tubu, 250 Punan Malinau, and 250 Merap (Machfudh 2002). Shiel conducted a long-term study in and around seven Malinau villages where he found 1,083 individuals in 248 households, with a population density of 0.49/km² (Shiel 2002). Another researcher working with Punan who live in very inaccessible areas far up the Tubu River noted villages hosting about 20 families each, with unusually high infant mortality rates, ranging from 17–46/1,000 births (Levang 2002, 123).

Management Goals—Large-Scale Timber

Out of Baru Pelepat's 7,265-ha territory, 4,750 ha are classified as state production forest while the rest is set aside for agriculture and settlement (Indriatmoko and Kusumanto 2001). Four logging companies have operated in the area. The first two operated from 1975 to 1995, and a third was in full operation until 2002. The latter reportedly did not rehabilitate logged areas, in violation of regulations under the TPTJ (*Tebang Pelih Tanam Jarak* or Selective Logging and Line Planting) system, and it is highly probable that the first two did not either (Hartanto et al. 2001).

To meet the new national autonomy regulations, large- and small-scale loggers were to stop their operations in 2001, and heads of the districts were to give no new destructive logging permits such as the *Izin Pemungutan Hasil Hutan* (Forest Products Utilization Permits or IPHH)—a controversial practice with negative environmental impacts. One timber company nevertheless continued to operate until 2002 (Marzoni 2004).

A prominent development over the last two decades in Baru Pelepat has been the establishment of transmigration settlements, which have increased the population and pressure on available land and natural resources. Another important development in Pelepat since the late 1970s has been governmentally instituted plantations, especially of oil palm (Kusumanto et al. 2002a).

Under Soeharto, concessions consisting of hundreds of thousands of hectares were routinely granted to the president's cronies for large-scale logging. In Kalimantan, the military had similar, lucrative timber concessions all along the Malaysian border, in disregard of pre-existing traditional claims to land and forests. Although the contracts between the government and logging companies included reasonable stipulations pertaining to forest sustainability, the regulations were in fact rarely enforced. The timber business was marked by cronyism, corruption, and general disregard for both local communities and good management practices. Forest quantity and quality have deteriorated continuously for the past 30 years at least, to the point where some studies suggest that large-scale timber is no longer commercially viable (Barr 2002). Since decentralization, in 2001, while the legitimacy of large-scale companies has been increasingly questioned and rights to govern and manage Indonesia's forests are being debated between the center and the districts, small-scale timber extraction has dramatically increased.

Management Goals—Small-Scale Timber

Once completely illegal, the status of small-scale logging is now unclear, because both central and regional governmental agencies claim the right to determine policy. The hegemony of the large companies has been severely restricted legally, but small-scale logging has seen an upsurge.

On July 31, 2000, the Pasir District government—where Rantau Layung and Rantau Buta are located—passed Regulation No. 16/2000 concerning "Community Forests," allowing communities with ownership rights to forests to profit through a mechanism called a Wood Utilization Harvesting Permit (*Izin Pemungutan dan Pemanfaatan Kayu* or IPPK). IPPK and other similar mechanisms were ostensibly designed to allow communities to profit from managing the forests in their traditional territories. Such "Community Forests" had to be on land owned by the community or land that fit the following criteria: at least 0.25 ha, with over 50 percent crown cover, or a minimum of 500 first-year trees not growing in formally designated state production forest (Article 1, Point 9). Until this regulation was temporarily suspended in 2001, the Pasir District government had issued 265 permits in 38 villages, covering 25,957 ha producing or expected to produce around 1,668,655 m³ of wood. Even after the suspension of the regulation, several companies continued to operate. In 2001 revenue deriving from IPPKs provided Pasir District with Rp12,893,817,000 (Rp8,500 = U.S.$1, May 2003). Almost 87 percent of the district's total revenue came from these permits (Adnan et al. 2002).

Community members have little access to information about new legislation and even less experience navigating bureaucracies, a situation that encouraged third parties to become mediators or brokers between the communities and commercial companies with the wherewithal, experience, contacts, and marketing channels to make the most efficient—if inequitable—use of local forests. Such brokers often became the authorized IPPK holders, handling administration at the district offices and with the companies. Often these brokers were government officials, further simplifying the process of dealing with the government. In the case of our research area, some of the brokers were officials from the subdistrict office. They often obtained permission which did not, in fact, comply with the law (Adnan et al. 2002). Community members and others participate in the actual logging using chainsaws.

In Baru Pelepat, some small-scale, illegal logging by outsiders began in the 1980s, relying on a traditional modern mix of buffaloes, chainsaws, and trucks. Since decentralization, nine small sawmills have opened up in the Baru Pelepat area (only two of which are legal). Now, villagers and outsiders both do the logging, with funding from a timber trader or sawmill owner. Villagers survey for commercial species, cut the trees, and carry them to the sawmill (see Indriatmoko's observations on illegal logging below).

In Bulungan, the district government began issuing IPPKs in April 2000, based on District Decision No. 19/1999. IPPKs in Bulungan could be assigned to individual landowners, village and government cooperatives, community conservation groups, or companies. Permit holders could clear-fell if it did not "have a negative effect" for up to six months, the permit being renewable up to three times. Bulungan District issued 75 IPPKs, covering almost 10,000 ha (Barr et al. 2001).

Malinau District (previously within Bulungan) issued at least 39 IPPKs, covering a total of 56,000 ha. As in Lumut Mountain, Malinau District expects significant income from IPPKs. IPPK holders pay a "third-party donation" of Rp200,000/ha and a production fee of Rp15,000/m³ of timber harvested. Barr et al. project that with these fees, the district government can collect around Rp53 billion from the 39 IPPKs issued through February 2001—a total that is nine times the district budget planned for 2000 (Barr et al. 2001).

Timber companies that form partnerships with local communities in the Bulungan area typically offer a variety of benefits such as employment, food, cash crop establishment, machinery, and fees based on volume of timber harvested. Barr et al. (2001) document some of the many cases where such promises have not been fulfilled, citing the lack of sophistication of local people in making such agreements as one important reason.

Management Goals—Large-Scale Conservation

Baru Pelepat is in the buffer zone of Kerinci Seblat National Park, which reaches 350 km along the Barisan Mountain range in southern Sumatra and is part of four provinces: West Sumatra, Jambi, Bengkulu, and South Sumatra. Covering an area of about 1,375,000 ha, it is one of the largest conservation areas in Southeast Asia (Kusumanto 2000, 4). The park, which was gazetted in 1995, includes 436,000 ha in Jambi Province.

Adjacent to the Bulungan Research Forest, in the northern part of East Kalimantan, is Kayan Mentarang. In the 1980s, a team sponsored by the Worldwide Fund for Nature (WWF) surveyed Kayan Mentarang and recommended that it be gazetted as a nature reserve. In 1996, the status of the reserve was changed to national park to allow residency by local people (Wollenberg et al. 2001c). WWF-Indonesia, the Indonesian Department of Environment, and CIFOR have supported local conservation and local people's rights to use forestland. Since 1996, WWF has been implementing a management plan that includes zones for biodiversity protection and "traditional" use by local villagers. One goal is for Kayan Mentarang to serve as a model for integrating conservation and development aims in remote areas (Wollenberg et al. 2001c).

External Diversity—High

In Baru Pelepat, Kusumanto's (2001b) list of external stakeholders includes more than five neighboring communities, plus two groups of hunter-gatherers (*Orang Rimba*), various groups involved in private timber harvesting (sawmill owners, small-scale investors, entrepreneurs, loggers), government logging companies, local officials (transmigration, agriculture, conservation agencies), and community facilitators from the Integrated Conservation and Development Project connected to the nearby Kerinci Seblat National Park and from WARSI, an NGO network. With the recent decentralization in Indonesia, involvement by government agencies has increased in issues such as resettlement, agriculture, forestry, Kerinci Seblat National Park, and the Integrated Conservation and Development Project—all of which now have district-level representation.

External Diversity—Medium

For Rantau Layung and Rantau Buta, Hakim (2001b) lists neighboring villages and forest workers from outside the communities as important stakeholders, along with small-scale sawmill owners. The governmental actors include the Regional Planning Agency (*Bappeda*), District Forestry and Soil Conservation Service (*Dinas*

Perhutanan dan Konservasi Tanah), District Forestry Service (*Dinas Kehutanan*), and District Environmental Impact Agency (*Bapedalda*). The subdistrict (*kecamatan*) government also plays a role. One forest concessionaire and four smaller logging companies operate in the area. An NGO, Padi, has been active locally. Some research has been done by the Environmental Study Center (*Pusat Penelitian Lingkungan Hidup*) and the Center for Social Forestry (CSF), both at Mulawarman University in Samarinda.

In Bulungan Research Forest, Purnomo et al. (2000) examined stakeholders in the Inhutani II concession. (Inhutani is an Indonesian parastatal timber company that operates in provinces outside of Java.) Bulungan's 17 identified stakeholders include neighboring villages, a couple of branches of Inhutani, the Forestry Department, the Malinau Forest Management Office (*Resor Pemangkuan Hutan*, RPH Malinau), several NGOs (Plasma;YAP, *Yayasan Adat Punan*; BIOMA, *Yayasan Biosfer-Manusia*), the subdistrict leader (*camat*), and university students and lecturers.

Age

In Borneo and Sumatra, Indonesia, although perspectives differ between the young and the old (cf. Colfer 1981; Hakim et al. 2001b), the relationship is generally amicable. The young are expected to respect the old, but both also expect the young to take on responsibilities for decisionmaking as they grow into mature adulthood. Parents expect to make significant sacrifices for their children but also to relinquish control as their own labor contribution decreases. The idea that the old might harm their own young (e.g., cursing them) is inconceivable. (Among the Javanese, from another Indonesian ethnic group, dead ancestors can curse their descendants for failing to honor them properly or observe tradition correctly, as 1980s field observations of Sitiung transmigration site near Baru Pelepat, in Jambi, showed.)

Internal Diversity—Medium

Baru Pelepat includes indigenous Minangkabau people, along with in-migrants primarily from the island of Java and other parts of Jambi province. Relations among the groups have been strained. The in-migrating Javanese are reputed to be hard-working farmers; the local people have a way of life that is more integrated with, and dependent upon, the forest. (See Case 9 for some parallels in Madagascar that are fascinating, given the historical links between the two countries.) The Orang Rimba, a hunter-gatherer group that has patron–client relations with the villagers (Kusumanto et al. 2002a) are also present.

Of Bulungan's four major Dayak groups in the area—Kenyah, Lundaye, Tidung, and Merap—most live in communities dominated by one or two ethnic groups (for details see Anau et al. 2004). The Kenyah have a traditional aristocracy, which retains some privilege vis-à-vis other community members. Punan hunter-gatherers maintain patron–client and other ties with settled communities, in some cases residing in settled communities part of the time. Large numbers of in-migrants connected with logging, mining, and plantation development inhabit Bulungan's more easterly and southerly regions.

Internal Diversity—Low

In Rantau Layung and Rantau Buta, indigenous Adang Dayaks comprise 98 percent of the population. (The Adang are Muslim, and their self-designation as "dayaks" is not consistent with anthropological usage.) The remainder are *Banjar* and Javanese (Hakim et al. 2001b).

Level of Conflict—High

Rantau Layung and Rantau Buta. Although this single microwatershed covers far less territory than the Bulungan Research Forest, both share many of the same conflicts. In the not-too-distant past, the communities dealt with a logging company and—because it was supported by a strong central government—therefore with government officials. Power was clearly in the hands of the company and the government, and conflicts were played down for pragmatic reasons. Again, the decentralization process has changed the power dynamics considerably.

In 2000, the communities claimed that the government boundary of the Lumut Mountain Protected Area incorporated part of their traditional territory. They were unhappy first because they had not been consulted during the determination of the boundaries, and second because they feared that in the next determination of protected area boundaries, their homes and yards might be included. The district head (*bupati*) initially understood them to be rejecting the entire concept of a protected area. The ACM team participated in clarifying the community's view to the government officials, and the problem was resolved (ACM PAR Team, Indonesia 2002, 41).

More potentially serious conflicts occurred relating to IPPKs. Two cases involved the ACM communities in disagreements with their neighbors. First, Rantau Layung and the neighboring community of Brewe agreed upon the boundary between their two villages. Five days later, a logging company we will call "PT-TMP" began logging in Jaung, ostensibly in accordance with an IPPK in Brewe's territory. Rantau Layung and Brewe each claimed Jaung as part of its own area. District Forest Service personnel met with PT-TMP, and the two communities discussed the issue, with ACM facilitation. Eventually all agreed that Jaung was traditionally owned by a Brewe resident but was administratively part of Rantau Layung. The fees to be paid by the company would be shared equally between the traditional and administrative owners. Villagers agreed not to log the protected forest any more and the district forest officer checked the IPPK holder's legitimacy. The officer's investigation revealed some irregularities. For example, community members whose signatures were on the application for an IPPK knew nothing about it (ACM PAR Team, Indonesia 2002).

The second IPPK-related problem was between Rantau Buta and neighboring Kesunge. The problem began when a logging company we will call PTW, operating in Rantau Buta, accused a company we will call PTM of logging in PTW's

IPPK area. PTM, operating in Kesunge's territory, then reciprocated the accusation. A meeting was called to convene the Forest Service Office, the community representatives from each village, and a member of the regional House of Representatives. A team of people from both villages and staff of both companies was organized, with the rationale that the local people knew their boundaries better than outsiders. The villagers formalized an agreement on the location of the boundary; the location of PTW's IPPK area in the Kesunge village; and that PTW should compensate Kesunge village. However, the disagreement between the two companies persisted and was still awaiting trial in late 2002 (Adnan et al. 2002, 6–7).

Baru Pelepat. Lying in the buffer zone of a national park, Baru Pelepat has production forest, plantation agriculture, and mining operations; it has been "invaded" by settlers, both spontaneous and government-sponsored, from nearby and from more distant origins. Numerous conflicts are under way.

Although the local community formally requested one group of settlers (see Kusumanto and Sari 2001), the underlying story appears more complex. When the ACM team began working in Baru Pelepat, the village headman was used to making rather unilateral decisions routinely and was widely considered corrupt. Others who supported the decision to invite the settlers included another powerful local man who was a member of the regional legislature and in the lucrative timber business, a mining company, and the district government. Such resettlement would provide a convenient solution to the mining company, which wanted the removal of inconvenient villagers from a potential mining area (and probably extra cash for low-salaried officials); meanwhile, it was defensible as "development." Although some villagers were opposed to this plan, including one community elder who has boycotted the group of elders ever since, the settlers were invited and came. Interactions between the local and settler communities have been marked by distrust and distance, if not outright hostility. Many local people are dismayed at the loss of land held by their ancestors (ACM PAR Team, Indonesia 2002, 63; Indriatmoko 2003b).

Illegal logging, which community members consider out of hand, has also sparked conflict. Indriatmoko (2002, 9–10) identifies several contributing factors:

- The nine sawmills in the area, some illegal, have a high demand for logs.
- Outsiders perceive that forests belong to all humans, who should have free access to forest products, including timber; locals, on the other hand, see the forest as communal property, governed by customary regulations.
- Customary (*adat*) regulations relating to forest management have not proved resilient or strong enough to cope with changing conditions.
- The powerful legislator/timber tycoon mentioned above has played a role.

Village concerns about illegal logging spurred their interest in clarifying their community territory by means of participatory mapping. The mapping process involved interactions with neighboring communities; additional conflicts came to the fore. In the course of the discussion, internal conflicts between two long-time rivals in the village initially weakened the negotiating position of Baru Pelepat (ACM PAR Team, Indonesia 2002, 70). See Chapter 10, for more on internal conflicts.

Bulungan. Like Campo Ma'an in Cameroon, Bulungan covers a large area and hosts an equally complex set of actors, issues, and desires for the future. The ACM team has been working with some 27 villages, composed of hunter-gatherer Punan, swidden farming Dayaks, and in-migrants of diverse ethnic backgrounds. Conflicts are more obvious to the south and east, where mining, logging, and oil palm development are intensive.

Until national decentralization began in earnest, conflicts were most obvious in relations between extractive industry and local communities. Since decentralization, the level of conflict has risen sharply from an already high level. In the absence of effective regulation of natural resource use, and realizing a possible opportunity to legalize their customary rights to land, communities began making serious efforts to clarify their territories (Anau et al. 2004). In May 2000, the governor of East Kalimantan instructed concession holders to compensate communities Rp1,000/m^3 to Rp3,000/m^3 for timber cut since the mid-1990s (Barr et al. 2001, 27). Private logging, mining, and agricultural companies began to take community rights somewhat more seriously and began negotiating directly with community representatives (Levang 2002). Intense competition within and between communities was spawned, both to clarify their boundaries and to gain access to benefits offered by these companies. At the same time, the companies— aware of the lack of sophistication and political savvy in these communities— routinely renege on promises and negotiate agreements that are patently unfair. See Box 10-4 in this book.

With decentralization, district governments also began giving out large numbers of IPPKs in 100-ha blocks. Unprecedented levels of logging resulted, as did increasing conflict among small-scale loggers, concession holders with permits from the central government, villagers (themselves often deeply divided), and mining interests (Barr et al. 2001). The Bulungan ACM team (CIFOR Bulungan Research Forest ACM Team 2001, 3) conclude that "A laissez-faire, frontier atmosphere has emerged in which making money has become more important than always being lawful or fair."

The backdrop for these complex machinations is the ethnic diversity of the region. Certain groups have inhabited the area for longer periods and have traditional power and prestige, as well as stronger claims to land. Others were moved to their present location by government-sponsored resettlement programs. Still others are hunter-gatherers who have traditionally used the forest but not in ways that are recognized by outsiders as indicating ownership. Conflicts among these groups center around overlapping ownership or use of agricultural fields or land containing coal deposits, with timber and valuable NTFPs like birds' nests and *gaharu* also important. Traditional and recent enmities have been exacerbated in this competitive context.

Although communities are enthusiastic about gaining access to resources long denied them, serious indications of trouble ahead have appeared. The Bulungan ACM team notes "community protests against the 'investors' for not paying expected fees or wages to local harvesters; complaints among villagers about opaque deals struck between leaders and investors; and forest logged in areas where permission was not granted by villagers" (CIFOR Bulungan Research Forest ACM Team 2001, 1).

Levang and team (2002, 124) are pessimistic: "Mutual aid among villagers gives way to individual enrichment. Competition for natural resources leads to conflicts with outsiders and among villagers, economic differentiation leads to *nouveau riche* behaviour and to jealousy, not to mention social pathologies such as alcoholism, gambling, and prostitution." He worries that the resentments spawned by marginalization of one ethnic group by another will fuel ethnic conflicts.

Consistent with others' observations on the dangers of natural wealth (Dove 1993; Peluso 2002), Barr et al. (2001, 33) note "The increased economic gains made possible by decentralization have therefore also helped to escalate conflict over land and forest claims."

Social Capital—Low

Rantau Layung and Rantau Buta. Lack of trust evident in Rantau Layung led to its characterization as having low to medium internal social capital, because efforts to negotiate agreements with the neighboring timber company had involved trickery by the local community leader in 2001 and factionalism within the community. However, they had been able to negotiate payments of Rp1,000,000 per household head from a timber company in return for rights to harvest in their territory, and they successfully removed their dishonest leader.

Rantau Buta was also seen by the ACM facilitators to have low to medium levels of social capital, though overt distrust was less evident. Meetings among the men in the evening were common, and community affairs were discussed at great length.

In both communities, these evening meetings of men tended to involve a considerable amount of information giving by those in charge, rather than genuine sharing of perspectives or joint analysis. Both communities had longstanding traditions of rotating labor parties called *sempolo*, which were important for their swidden cultivation system, suggesting a basis on which more effective social capital could be built. They also collaborated in Muslim religious observances (ACM PAR Team Indonesia, 2002).

Relations between the communities and others were more problematic. Problems between other communities and logging and plantation companies in the area were well known, widely discussed, and marked by "bad blood." Violent confrontations occurred periodically, and deception seemed to be the order of the day. In both communities, inhabitants expressed significant concern about their dealings with other more powerful stakeholders.

The information flow between the local government and the communities was felt to be inadequate, partly because of difficulties of access. Local government agencies like the District Forest Service and the District Environmental Impact Agency were very concerned about maintaining the conservation function of the surrounding Lumut Mountain Forest, while local communities were concerned about their continued livelihoods. Openness between communities and with other stakeholders was low, resulting in low levels of trust and ineffective partnerships (ACM PAR Team Indonesia, 2002).

Baru Pelepat, Jambi, Sumatra. Local institutions were seen as weak (Jambi ACM-PAR Team 2001a) when the ACM research began:

> It remains an open question to what extent the 13 [new] district regulations could foster true devolution to village-level governance as well as a democratic life in such villages as Baru Pelepat where social capital is low, the relationship between the original community and the settlers from elsewhere is poor, village decisionmaking is monopolized by a handful of village elites with poor representation of community groups, and the *adat* [customary law] institution is likely to discriminate rights and responsibilities between the original people and settlers, for the latter are no part of the customary community. (Kusumanto et al. 2002a, 6)

Kusumanto and Indriatmoko (2001, 12) further describe the situation as generally characterized by "… much distrust, little reciprocity and weak social networks." *Turun betahun,* a traditional yearly planning activity among the Minangkabau, has not been practiced for many years, and formal farmer groups, once set up, have dissolved. Women's rotating credit groups and Koran reading groups do function, however (ACM PAR Team, Indonesia 2002).

This state of affairs derives partly from a loss of respect for adat leaders who have failed to comply with social norms. They are reported to have communicated badly with the community at large, with other leaders, and with outsiders and neighboring communities. Local leaders are also seen as deficient both in personal capacity and in time devoted to village leadership (ACM PAR Team, Indonesia 2002, 32). For more on this dynamic, see Krishna's (2002a) finding that in India the presence of leaders capable of representing the community vis-à-vis other powerful groups was critical for making good use of social capital for effective development.

Despite the fact that the Baru Pelepat site is much smaller in area than the other Indonesian sites, it is characterized by the same complex mix of stakeholders. The potential for high levels of social capital is much higher within a given ethnic group, where kinship, common language, and traditions make communication easier. This is evident in the often-strained relations among these communities (ACM PAR Team, Indonesia 2002, 28).

A bit of light in this bleak portrayal relates to relations between the Orang Rimba and the inhabitants of Baru Pelepat, despite the fact that the former's territory overlaps with the Minangkabau territory of Baru Pelepat. Each party has traditionally recognized the rights of the other. The Orang Rimba have tenure over forest products, and the Minangkabau community has customary land tenure in this shared territory, with little conflict. Each Orang Rimba group has a patron–client relationship with a particular person in the village, called a *jenang,* who may help resolve conflicts within the Orang Rimba community. The villagers requested ACM help in recognizing the Orang Rimba as part of their formal administrative structure (Indriatmoko 2002, 10).

Local people also interact with a variety of organizations, governmental entities, and private enterprises. Traditional linkages are well established but fragile in terms of trust and mutual respect, reportedly due to their hierarchical, authoritarian nature. The ACM-Jambi Team (ACM PAR Team, Indonesia 2002, 28) writes "Different views on human and forest development objectives, distrust and lack of knowledge and information, have generated poor relations between the government, community stakeholders, and private enterprises."

Bulungan Research Forest. Bulungan's social capital context is complex. The area has historical enmities among ethnic groups, new arenas of interaction among in-migrant ethnic groups, and a whole host of new or at least more recent players like timber, plantation, and mining companies, district-level government agencies, and even NGOs, universities, and research organizations.

Anau et al. (2004) found the whole gamut of what they called community cohesion, measured by low levels of factionalism, cooperative efforts at the village level, and support for the village leader, among the 27 villages they studied in Bulungan.

Levang (2002) found that the Punan carried out their daily work on a mutual aid basis among households organized into small neighborhood groups. He concluded that confidence and trust depended on family links or ethnicity. The Punan are in patron–client relationships with traders, with the usual problems such relationships suggest strong dependency ties, opaqueness of the market, and some deception. However, even traders must take some care not to go too far and lose their clients' confidence.

Levang also talks at some length about the disruptions brought about by economic change and the introduction of outsiders. He sees family ties becoming looser, self-help losing popularity, and greed gaining ground ("Everybody is looking for big and easy money." 2002, 121) The Punan bride price, once a mechanism for strengthening links among groups, has become a monetary transaction. See also Barr et al. (2001) for a good discussion of the chaos that currently characterizes land claims and governance in the area, with obvious and adverse impacts on social capital among groups.

Each ethnic group complains about the preferential treatment given to others. The local people see the concessionaires favoring the in-migrant Javanese, Bugis, and Batak. And in the local civil service, the Tidung, Kenyah, and Lundayeh take the best spots, with the Punan the most marginalized of all. Levang (2002, 126) concludes that significant social capital exists within groups but little between them.

CASE 8

Kyrgyzstan

Contributing Author
Kaspar Schmidt

Sites in Kyrgyzstan
Achy, Arstanbap-Ata, Ortok, and Uzgen

Devolution Status—Intermediate

Kyrgyzstan's state forest farms (Russian *"leshoz"*) in the ACM study area include natural and planted stands of fruit and walnut trees. As with all natural resources, forests were subject to strong state control under the Soviets. With financial support unavailable after the breakdown of the Soviet Union, interest in involving private stakeholders in forest management grew. In recent years, there has been increasing interest in negotiated agreements that specify the responsibilities a tenant will undertake in return for rights to use the forest areas. Yet decisionmaking power and ownership remained in the hands of the state. A new

Kyrgyzstan Key ACM Characteristics	
Devolution status	Intermediate
Forest type	Temperate
Population pressure	High
Management goals	Multipurpose reserve
External diversity	Medium
Internal diversity	Medium: Arstanbap-Ata, Achy
	Low: Ortok, Uzgen
National conflict	Low
Local conflict	Medium: Arstanbap-Ata, Achy, Uzgen
	Low: Ortok
Social capital	High: Ortok, Uzgen
	Medium: Arstanbap-Ata, Achy

forestry code, passed in 1999, recognized collaborative forest management through leasing arrangements, among other things (Carter et al. 2003). But real authority for forest management has largely remained with the *leshoz*. Local people have become increasingly involved in forestry activities beyond their traditional harvesting of walnuts (Schmidt 2002b, 7–8).

The Arstanbap-Ata *leshoz* (not atypically) applies a system of paid leases, with the lease payment (paid in cash or in kind) dependent on the area and production potential of the plot. Most leases are for horticultural plantations, with only a few examples of natural forest stands. Leases are renewed annually, with some plots having multiple leaseholders (depending on the crop, which may be fruit, hay, fuelwood, and others). Arstanbap-Ata *leshoz* began leasing forest plots near the village to private households in the late 1980s as a means to ensure forest protection and improve the forest belt around the village (Schmidt 2002b).

The Achy *leshoz* applies a similar system of leases, with the payment a function of area and production potential. Many households—often former *leshoz* workers—are involved in forest management through this system, with leases typically lasting five to ten years. Achy *leshoz* began leasing out plots in the 1980s partly to ensure the fulfillment of the prescribed state production plan, despite cuts in state forestry funding, and as a means of motivating its staff. More recently, in autumn 2002, interested households in some villages in Achy *leshoz* were assigned a number of walnut trees, fixed as a function of the size of the household, for the walnut harvest in coming years. This was often at the expense of other long-term leaseholders, who saw the size of their leased plots considerably reduced. It seems that no precise information as to how many years the households will be entitled to harvest walnuts from these trees has been given to them. A written contract fixing the amount of walnuts to be delivered to the *leshoz* by these households is made in years when the walnut yield is deemed worth the administrative effort (Schmidt 2003e, 6).

The situation on the two other Kyrgyz sites, Ortok and Uzgen, is quite different. New, collaborative approaches to forest management (CFM) have been explored in these two *leshozes* since 1998, in collaboration with the Kyrgyz-Swiss

Forestry Support Programme (KIRFOR) and the State Forest Service (see Fisher 1999; Carter et al. 2003; or Fisher 2003b). In an early CFM workshop in Kyrgyzstan it was decided that a system of leasing of plots of mature forest would be the appropriate approach for the beginning (Carter 1997). The tenants of the plots have the right to harvest all fruits growing on the leased plots and to use or to sell the fruits and a defined amount of dead branches. In return, they are responsible for forest protection; they also must agree to other forestry work, such as the establishment of nurseries or the maintenance of orchards.

Under this system, no money is expected to exchange hands—an innovation when compared with other local lease systems such as in Arstanbap-Ata or Achy. Additionally, the length of leases has been increased in the CFM areas to 49 years, although in most cases a probationary period of five years applies. In Arstanbap-Ata, leases are formally only for one year, but are generally renewed; in Achy they are usually for 10–15 years. The Kyrgyz CFM has grown continuously in Ortok and Uzgen, both in the number of households involved and the area under CFM lease. CFM has now been extended to other *leshozes* in the walnut-fruit forests and other regions of the country (Fisher 2003b).

In Ortok, in March 2003, 127 CFM leaseholders were working on 500 ha of walnut forests and 427 ha of plantations (Fisher 2003b). At this time, leaseholders remain in a five-year probationary period, prior to the issuance of the planned, long-term leases. Seedlings for the plantations, which the CFM tenants have to establish as part of their contractual duties, are provided by the *leshoz,* and the tenant must ensure a 70–90 percent survival rate at the end of the lease (Messerli 2000).

As in Ortok *leshoz,* some management responsibility has been given over to local Uzgen farmers since 1998, also within the CFM framework. In Uzgen, in March 2003, there were 92 CFM leases (Fisher 2003b), including some group leases, covering a total of 729 ha. People had agreements with the government to use that land or forest for 49 years. Again, leaseholders remain in a five-year probationary period.

Forest Type—Temperate

All of the Kyrgyzstan ACM sites were in temperate zones with similar climatic conditions (see Schmidt 2003e).

Climate Conditions. Summer temperatures in the walnut-fruit forest belt are moderate and winters comparatively mild from a central Asian perspective, with this zone protected from the intrusion of cold northern air by high mountain ranges (see Table A-4). Abundant precipitation in early spring and short bursts of rain in early summer are followed by a dry July, August, and September. A second rise of precipitation occurs from October to December, never reaching the spring level (Academy of Science of the Kyrgyz Republic 1992).

Precipitation depends strongly on altitude. For areas such as Achy, lying between 900 and 1,100 meters above sea level (m asl), precipitation averages 800 mm, with

Table A-4. *Climatic Conditions on Four Kyrgyz ACM Sites*

Sites	Altitude (m)	Mean air temperature (°C)	Mean air humidity (%)	Annual precipitation (mm)
Arstanbap-Ata	1,520	9.2	55	1,000
Achy	1,050	11	55	730
Ortok	1,500	9	57	1,000
Uzgen	1,300	10.3	60	800

Sources: Academy of Science of the Kyrgyz SSR Geography Section 1965; Academy of Science of the Kyrgyz Republic 1992.

a rain-free period of 80 days. Above 1,100 m asl (e.g., Uzgen), precipitation increases and the rain-free period decreases to 35 days, on average. In the range from 1,400 to 1,800 m asl (Arstanbap-Ata and Ortok), precipitation increases about 60 mm for every 100 m of altitude (Academy of Science of the Kyrgyz Republic 1992).

The mean absolute minimum temperature in the walnut-fruit forest belt is −22°C; the absolute maximum temperature 38°C (Bulychev and Venglovsky 1978). Snow cover, of about 20–60 cm in the forest belt, lasts about four and a half months in winter and early spring (Academy of Science of the Kyrgyz SSR 1965). The vegetation period in the walnut-fruit forest belt ranges from 160 to 180 days. The frost-free period is about 198 days, with first frost on average on September 15 and the last frosts on April 20 (Academy of Science of the Kyrgyz Republic 1992; summarized from Schmidt 2003e).

Forest Characteristics. Current forest cover in Kyrgyzstan is estimated at 4 percent of the country, or 700,000–800,000 ha. Small but highly diverse, Kyrgyz forests are all state-owned. All four ACM research sites lie in the walnut-fruit forest belt in the south of the country. Fruit-bearing woody species such as walnut (*Juglans regia*), apple (*Malus* spp.), hawthorn (*Crataegus* spp.), plum (*Prunus* spp.), and rose species (*Rosa* spp.) occupy hills and mountain slopes at altitudes between 800 and 2,400 m asl. The walnut-fruit forests of Kyrgyzstan are considered the largest of their kind in the world and therefore of global significance for biodiversity conservation (Blaser et al. 1998; Hemery and Popov 1998; Venglovsky 1998). Humans have influenced and modified a considerable part of this naturally occurring forest type by planting selected species and varieties, grafting, and by using some of the forest for grazing and tillage.

Other important forest types in Kyrgyzstan are spruce (*Picea schrenkiana*) and juniper (*Juniperus* spp.). Juniper grows under arid conditions or in very high altitudes up to 3,500 m asl. Riverside forests, typically of willow (*Salix*), poplar (*Populus*), and birch (*Betula*) species, also exist. Additionally, there are plantations of naturally occurring and introduced species as well as of poplar, the latter mostly near or within settlements (Müller and Venglovsky 1998).

ACM *Leshozes*. The *leshoz* of Arstanbap-Ata comprises 7,489 ha of mainly broadleaf forests in an upper valley in the mountains of southern Kyrgyzstan (with 0.6 ha of forest per person in the area). About 3,000 ha of the forest are dominated by

walnut trees (*Juglans regia*). Apple (*Malus* spp.) and juniper (*Juniperus* spp.) also occupy substantial areas. The forests of Arstanbap-Ata *leshoz* are under fairly heavy
human pressure. Signs of overuse are evident, especially near villages, mainly due to
increased firewood use since independence and to unregulated grazing. Illegal felling for commercial purposes—walnut timber and burls are exported—has been
reported, especially in the late 1990s (Schmidt 2003e).

The forests of Achy *leshoz*, which borders Arstanbap-Ata, lie in the hilly flanks
of the valley of the Kara-Unkur River in southern Kyrgyzstan. In total, 6,996 ha of
the *leshoz* territory are forested; the remainder is made up of haymaking plots,
pasture land, and, closer to the villages, fields. Some of the summer pastures are
located in Arstanbap-Ata *leshoz* territory. The most widespread type of forest is
open hawthorn (*Crataegus* spp.) woodlands (~1,750 ha), followed by stands dominated by apple and dog rose. Walnut stands occupy only about 760 ha. Other
important species include pistachio, maple, almond, and cherry plum (*Prunus sogdi-
ana*). Of the four Kyrgyz sites, Achy has the greatest pressure on the forest resources,
especially for fuelwood and to some extent also for precious walnut burls and
timber. The forest is relatively accessible, including to outsiders from the lower part
of the district (Schmidt 2003e).

Ortok *leshoz* comprises 10,282 ha of forests, and is the richest *leshoz* of the Kyrgyz sites, with 1.8 ha/person and a high percentage (47 percent) of walnut stands.
Forest areas are near the settlements, and some crops are planted within the forest.

The landscape in Uzgen is dominated by strips of natural and planted *leshoz*-
managed forests, pastures, and agricultural land. The latter two are mostly in the
hands of the municipal administration and leased to local farmers. Out of a total of
21,777 ha, 5,444 ha are walnut stands (*Juglans regia*), partly mixed with other species. Other significant tree species include maple (*Acer* sp.), hawthorn (*Crataegus*
spp.), juniper (*Juniperus* spp.), poplar (*Populus* spp.), and typical species for the widespread riverside forests, such as willow (*Salix* spp.) and birch (*Betula* spp.).

While somewhat restricting livestock and land use, the forest provides an important economic resource for the villages, especially walnuts. Other main forest
products are firewood (key for cooking and winter heating); wild fruits (apples,
haws or hawthorn, and rose hips), medicinal herbs, mushrooms and edible fungi,
cherry plums, and almonds (*Prunus* spp.).

Much of the land is poor and stony or sandy. A number of rivers run through
the area, but water access for irrigation is a common problem. Agriculture and
pasture are also important land uses (summarized from DFID 2001b, Messerli 2000;
Schmidt 2003e).

Population Pressure—High

The calculation of hectares per person was computed based on the forested area
available in the *leshoz* and the population living in villages within the *leshoz* boundaries using census data from 1999 (Abdymomunov 2001). The population from
adjacent areas beyond the boundaries of the respective *leshoz* was not taken into
account. Thus, Uzgen presents an apparent population pressure anomaly when
compared to other sites.

In the Arstanbap-Ata *leshoz*, 7,489 ha of forested area was available to 11,724 people in 1999, yielding 0.6 ha per person. The population growth rate is 3.3 percent per year. Achy *leshoz* (6,996 ha of forested area) includes five villages, with about 12,189 people in 1999 (2,300 families), with a density of 0.6 ha of forested area per person. The population growth rate is 3 percent per year. Ortok *leshoz* (10,282 ha of forested area) supported a population of about 5,621 people in 1999, yielding 1.8 ha of forest per person. Population growth in the district at large is 2.5 percent per year, and the population is virtually all Kyrgyz (Schmidt 2003e).

In Uzgen *leshoz*, there are around 22,000 people in the roughly 21,777 ha of forest, resulting in about one ha of forested area per person. This means that as far as human pressure on forests is concerned, Uzgen occupies an intermediate position between Ortok, with more forests per inhabitant on the one hand, and Arstanbap-Ata and Achy, where forest resources are particularly limited, on the other hand. The annual population growth rate was 2.6 percent per year between 1989 and 1999 (Schmidt 2003e). The population is almost 100 percent Kyrgyz, with less than one percent "other." Marti (2000) reports that compared to other *leshozes* within the walnut-fruit forest belt, Uzgen has more heterogeneous income-generating options.

A rapid rural appraisal (RRA) conducted in one Uzgen village, Ak Terek (DFID 2001b), found 150 households containing 880 people, with an adult sex ratio of 0.92 (202 men and 218 women). The number of children under 16 years old was 408, or nearly half the population. The village's 28 male pensioners and 47 women made for a typically unbalanced old-age sex ratio: 0.59.

Population growth rates appear to be high but declining. Villagers reported that most families had about eight children during Soviet times, but now have four to five. Important illnesses affecting the population include anemia, goiter, influenza, tuberculosis, gastritis, and toothache. The sex ratio is fairly even in all four sites.

Management Goals—Multipurpose Reserve

Kyrgyzstan is currently redefining its forest policy, widening its former narrow focus on forest conservation to include broader objectives. Soviet forest policy from the 1950s up through Kyrgyzstan's independence targeted conservation of the remaining forests and increase of forest cover. Today, the concept of multipurpose forest management has been introduced. Attempts to promote sustainable forest management are currently being made. These include ecological, economic, and social aspects (Cornet and Rajapbaev 2004; Yunusova 1999), but they have yet to be made operational and taken up fully by foresters on the ground. Thus, the current forest policy regarding the walnut-fruit forest is still rather conservative, strongly emphasizing conservation of existing forests and re- and afforestation to increase the forested area (Musuraliev 1998).

The local population and the *leshozes* use walnut-fruit forests for fuelwood and a range of non-timber forest products (NTFPs), most prominently walnuts. Timber production plays only a very limited role; harvesting is legally restricted to sanitary felling, thinning, or other silvicultural interventions undertaken by the *leshozes*. Much of the walnut-fruit forests is used for subsistence agriculture: as pasture for livestock, as haymaking plots, or for farming in openings and in open stands. The predominant use of much of these forests is therefore agroforestry in the form of an extensive sylvopastoral system. The forest is part of a wider cultural landscape, a patchy mosaic of natural forests—often over-aged stands with insufficient regeneration—plantations, orchards, small farming plots, pastures, and drier open areas.

External Diversity—Medium

All four Kyrgyz sites share a number of common external stakeholders (Schmidt 2003e): local and neighboring communities; the *Aiyl Okmot* (municipality); Regional Forest Administration; State Control for Bioresources and Wildlife; Ministry of Emergency and Ecology; Institute for Forest and Walnut Research of the Kyrgyz Academy of Science; walnut buyers who come to the villages during the harvest season; and companies buying NTFPs (walnuts, but also medicinal herbs, hawthorn fruits, rose hips, mushrooms, wild apples, and jam). In addition, each *leshoz* has some stakeholders unique to itself.

Arstanbap-Ata *leshoz* has the following additional stakeholders: Helvetas, a Swiss NGO supporting community-based tourism initiatives; a United Nations Development Programme (UNDP) conflict prevention project; the Biosphere Institute of the Kyrgyz Academy of Science; fruit processing factories (processing wild apples gathered in the forest); and a Turkish timber exporting and processing company. Achy shares some of these same stakeholders: the UNDP project, fruit processing factories, and the Turkish company.

In Ortok, a central stakeholder is the Kyrgyz-Swiss Forestry Support Programme KIRFOR (with whom Schmidt has collaborated). In particular, the KIRFOR project supporting the implementation of Collaborative Forest Management (CFM) plays a role. Helvetas and timber processing units also work in Ortok. In Uzgen, KIRFOR is similarly important, as is a UNDP microcredit program, a DFID-funded livestock project (pilot phase 2001), and timber processing units.

Age

In Kyrgyzstan, the young depend financially on the older generation. Only as they grow older do they obtain the resources—animals, land, labor—they need to prosper. In their youth and young adulthood, they remain part of their parents' households, subject to their parents' decisions. Parents are expected to provide a home for each married son and find him employment. Opportunities for the young to make money thus depend on peripheral forest resources, with other economic activities in the area limited. A fair number of young Kyrgyz and Uzbek men and women migrate to cities or other countries, mainly Russia and to a lesser extent Kazakhstan, to find work (Marti 2000; Schmidt 2002).

Internal Diversity—Medium to Low

Arstanbap-Ata is 77 percent Uzbek, 22 percent Kyrgyz, and less than 1 percent Russians, Tatars, Tajiks, and Chechens. Achy is 55 percent Kyrgyz, 45 percent Uzbek, and less than 1 percent other ethnic groups (Schmidt 2003e). Ortok and Uzgen are nearly 100 percent Kyrgyz (Schmidt 2002).

Level of Conflict—Medium

Tension characterizes all the Kyrgyzstan sites during the time immediately preceding the walnut harvest. Local people and outsiders—some invited as helpers, some spontaneous arrivals—jostle to secure their share of the harvest. Conflicts also blossom between local people and *leshoz* officials, who are accused of distributing collection rights for the same plot twice. Fistfights are quite common, and a forest ranger and a forester were killed in the late 1990s, both apparently following disputes regarding walnut. The atmosphere is explosive at harvest time (Schmidt 2003e).

Achy. Achy is the most densely populated of the Kyrgyz sites and generally perceived to suffer from too few walnut trees for too many people. Given the land/person ratio, it would be impossible for each family to have access to a sufficient number of walnut trees (in the form of a lease), to make income from walnut harvest worth the effort. Additionally, villagers have a litany of complaints against the *leshoz* authorities, who are said to be corrupt, to allocate leases in an arbitrary and inequitable manner, to make decisions based on their own personal interests, and to have abdicated their responsibility to protect the environment. The authorities are also accused of clearing forest plots to plant potatoes or to sell fuelwood, accepting bribes for leases, and failing to treat forest diseases (Marti 2000, 55). Until 2002, access to walnuts at the time of harvest was very uneven, with some leaseholders having huge groves of up to 25 ha and others having legal access to only ten walnut trees—or even no trees at all. All guard their plots assiduously. Nontenants complained that they had no access to walnuts, vital to their livelihood; some were reduced to stealing walnuts and firewood (Messerli 2000, 23–30). The redistribution of harvest rights in autumn 2002 in some villages reduced such tensions and made the distribution more equal. Messerli mentions other important conflict issues: theft, corruption, illegal cutting of trees, and reimbursement in kind rather than cash from the *leshoz*. Firewood prices that vary without apparent reason and that must be paid in cash to the *leshoz* are other sources of community antagonism.

Arstanbap-Ata. The valley of Arstanbap-Ata is also very densely populated, resulting in considerable pressure on natural resources and forests in particular (Schmidt 2003e). Local people have a widespread mistrust of the *leshoz*; corruption is alleged. *Leshoz* staff are said to be more interested in earning money than in forest conservation, and some people question staff competence. Local informants report a quantitative and qualitative decline of forest resources over the last ten years. The *leshoz* controls most land in the valley, including pastures and agricultural land, while the municipality (*Aiyl Okmot*) has virtually no access to land—an uneven distribution of land resulting in some tension between these two key institutions. The head of the municipality seems in favor of models giving local residents an improved say in land use, whereas the *leshoz* tries to keep control over the use of its land.

Local communities not only fear that equal access to forest resources will not be granted, they experience conflicts in forest and land use among themselves. In some instances people taking care of and planting on forest plots were forced by their neighbors to open up the fenced forest plots, as the neighbors feared that there might not be enough land for their cattle to browse. Such land-use conflicts can partly be explained by general changes in land use since the breakup of the Soviet Union: people tend to keep more cattle at home, which then browse in the vicinity of the villages. Also, the population is rapidly growing, at 3.3 percent per year for the period 1989–1999, as calculated on data given in Abdymomunov (2001). On the other hand, examples exist where several households collaborate and take different forest products from the same forest plots, a system which seems to work well in many cases.

Uzgen. The walnut forests, often at some distance from the settlements, are less central to people's livelihoods in Uzgen than in Achy, for instance. Nevertheless, *leshoz* corruption remains a problem. Uzgen residents accuse the *leshoz* authorities of illegally selling the valuable walnut burls. Though less significant than in Achy, theft of walnuts is also a problem in Uzgen (Messerli 2000, 32).

The first CFM lease contracts made in 1998 were unrealistic about the amount of work required of the leaseholder. The obligations included preventing fires, detecting pest outbreaks, excluding livestock, establishing nurseries, collecting various kinds of seeds, providing propagation material for poplars, planting walnut seedlings in degraded areas, and more. When the CFM tenants could not fulfill all these obligations, they had to pay fines to the *leshoz*, angering the leaseholders, who felt the agreements had been unfair (Messerli 2000, 16). The unfairness of the leases was also recognized by some of the authorities (Fisher 1999; Carter et al. 2003). Additionally, some new leases have overlapped pre-existing land uses. Such conflicts seem to have been resolved reasonably amicably, but improved lease allocation procedures could avoid them (Fisher 1999).

One Rapid Rural Appraisal (RRA) in the village of Ak Terek also found significant disagreements between the populace and the administrative center (*Aiyl Okmot*). People accused the *Aiyl Okmot,* the leader of which keeps changing, of blackmailing people in need of various kinds of documents. One conflict involved four or five families who were not given land during privatization. Another revolves around the post office in Salam Alik where pensions and child benefits are supposed to be paid; the post office staff is accused of keeping the funds and using them for their own purposes (DFID 2001b, 10).

Level of Conflict—Low

In the fourth ACM site, Ortok, residents expressed only a general distaste for the corruption that is said to characterize *leshoz* authorities. Illegal sale of the valuable walnut burls was alleged.

Ortok. Messerli (2000) found only a few complaints about inequity in the distribution of leases in Ortok. Conflicts revolving around the walnut harvest do exist among community members. School is closed during the walnut harvest for a couple of weeks in September, and people even climb walnut trees to harvest them green, in an attempt to forestall theft. Messerli (2000, 33) notes that

> [c]hildren can be seen up in the trees shouting to one another to guard the territory and women, sleeping up in the tents, are fighting other people off in an aggressive and sometimes even violent fashion. Thus the principle of whoever is the quickest can collect most walnuts, applies with or without *arenda* [lease] plots in the whole forest area of Ortok. Since misunderstandings about the sizes and demarcations of the *arenda* plots … have also existed up to now, disputes between tenants and nontenants could not be excluded over territory.

Social Capital—High

Kyrgyzstan has relatively strong traditional institutions with high levels of distrust of unconnected households and of the state. The traditional organization of Kyrgyz, and to a lesser extent Uzbek, society in clan groups—along with the experience of forced Soviet collectivization—are key to understanding levels of social capital. As so often happens with this issue, increasing knowledge of the sites reveals different understandings of local conditions. The Kyrgyz sites were initially assessed as high on social capital for Ortok and Uzgen, medium for Arstanbap-Ata, and low for Achy. More recent assessments are reflected here.

Although internal social capital among community members appears to be fairly high on all four sites, this is more true in the more homogeneous communities of Uzgen and Ortok. (However, see Messerli 2000 on the lack of interest in working collaboratively on leases, as well as walnut theft problems.) High levels of participation in community ceremonies and mutual aid, as well as trust among family, neighbors, and friends are present. Social networks are seen as increasingly important on all sites in Kyrgyzstan because of the disintegration of formal institutions that functioned better during Soviet times (see Box 10-7. Fisher (1999) notes with approval the signing of two group leases in Uzgen, each involving four families. A group in Ak Terek Range formed on the basis of friendship rather than kinship, simply liking to work together and support each other.

Comparable levels of social capital linking local communities with the state are lacking. The women of Uzgen have felt exploited by *leshoz* managers, who see community forest management activities as a means of mobilizing free labor for tree planting and forest protection (see also Fisher 2003b). Tenants agreed to produce specified amounts of apple seeds and other NTFPs, for instance, without having had the experience to estimate the reasonableness of these contractual demands. Some had to pay a small fee in compensation for their inability to fulfill their contractual obligations. These fees, the inappropriate workloads, and the lack of walnut harvests for the late 1990s (combined with the more general complaints listed in Boxes 10-7 and 10-8) resulted in considerable dissatisfaction with the leasing arrangements and the state (Messerli 2000).

Although in Ortok, Uzgen, and Achy, a general dislike for working in groups has been noted (Messerli 2000), this is also a widespread attitude in Arstanbap-Ata. Fisher (1999, 12–13) has identified several Kyrgyz institutions of interest from the point of view of social capital. These are

1. the Ak Sakals—literally "white beards"—the village elders who have had a traditional role in leadership, decisionmaking, and conflict resolution;
2. Ashar, a form of voluntary group labor within extended families or a group of friends sharing a common interest; and
3. Sherine, a rotating system of responsibility for hospitality.

The governmental efforts to encourage group leasing have been more successful in Ortok than in the other study sites. In Baka-Chardak (in Ortok), a large area of contiguous leases has been arranged, allowing for labor- and cost-sharing. Tenants have fenced the entire area, rather than individual plots, and protection tasks are being shared (Fisher 1999, 12). The fact that Fisher notes the *potential* of such cooperation to contribute to improved local institutional and organizational capacity within the local populace acknowledges a need for improvement.

Messerli (2000, 28) also finds that the links between the communities and *leshoz* are better in Ortok than in Uzgen or Achy. One factor may be that Ortok tenants have written contracts with more reasonable requirements.

Social Capital—Medium

In a gender study in four *leshoz*es, including Achy, Uzgen, and Ortok, Messerli (2000) found that almost all the women questioned had difficulty imagining working in a collective way with other families on their leased plots. They preferred working individually or in their family groups, all the while complaining about the need to organize transportation, sale, and so on by themselves. They see others as "uncooperative" or "unreliable," and they had little trust and much fear of being exploited by others. A similar picture emerged for Arstanbap-Ata from interviews conducted by the ACM team in summer 2003 with a representative of the local women's council and other stakeholders. Inequity in the distribution of lease plots, with some community members receiving much bigger, more productive walnut forest plots than others, is another source of intracommunity friction, as is theft of walnuts and firewood (Messerli 2000).

Marti (2000) reports similar distrust in the community of Oogon Talaa, in Achy, toward the *leshoz* staff. Foresters are accused of allowing unregulated activities in the forest, being careless with forest resources, and inappropriately profiting from the sale of firewood, and other resource appropriation. The people see the state as having reneged on its responsibilities for forest conservation in general. This is confirmed by Messerli (2000, 28) who reports rumors of illegal sale of burls by *leshoz* personnel and misunderstandings about the allocation of lease plots. Such accusations are quite common on all sites in Kyrgyzstan, but were voiced especially frequently in Arstanbap-Ata and Achy by local residents in general, as well as by representatives of different stakeholder groups in the course of the research process.

CASE 9

Madagascar

Contributing Author
Louise Buck

Sites in Madagascar
Ranomafana National Park

Devolution Status—Intermediate

In 1996, the National Law on Community Based Management of Natural Resources was passed, as part of a national devolution policy that began in 1993. The National Association for the Management of Protected Areas (ANGAP), a multi-state organization responsible for national parks in the country, was formed as part of the national development strategy facilitated by the World Bank. With the support of donors, ANGAP was able to garner the services of the "best and brightest" in the Forest Department—people with vision, legitimacy, and authority. Another important set of actors included the National Center for the Environment (later, the Ministry of Environment), which took responsibility for facilitating the "GELOSE process" (*Gestion Locale et Securisé des Ressources Renouvelables*, Local Management and Security [of tenure] of Renewable Resources). The GELOSE process focused on bringing about greater tenure security to local communities (see Cowles et al. 2001; or Carter 2000).

CIFOR, with help from Cornell University, was involved in developing management agreements among stakeholders for more pluralistic management of Ranomafana National Park (Buck 2002; CIIFAD 1997). Partnerships were initiated that would form "models" for developing the Implementation Decrees for the new law. A simplified ACM process was carried out in three communities, with the intent of developing partnership agreements between the communities and the government (see also Ford and McConnell 2001, who initiated a similar process in another park). Activity centered around a particular economic activity, such as ecotourism or extraction of reeds. Traditional management rules (*dina*) could be legally acknowledged if written down. One focus of this research was trying to integrate these rules into the partnership agreements.

Forest Type—Humid Tropical

Ranomafana National Park (RNP) covers 41,600 ha and lies in Fianarantsoa Province of southeastern Madagascar, some 90 km west of the Indian Ocean on the east-facing escarpment of Madagascar's central high plateau (IDPAS 2003). The nearly completely forested area has an exceptionally high level of biological diversity, which is considered by many to be under threat from human activity. During the 1970s and early 1980s, when logging was common, *Dalbergia baroni*, a valuable hardwood, was a principal product, with 11 active concessions (Peters 1994). The forest's preservation received the highest priority in Madagascar's Environmental

Madagascar Key ACM Characteristics	
Devolution status	Intermediate
Forest type	Humid tropical
Population pressure	High
Management goals	Large-scale conservation
External diversity	High
Internal diversity	Low
National conflict	High
Social capital	Medium

Action Plan. RNP is surrounded by a mosaic of paddy fields, upland agriculture, regenerating forest fallows, and some intact rainforest (Ferraro 2002, 263). Its steep slopes are considered to have been important in protecting it somewhat from logging (Ryan 2003). Elevation in the park ranges between 400 and 1,400 m. Annual rainfall ranges from 2,300 to 3,000 m, and average temperature is 21°C (Styger et al. 1999). Soils are among the poorest in the world (Peters 1994). Lowland rainforest, cloud forest, and high plateau chaparral can be found within the park (IDPAS 2003).

While logging is no longer important, the major threat to the integrity of the park is shifting cultivation. Shifting cultivation inside the park was banned officially when the park was created, and later it was banned in parts of the park periphery. Enforcement is difficult and uneven, but it has nonetheless stimulated local people to search for alternative economic strategies. Unsustainable extraction of certain forest products such as ferns, ebony, reeds, and crawfish is the second most disruptive type of activity.

Fernandes (1997) identified separate development strategies, designed to improve local livelihoods while protecting the park. These focused on

- home gardens, where a multistrata agroforestry system was envisioned (including fruit trees, living fences, fuelwood trees, and leguminous hedgerows around a compost heap);
- lowlands, where flood-tolerant legumes might be added to paddy rice areas;
- middle slopes, where leguminous cover crops and hedgerows were envisioned to enhance upland rice and cassava production; and
- forest margins (upper slopes), which are suitable for a multistrata banana–coffee farming system that would curtail the growth of the weed *Lantana camara* and would also protect the park from accidental fire.

The most intensive action research was undertaken in Sahavoemba (a Tanala village), Bevoahazo (a Beteseleo village), and Anjamba (a mixed village about six km from RNP in the buffer zone). Additional work was conducted in four more villages (Andafiatsimo, Andranomakoko, Riambondrona, and Vohipeno; Rakotondrabe 1999; Rakotoson 1999; Buck 2003).

Population Pressure—High

For Madagascar as a whole, the population growth rate is 3 percent, with an average density of 26 people/km^2 (Pollini 2000). The population data available on Ranomafana pertain to the whole park. At the time of its establishment there were 26,000 people living in over 100 villages within a 5-km radius of the park's boundaries; about half of the villages are small, with 150 people or less (Ferraro 2002, 263).

Freudenberger's (1999) data from the "forest corridor" of which Ranomafana is a part suggest high rates of population growth (families with typically 8–10 children per family, marriage for girls at age 12–13, half the population 18 or under). These data were confirmed by Ferraro (2002, 268). A high growth rate is compounded by dramatic rates of in-migration. The Tanala village of Sahavoemba, one of our study villages, has 184 inhabitants in 34 households (Rakotoson et al. 2000).

Management Goals—Large-Scale Conservation

RNP—established in May 1991, its boundaries redefined in 1996—was created partially in response to suggestions made by the American lemur researcher Patricia Wright within a political and economic context that supported park formation (see Cowles et al. 2001) but with virtually no attention to existing uses of the area. Indeed, some communities were removed (Buck 2002). Despite large expenditures of money to improve conditions for local people, the results have been disappointing.

External Diversity—High

External stakeholders include RNP managers; agents for conservation and development; ANGAP; the Forest Department; the National Center for the Environment; Antananarivo University; individuals involved in the various processes such as GELOSE and AGERAS (*Appui à la Gestion Environnementale Régionalisée et à l'Approche Spatial* or Aid in Regional Environmental Management and Spatial Approaches); the mayor (head of the *commune*); NGOs (Tefy Saina and the National Association for Environmental Action, ANAE); and various donors and project administrators (USAID, Chemonics, Cornell University's CIIFAD). See Buck (1998, 2002) and Rakotoson et al. (1999). Locally, villagers of Anjamba created an association called MANIRISOA to act on their behalf in the process of developing "forest use agreements" with the government (Rakotoson 1999).

Age

Relations between the young and the old in Madagascar, at least among the Tanala, seem to be marked by considerable antipathy. Freudenberger (1999) recounts a "revolution" of the young against the old in 1990 in a number of communities in the forest corridor of which Ranomafana is a part. The youths' primary complaint was the harshness of traditional justice, including the possibility of complete social ostracism—known as becoming a "black banana"—and higher local fines than those imposed by the state. Rakotoson et al. (1999) describe the unwillingness of elders to listen to youth, even when the young dare to speak up in public meetings in Sahavoemba.

Internal Diversity—Low

Ranomafana sites include the local Tanala in Sahavoemba, the in-migrating Bete-seleo who practice lowland rice cultivation in Bevoahazo, and both groups in

Anjamba. The Tanala (meaning forest people) are the most forest-dependent of the ethnic groups. Conflict levels are high. The Beteseleo are considered hardworking, prolific, and prone to migration and wage labor. The Tanala have a much stronger traditional forest management base on which to build in joint management efforts (Buck 2003). Peters (1997), talking about Ranomafana National Park overall, documents considerable gender and ethnic inequity in participation in formal project activities, with men and Beteseleo dominance in hiring and training. Rakotoson et al. (1999, 12) briefly mention the importance of involving the "kings, nobles and villagers" in the development of an association in Sahavoemba.

Level of Conflict—High

Madagascar has been in turmoil at least since December 2001, when the presidential election results were disputed by Marc Ravalomanana, who argued that President Didier Ratsiraka and his Malagasy government rigged the election. In February of 2002, demonstrations in the capital, Antananarivo (of which Ravalomanana was mayor), led to violence. Subsequently further demonstrations were outlawed under new emergency powers (Quist-Arcton 2002). In May 2002, a court-ordered recount found Ranalomanana to have won, but human rights abuses and armed conflict continued between the two camps (Amnesty International 2002). This level of turmoil, however, began after the research reported here had drawn to a close.

Although intravillage conflicts appear to be minimal in the Ranomafana study villages and are related primarily to age, the long history of fear toward outsiders suggests a more significant level of midlevel conflict. Local dependence on shifting cultivation and its close connection with ethnic identity and freedom (see Peters 1994), coupled with the antagonism of outsiders to this practice, inevitably involves conflict. Peters recounts a variety of conflicts between local communities and park management, noting that greater trust and improved communication are essential if better management can be hoped for. Rakotoson et al. (1999) report that many promises were made and not fulfilled by external agents, resulting in a lack of trust and an abundance of worrying rumors in ACM villages. Sahavoemba villagers feared forest guards (*Lehibeniala*), lacked an understanding of the roles and responsibilities of various outside agents (such as CANFORET and park conservation agents), displayed fear-based obedience to administrative authorities, and expressed jealousies of neighboring communities. All of these observations suggest that conflict is latent at the very least.

Conflict in Madagascar, though not violent, persists in the region, particularly between the Beteseleo and Tanala peoples. The potential for violence may be mitigated due to relationships of mutual economic dependency between the groups, particularly regarding methods of upland and lowland rice cultivation. Rice is extremely significant materially and symbolically (Buck 2003).

Social Capital—Medium

Ferraro (2002, 270–271) found the residents of RNP to be very concerned with "… the fostering of *firaisina* (union), *fihavanana* (family), and *firaisankina* (solidarity). Community harmony is extremely important too because it fosters mutual assistance and security." Ferraro points out that breakdowns in social relations can lead to economic losses if household members become unable to work. In semistructured interviews, residents explained that disruptions to traditional patterns of authority, reciprocity, and social bonds were already beginning to occur even before the establishment of RNP. The specific examples mentioned were disregard of taboos on commercial sale of aquatic species and frequent borrowing from richer neighbors that disrupted community harmony.

Rakotoson et al. (1999), writing specifically about the Tanala community of Sahavoemba (where ACM activities took place), note the necessity for consensus—from the king to the royalty to the counselors—in community decisionmaking. Unwillingness to make independent decisions was observed (cf. Hanson 1995). Moreover, Sahavoemba was frequently chosen for development and conservation efforts, due to various factors, including the "honesty of its population and its tradition" (Rakotoson et al. 1999, 15). Rakotoson et al. (2000, 39) later note that in Sahavoemba "... the traditional structure is intact, which favors the transmission of messages via the traditional chiefs (*mpanjaka*) and notables (*Raiamandreny*)." See also Hanson's 1995 discussion of the dual functions of *fokonolona* meetings for transmission of state messages and managing internal issues.

CASE 10

Malawi

Contributing Author
Judith Kamoto

Sites in Malawi
Chimaliro and Ntonya Hill

Devolution Status—Intermediate

In Malawi, several laws were promulgated in the late 1990s that effectively decentralized much forestry-related decisionmaking to the district level: The Decentralization Policy (1998) and the Local Government Act (1998) were adopted after many relevant sectoral policies had already undergone a reform process. A significant amount of uncertainty about who has which responsibilities and rights remains (Ngulube 2000). However, compared to many southern African nations, Malawi's devolution process has been effective, in, for instance, devolving rights to Village Natural Resource Management Committees (VNRMC) to formulate their own bylaws (Shackleton et al. 2001, 7). Partnerships with local communities are theoretically part of the new modus operandi, but their implementation has so far been marred by uncertainty about rules, slow-moving bureaucracy, lack of representation of intracommunity variation, and near absence of skills relating to participatory processes within the Forestry Department (Ngulube 2000).

In line with government interest in comanagement of forests, the Forestry Research Institute of Malawi (FRIM) initiated a research project in Chimaliro Reserve (an ACM site). The first phase (1992–1994) explored appropriate management plans and silvicultural systems for the woodlands. The second phase (1996–1999) involved the communities in the management of the reserve. On a pilot basis, in 1997, the Forest Department (via FRIM) divided part of the Chimaliro Reserve into blocks for comanagement with communities from the surrounding communal lands (Village Forest Areas). Three blocks (I, II, and III) were allocated to three "communities," representing 11 villages. Communities were allowed to collect nontimber forest products (NTFPs) and were given modern beehives (Kamoto 2002g). Ethnic groups include the Tumbuka and Chewa tribes, with most householders being smallholder commercial and subsistence farmers (Shackleton et al. 2001, 14).

The Village Natural Resource Management Committees have been comparatively successful here, reportedly because of the deep respect with which villagers view their traditional leader. This leader has supervisory control over Village Forest Areas, as well as the Forest Reserve comanagement blocks, and has the final say about management rules (Shackleton et al. 2001, 35).

According to Shackleton et al. (2001, 56),

> ... in Chimaliro the state takes a proportion of the benefits. Under the comanagement arrangements for Chimaliro Forest Reserve, the Forestry Department and the VNRMC agreed that 70% of the revenue from forest products collection would go to government and 30% to the community. On customary land this is reversed, with 80% and 20% of the revenue accruing to the community and government, respectively. The VNRMC, with approval of the village head, decides how the income should be used, based on development needs in the community. Only when these needs are satisfied can money can be paid out as household dividends.

Kamoto (2003a) sees the current decentralization policy as advocating local governance, which will in turn enhance interaction and involvement in the decisionmaking process by all stakeholders at the district level. She anticipates that all development activities in the district will be discussed in the district assembly, where NGOs and government sectors will sit together. Line ministries, NGOs, and donors are intended to sit together at meetings and workshops to plan their interventions in the district. The degree to which this has actually occurred to date is debatable.

The second ACM site, Ntonya Hill Plantation in Zomba District in southern Malawi, is a periurban plantation site consisting of 11 villages under the control of a management committee, in an area previously receiving technical assistance from the Blantyre City Fuelwood Project. The fact that the traditional authority (chief) disbanded the management committee—because it "was not representative and some of its members were flouting rules governing the management of the plantation" (Milner 2002c)—gives some indication of the unusual level of real devolution in that country. The ACM team has worked primarily with two communities: Katunga and Kalimbuka.

Forest Type—Dry Tropical

Chimaliro Forest Reserve is located in Kasungu district: 330°30′E, 120°30′S, midway between Lilongwe and Mzuzu. It is a typical semideciduous, wet plateau miombo woodland. It covers 160 km² of hills. Annual rainfall ranges between 800 and 1,200 mm/year. The mean annual maximum and minimum temperatures range from 22°C to 24°C and 12°C to 14°C, respectively. The forest reserve lies on precambrian metamorphic rock characterized by gneiss. The ferruginous soils at the base of the range are generally of poor

Malawi Key ACM Characteristics	
Devolution status	Intermediate
Forest Type	Dry tropical
Population pressure	High
Management goals	Multipurpose reserve
External diversity	Medium: Ntonya Hill Low: Chimaliro
Internal diversity	Low: Ntonya Hill Low: Chimaliro
National conflict	Low
Local conflict	Medium
Social capital	High: Ntonya Hill Medium: Chimaliro

sand clay type, with pockets of better alluvial soils. The steeper upper soils are characterized by shallow strong lithosols. The reserve was initially gazetted for soil and water conservation purposes.

Forest fires have probably had the biggest impact on these miombo woodlands. Although fires do not cause high mortality among most species of trees and grasses, they have large effects on size and structure of woody plants (replacing small numbers of large stems with large numbers of small stems). They also alter species composition of trees and of the grass and herb layer. The loss of animal life is probably large, but has not been evaluated.

The most dominant canopy species in Chimaliro were *Julbernadia paniculata, Brachystegia floribunda, Brachystegia utilis,* and *Brachystegia boehmii.* The highest stocking density and basal area per hectare found in Chimaliro was 932 stems and 16.33 m³ per hectare, respectively. Sixty-one species were recorded in Chimaliro. Surveys revealed a variety NTFPs: mushrooms, fruits, insects, medicines, tubers, honey, thatch grass reeds, and poles (summarized from Kamoto 2002e). Mushrooms, of which there are 23 edible varieties, constitute an important NTFP (Ngulube et al. 2001, 34).

The Malawian government's allocation of the three blocks of forest to the three communities was based on proximity and historical attachment to the area of woodland to be managed by the community. Blocks I, II, and III were 18, 74, and 118 ha, respectively (Kamoto 2002a).

Ntonya Hill is a forest plantation of 370 ha of *Eucalyptus camadulensis,* located in Zomba District in southern Malawi at 15°27′S and 35°19′E. This periurban area was originally miombo woodland, at an altitude of 1,075 meters above sea level. Soils are skeletal lithosols with many rocky outcrops. Mean annual rainfall is 780 mm, most of which falls between November and April. Mean temperatures range from 16°C to 28°C.

Prior to the establishment of the plantation (from which local communities were excluded until recently), 11 communities used the woodlands for fuelwood, fruits, medicines, and for livestock grazing and browsing. In 2000, when the ACM work began, management of the plantation had just been turned over to the local communities, after a period of serious conflict and encroachment (Milner 2002b). Ntonya Hill can be divided into three ecological categories: areas where the eucalypts continue to dominate; coppiced areas that are mixtures of eucalypts and indigenous species; and seriously degraded areas. The northern part of the area is in worse condition than the southern and eastern parts. Sixty-seven species were recorded in the context study, with the greatest diversity occurring in the coppiced area. Milner (2002a, 7) reports

> The basal area for all species was greatest in the stand of eucalypts with 7.8 m³/ha. Dominant height of the eucalypt trees in this category was 16.4 m. Basal area for non-eucalypts was greatest for the coppiced areas. The coppiced areas had a significantly greater density of trees and shrubs other than eucalypts in the diameter class 1, >5 cm.

Fires are a problem, sometimes started deliberately to enhance grazing (Milner 2002a).

Population Pressure—High

Shackleton et al. (2001) report the population density in Chimaliro as 74 persons/km². Kamoto (2002b) describes three communities of 11 villages, each attached to one of the three forest management blocks, with 688 households overall. Female-headed households average 34 percent, ranging from 16 percent in Block I to 56 percent in Block III. Overall, there are 457 male-headed households and 231 female-headed households. Average household size varies from five in Blocks I and II to seven in Block III (Kamoto 2002b).

Kamoto (2003d) quotes 1998 government census figures for the Kaluluma Extension Planning Area of the Ministry of Agriculture, in which Chimaliro is found. For the total population of 29,660, the sex ratio is 0.97 (more women than men). Forty-five percent of the population is under the age of 15. Intercensal population increase was reported as 37.3 percent.

The ACM team has been working in 2 of the 11 villages (Katunga and Kalimbuka) adjacent to the Ntonya Hill plantation area (Milner 2002c). No information is available on the population of Ntonya Hill.

Management Goals—Multipurpose Reserve

Government protection of Chimaliro Forest Reserve began in 1924, and the area was gazetted in 1926. At the time, people were moved out of the reserve and settled on the surrounding plains, with the exception of one village, which became a caretaker with rights to cultivate in specific areas in the reserve until 1945. There is a tradition of headmen having some responsibility for protecting the reserve, and this worked well in the past when there were sufficient off-reserve trees. By 1952, encroachment had become a problem, and government employees were assigned to manage the reserve. In 1990 approximately 60 ha of the reserve adjacent to what would become a forest compound for forestry staff in 1992, was cleared and planted with *Eucalyptus* (Kamoto 2003c, 3).

FRIM's research project, focused on the management of miombo woodlands by local people, has aimed for comanagement with the forestry department of three forest blocks, with three block natural resource management committees. Kamoto (2003c, 2) says,

> As part of the agreement, the communities harvest products from the reserve in the blocks they manage. These include dead wood for firewood, thatch grass, mushrooms, reeds, bamboo, caterpillars, cattle grazing and wild fruits, but harvesting of live trees for firewood or construction purposes is prohibited.

Although considerable ACM activity has occurred on the site, the involvement of government personnel has not always been helpful. Local forestry officials have resisted greater community involvement in forest management and also exhibit a lack of skills in dealing with communities.

The Ntonya Hill Plantation is connected with the Blantyre City Fuelwood Project (BCFP), which was launched in 1986 and has since been phased out. Its primary objective was growing and supplying fuelwood requirements to the urban dwellers in Blantyre and Zomba District. The Department of Forestry was the implementing agency for this Southern African Development Community (SADC) project funded by the Norwegian Agency for International Development. The overall project comprises 5,000 ha of *Eucalyptus*, of which 4,700 is considered productive. There are also 1,200 ha of indigenous woodlands (Ngulube 2000).

Phase 1 of the project focused on plantation establishment and management; Phase 2 concentrated on developing cost-effective ways to harvest, transport, distribute, and market the fuelwood. Phase 3, which began in 1997, involved handing over the forests to local communities for sustainable management, to contribute to improving their standard of living.

The plan involved management by traditional authorities and Village and Area Natural Resource Management Committees (VNRMC and ANRMC, respectively). Although Ngulube (2000) describes the training received by communities (in leadership skills and forest management) and their enthusiasm for the initiative, Milner (2002c) stresses the problems. Milner, the ACM facilitator, worked in 2 of the 11 villages of the Ntonya Hill Plantation (Katunga and Kalimbuka). As mentioned previously, the traditional authority or chief had disbanded the ANRMC on the basis that it was nonrepresentative. No new committee had been elected. Ngulube found such problems to be common throughout the project area. He noted strained relationships between the VNRMCs and local leaders, plus power struggles between VNRMCs and ANRMCs pertaining to training opportunities and control of funds.

Ngulube also notes that the process of allocating lands to communities had not followed traditional boundaries, but rather proximity of the communities to the plantations. He suspects that this approach may complicate the issue of land tenure, with the original owners wanting to establish gardens. Infrastructure, training, and technical support are donor dependent, and communities express an unwillingness to share revenue from forest products with enabling agencies. Nor are plans for revenue distribution within the communities from the project at all clear (Ngulube 2000). The lack of ACM progress may be related to these problems, though the lack of documentation makes this difficult to assess.

External Diversity—Medium

In Ntonya Hill, Milner (2002b, 3) concluded, based on a stakeholder analysis, that the following external stakeholders were important. The Department of Forestry, through the Blantyre City Fuelwood Project, facilitates sustainable management and utilization of the forestry resources. Other stakeholders include (a) government ministries available at the local level (ministries of Agriculture, Education, Health, Community Development, and Social Welfare); (b) government departments available at the district level such as the District Commissioner, the Police, and courts for settling some disputes; (c) NGOs such as the Evangelical Lutheran Development Program, promoting agroforestry in the area; (d) buyers of forestry resources such as poles and firewood, including brick burners, tobacco farmers, and others; and (e) research organizations such as FRIM and university research institutions.

External Diversity—Low

In Chimaliro, Kamoto (2002b) listed the following groups as most important: local communities; FRIM; Kasungu District Office; Chimaliro Forestry Compound staff; and PLAN International. Others included the primary schools, members of parliament, the Department of Environmental Affairs, and the councilor (a politician).

Internal Diversity—Low

In Chimaliro, most community members descend from the Ngoni group and follow their traditions, but having intermarried with the indigenous Tumbuka ethnic group, people now speak the Tumbuka language. Chewas from south of Chimaliro, employed as cattle herders for the Tumbuka, come and go and do not spend a great deal of time in the area (Kamoto 2002b; 2002f). In studies produced by FRIM, status, wealth, farm size, and crop income were used as differentiating variables. These were negatively related to the intensity of men's and women's collection of foods from the forest (Ngulube et al. 2001, 29). The Ntonya Hill site has two ethnic groups, Yao and Lomwe, with the former being numerically dominant.

Level of Conflict—Medium

Chimaliro's forests are comanaged by the communities and the state. Village and Block Natural Resource Management Committees (VNRMC and BNRMC, respectively) have been established for each of the three management blocks. In one block, the BNRMC members were appointed and not elected. This problem was further compounded by nepotism. Accusations were made that the committee was managing the forest for its own benefit, not the community's. In another block, alcoholism on the part of the chair led to intracommunity conflict. Also, traditional village heads sometimes show little respect for the committees, though the committees do use traditional conflict management mechanisms. Some decisions were reached about distributing forest benefits, but the communities dispute whether these decisions were made jointly or equitably (Kamoto 2002e).

Although the land was initially allocated to blocks based on population, the communities perceive the arrangement to be inequitable. No maps delineate the boundaries, which remain unclear (Kamoto 2003b, 1). Theft remains a problem, with villagers suspecting collusion between butchers and villagers who conspire to steal other people's cattle. Careless herdsmen are accused of grazing their cattle in other people's maize fields. Scarcity of land leads to unauthorized opening of lands for gardens, resulting in more conflicts. Some conflicts are attributed to jealousy (Kamoto 2002a). Additional conflicts revolve around access to forest products like honey, firewood, and construction poles (Kamoto 2003b).

Cattle grazing has been an issue of contention. All suitable grazing land lies in the much larger Block III, to which all three communities have historically had access. In 1980, paddocks were set up in the present Block III—at the time the most suitable place for grazing in the forest reserve. The paddocks allowed rotation to avoid overgrazing and kept animals off other parts of the reserve. When comanagement became a goal, however, the demarcation of the area into blocks occurred without taking the grazing issue into consideration. Block III residents considered that because NTFP collection and use were theirs, grazing rights belonged to the block as well. Residents of Blocks I and II were also uncertain whether grazing was allowed in their blocks at all.

For a time Block III refused access to the other blocks' cattle. Block I is distant from Block III, so its animals have occasionally had to graze in nearby permanently flooded areas (dambo). When the dambo was impassable after heavy rains during the 2000–2001 growing season, Block I animals had no place to graze, as the forest reserve was off-limits according to the comanagement agreement.

As part of the ACM initiatives, members of the three BNRMCs met to discuss land issues, along with the forest guards. In one meeting, the women committee members from Block III were charged with the responsibility to discuss grazing rights with village headman Ng'onomo of Block III—a remarkable turn of events, as no women had been allowed to discuss traditional issues with local leadership in the past. The BNRMC men's awareness, brought about by participating in ACM processes, has allowed women to be seen as more equal partners who must be involved in formal decisionmaking processes. Previously women only contributed informally, behind the scenes, but this particular incident was a true reflection of the changes that have taken place. The men are adapting to these changes, changing tradition and culture.

After many productive meetings, land ownership was clarified, with the land remaining public and grazing rights made more explicit within comanagement arrangements. Ultimately the Block III committee returned grazing access, after the Block II committee agreed to keep its people from stealing honey, bringing axes and pangas into Block III, and setting fires (Kamoto 2002e). All community members who followed the agreed-upon rules were welcome to use the grazing area (Kamoto 2003c).

In Ntonya Hill, due to the degraded condition of the landscape, a decision was made in 1986 to develop a eucalyptus plantation. This area was planted by 1988, under the Blantyre City Fuelwood Project, and villagers surrounding the plantation were prohibited from using forest products (particularly eucalypts, bamboo, and firewood). The demand for these products increased, and conflicts flared between the BCFP and the communities, with thefts and encroachment increasing dramatically. In response to this conflict, the BCFP management finally decided to hand management of the area to the communities, with BCFP staff available as facilitators, with the last area handed over in 1999 (Milner 2002a; 2002b). The conflicts appear to continue to this day, however.

Social Capital—High

The people of Ntonya Hill Plantation were seen to have high levels of social capital, though little information is available on the forms this took.

Social Capital—Medium

In Chimaliro, well-organized community groups were the indicator of high levels of social capital in the team's initial assessments. However, significant problems appear to be interfering with intracommunity cooperation. Kamoto notes the unwillingness of cattle owners to help each other when in difficulty. Previous customs involving the sharing of cattle for feasts and other special occasions are in decline. The severity of the HIV/AIDS epidemic may be a contributing factor. Deaths occur nearly every day, and each illness affects at least two or three other people directly, interfering with efforts to organize for improved forest management, among other things (Kamoto 2002a). Kamoto attributes the communities' inability to negotiate better access and rights to firewood and construction poles, or to resolve overlapping claims on land, to weak community organization (see Chapter 10; Kamoto 2003b).

On the other hand, during 2002, when Malawi suffered a serious drought during the growing season and famine resulted, with some households having no food, no seed, and increased malnutrition, wealthier families found piecework for the destitute in exchange for food or seed. Funerals, now very common, remain important cultural events, bringing 75–100 percent of the community together. All other work and activities come to a standstill while the bereaved family provides food to all guests. If the family does not have enough food, other well-wishers help (Kamoto 2003e).

Block I has demonstrated considerably more social capital than the other two blocks, having strong committee leadership and an active village headman. Committee members work cooperatively and have responded enthusiastically to the available income-generating opportunities; their attendance rate is 75–100 percent for meetings, whether organized by the ACM researcher, frontline forestry personnel, or outsiders (Kamoto 2003b, 5). The other committees have been plagued by drunkenness in one case and nepotism in the other.

Following the participatory forest resource assessment training in November 2002, however, Block II also demonstrated social capital when it instituted a collaborative monitoring arrangement that involved members of the resource management committees from all three blocks. In another case, the European Union initiated a food-for-work program aimed at women, focused on tree nursery establishment. In the past, men had distanced themselves from activities designed for women, such as a PLAN International maize mill project. But in this case they put up fences for the nursery without requesting payment from the women, suggesting a possible increase in cross-gender social capital (Kamoto 2003e).

Some challenges to building social capital probably derive from the changing context in which the communities find themselves—trying to develop a collaborative management regime with a forestry department long considered something of an enemy, under dramatically adverse economic and health conditions. The forestry personnel are not trained in the facilitation and extension skills needed for effective collaborative management, nor are they enthusiastic about the idea. Kamoto (2003b, 11) concludes "… backbiting, fear of the unknown, resentment, and mistrust engulfed the local level actors in their efforts to embrace the comanagement initiative."

Other problems are institutional, involving the links among the forestry personnel, the resource management committees, and the communities. The forestry personnel are not well integrated into the communities socially. Committee members receive benefits, such as regular training, tours, and field trips; they cling to their offices sometimes at the expense of representing their constituency (Kamoto 2003b, 5). Serious jealousy over subsistence allowances between the forest guards and the research collaborators has slowed down ACM progress (Kamoto 2003e). However, the fact that no honey thieves have been apprehended—despite continuing patrols and the lack of honey harvests for the community—may be a sign of strong intra-community social capital, functioning to protect the thieves and arrayed against the externally dominated plans of the forest department.

CASE 11

Nepal

Contributing Authors
Sushma Dangol, Chiranjeewee Khadka, Cynthia McDougall, Raj Kumar Pandey, Kalpana Sharma, Narayan Prasad Sitaula, Netra Tumbahangphe, and Laya Prasad Uprety

Sites in Nepal
Andheribhajana (Sangkhuwasaba District), Bamdibhir (Kaski District–also called Bamdi and Bamdibhirkhoria), Deurali-Baghedanda (Kaski District), and Manakamana (Sangkhuwasaba District)

Devolution Status—High

Nepal is one of the most well-known cases of early devolution of authority (not ownership) to communities (Hobley 1996b). Its history goes back at least 20 years. In multiple publications, the New Era Team has documented the important changes

Nepal Key ACM Characteristics	
Devolution status	High
Forest type	Submontane, subtropical
Population pressure	High
Management goals	Multipurpose reserve
External diversity	Low
Internal diversity	High
National conflict	High
Local conflict	High: Andheribhajana
	Medium: Bamdibhir, Deurali-
	Baghedanda, Manakamana
Social capital	Low

in Nepal's forest-related policies. (New Era is one of CIFOR's partner research organizations.) As early as 1976, the government realized its failure in managing Nepal's forests and adopted the National Forestry Plan of that year. The government embarked on a Community Forestry Development Programme (CFDP) in 1978 to encourage initiatives of local people in the management of the forest resources, supported by various donors (see Gilmour and Fisher 1991).

The stage was set for the development of forest user groups (FUGs), so crucial in current forest management, in the Decentralization Act of 1982 (Gilmour et al. 1992). This act was amplified in the Master Plan for the Forestry Sector in Nepal in 1988, which provided the policy context for community forestry. All accessible forests in the hills were to be handed over to communities. In the Forest Act of 1993, user groups were recognized as independent, autonomous, nongovernment institutions. Forest regulations introduced in 1995 provided guidelines on group formation and recognized users' group rights and responsibilities regarding the management and use of local forest products. Also see Malla (2001) for a discussion of persistent problems deriving from cultural characteristics, under various policy regimes.

ACM field staff have noted inordinate legal power on the part of the District Forestry Officer (DFO) over decisions and actions taken by FUGs (New Era Team 2001d). Examples include the DFO's rights to determine if conditions were met for a handover of a forest or whether a given action was environmentally destructive. The DFO can reject an operational plan or limit opportunities to sell produce. The team also saw the government's decision to retain ownership of forests as detrimental to local management and security; it called for better conflict resolution mechanisms. Further, additional attention should be paid to gender and women's important role in forest use and formal management.

ACM has four sites, two in the center and two in eastern Nepal. Our interest in selecting a site in the *terai*, where the availability of valuable timber would have made Nepal more comparable with our other sites, was discouraged by the Nepali government. The liberal rules that apply to user group formation and research in the hills do not apply where valuable timber remains.

Bamdibhir Community Forest (CF), formally established in 1994, is located in Chanpakot Village Development Committee (VDC) ward six, about 15 km west of Pokhara municipality in the Kaski district of central Nepal. The FUG is within the Pumdibhumdi Range Post of the Kaski District Forest Office, and consists of 128 member households. FUG members live in wards 3, 5, and 6 of the village.

Magars are settled in ward 3, the "untouchables" in ward 5 and Brahmins in ward 6. The group manages 48 ha (Dangol et al. 2001a).

The Deurali-Baghedanda FUG is situated in Hangsapur village on the eastern border of Kaski District (Dangol et al. 2001a). It includes three separate forest areas—Deurali, Baghedanda (both natural forest), and Hangsapur Pahiro (plantation)—handed over to the community in 1993. The FUG is comprised of members from ward four of Hangsapur village. The total number of FUG member households is 140. The total size of the three forest areas is 181 ha.

The New Era Team describes the Manakamana Community Forest User Group (CFUG) of Sankhuwasabha District as including six villages: Chilaune Chautara, Dhunge Bisaune, Deurali, Baidhyagaon, Gidde, and Dumribangi *tole*, or hamlet. The villages occupy nearly half the area of ward 12 of Khandbari municipality. The FUG also includes a few households of ward 13 of the same municipality. The community was given 132 ha of natural forest in May 1993. The forest borders the Sabha River in the east, the Arun River in the west, and the Dasmure Dhunga Bisaune community and Chisapani sandy land in the south. A one-hour flight from Kathmandu to Tumlingtar, followed by a 1.5- to 2-hour walk is required to reach Manakamana CFUG (New Era Team 2000).

Andheribhajana CF in Sankhuwasabha District covers 113 ha and serves ward 9 of Khandbari Municipality and six villages: Barajuthan, Agrakhe, Pulamidada, Dadatole, Gairigaun, and Khanidada. Handed over to the community in 1995, this forest borders the settlement area of Dare Dnada in the east, the Higuwa River and Tamang village in the west, Thale Dnada Khani Dnada in the north, and the Pangtha and Higuwa Rivers in the south. The trip from Kathmandu takes one hour by plane to Tumlingtar and another six hours of walking to reach Andheribhajana (New Era Team 2000).

Forest Type—Submontane

Bamdibhirkhoria CF covers 48 ha, with a relatively even-aged, mature forest, with more than 70 percent crown cover, and more than 60 percent ground cover. It is composed of 20- to 30-year old plantation mixed with natural regeneration (pole size). The upper part of the forest is natural regeneration. Primary species include *Katus* (*Castanopsis indica*), *Mahuwa* (*Engelhardia spicata*) and *Chilaune* (*Schima wallichii*). The lower part was planted to *Uttis* (*Alnus nepalensis*), but more recently the natural regeneration of *Schima wallichii* and *Engelhardia spicata* has begun to dominate.

From a biodiversity perspective, the following are important tree species: *Castonopsis indica; Schima wallichii; Engelhardia spicata; Alnus nepalensis; mallato* (*Macaranga postulata*); *tindu* (*Dyospyrus montana*); *angeri* (*Lyonia ovalifolia*); *kaphal* (*Myrica esculanta*); *semal* (*Bombax ceiba*); *painyu* (*Prunus cerasoides*); *kyamun* (*Syzygium operculata*); *Tooni* (*Toona serrata*); *ankhitare* (*Trichilia connaroides*); and *siris* (*Albizia julibrissin*). Important shrubs include *ghatu* (*Clerodendron viscosum*), *archale* (*Antidesma bunius*), *dhayero* (*Woodfordia fruticosa*) and *ainselu* (*Rubus ellipticus*).

Locally important herbs include *Dattiwan* (*Achyrium bidentata*), *ban tarul* (*Dioscorea alata*), *siru* (*Imperata cylindrica*), *pani amala* (*Nephrolepsis cordifolia*), *amliso* (*Thsanolaena maxima*), *rudhila* (*Pogostemon benghalensis*), *githa* (*Dioscorea bulbifera*), *guhe amala* (*Polys-*

tichum lentum), *seto kanda* (*Rubus rugusus*), *kukurdaino* (*Smilax aspera*), bamboo, and *gurju*. Common wildlife include deer, rabbit, *kalij*, jackal, *kokal*, leopard, monkey, squirrel, porcupine, jungle cat, and birds (parrot, *row, titra*, vulture), snakes, and butterflies (McDougall et al. 2002b).

Fodder, fuelwood, and timber (from dead, dying, or diseased trees) for community consumption are the primary forest products. Some coffee is being grown experimentally (Dangol et al. 2000). The community experienced a devastating landslide some years back that has strengthened its commitment to improving forest management in their community; the event brought in outside help as well.

The Deurali-Baghedanda site includes three community forests covering 181 ha, which are composed of natural, multistory, deciduous, broadleaf forest and plantation (at the pole stage). Regeneration is very good in Baghedanda and of medium quality in Deurali. Crown cover is 70–80 percent, and ground cover 60–70 percent. Deurali forest is dominated by *Castanopsis indica* and *Schima wallichii*. Stem volume averages 28.37 m³, with a tree density of 218/ha. Baghedanda forest is dominated by *sal* (*Shorea robusta*), associated with *Schima wallichii* and *Castanopsis indica*. Stem volume averages 53.06 m³, with a density of 362 stems/ha. The main species in the third area that is planted include *Schima wallichii, salla* (*Pinus roxburghii*), *chanp* (*Michelia champaca*), *Engelhardia spicata, Castanopsis indica, Alnus nepalensis,* and *lankuri* (*Fraxinus floribunda*). Stem volume averages 25.06 m³, with 310 stems/ha.

Important wildlife include porcupine, leopard, leopard cat, wildcat, jackal, bear, monkey, wild boar, jungle cat, rabbit, and squirrel. Birds of importance include the spotted dove, *kaliz*, cuckoo, long tailed bird, eagle, bulbul, sparrow, crow, *dhobini, dangre*, and more. Important nontimber forest products include *nigalo*, a small bamboo used for basket making; *dalchini*, used for flavoring and medicine; *tite aushadhi*, a local herb; and *Myrica esculanta* (Dangol et al. 2000; McDougall et al. 2002b).

The CF in Manakamana comprises 132 ha of natural forest of intermediate quality in terms of health and level of natural regeneration. The primary forest species include *Shorea robusta, Schima castanopsis,* and *Botdhairo* (*Lagerstomia parviflora*). Birds include jungle fowl, woodpecker, owl, wolf, myna, partridge, crane, quail, pigeon, and cuckoo. Other important animal species include monkey, bat, deer, fox, jungle cat, porcupine, rabbit, *Salak* (a wild stork), and wild dog. The forest is mainly used for fuelwood, fodder, grass, grazing animals, and timber for construction and furniture making. Forest products are not yet used for commercial purposes. Fire has been a major threat to the forest, occurring in 1989, 1993, 1995, and 1998 (New Era Team 2000; McDougall et al. 2002b).

The Andheribhajana CF covers 113 ha and includes the following main timber species: *Sal* (50 percent); *chilaune; katus; uttish; jamun; patle;* and *lampate*. The last two are nontimber trees. Important wildlife include jackal, various monkeys, rabbit, deer, porcupine, and birds (long-tailed bird, dove, wild chicken, and others). The forest is used primarily for subsistence purposes. People harvest firewood, grass, leaves, and medicinal plants. They graze animals and use timber for building construction and making agricultural tools, such as plows, poles, and pillars. None of the forest products are yet used for commercial purposes. Between 1958 and 1963, a large number of migrants came to the community, resulting in significant forest loss for agriculture and house construction. Landslides occurred in 1968 and 1998, and fire in 1998 (New Era Team 2000; McDougall et al. 2002b).

Population Pressure—High

Nepal's hills are well-known for high population levels. The apparently low population density in Bamdibhirkhoria relates to the comparatively large size of the community forest, vis-à-vis the population, rather than to the overall population density in the area.

The Bamdibhirkhoria Forest User Group has 134 households, with a population of 722. The sex ratio is 1.14, with more men than women (McDougall et al. 2002b). Average household size was 4.9 in 1991. The forest area is 48 ha, with a forest/household ratio of 0.32 ha/household, or 3.13 households/ha (Dangol et al. 2000). Based on these data, the calculated population per area of community forest is 16 people/km².

In Deurali-Baghedanda, there are 142 households, with a population of 864 in the forest user group, which includes three separate executive committees (McDougall et al. 2002b). Dangol et al. (in McDougall et al. 2002b, 51) compute the forest to household ratio in 2002 as 1.26 ha/household (total community forest area is 181 ha), or 0.79 households/ha. This figure only includes the forest, not the settlement area. Average household size was 4.9 in 1991 (Dangol et al. 2000). The calculated density of people per square kilometer of forest here is 40.

In Manakamana FUG, the 132 ha of community forest serves 164 households, totaling 879 individuals. This yields 0.8 ha available for use by each household (McDougall et al. 2002b). Manakamana then has 28 people/km² of community forest.

The 185 households in Andheribhajana FUG (1,020 persons) share a community forest of 113 ha, making 0.63 ha for the use of each household. The sex ratio is quite even at 1.02 (McDougall et al. 2002b). Andheribhajana is the most densely populated of the Nepali sites at 357 people/km² of community forest.

Management Goals—Multipurpose Reserve

The Nepali government's involvement in community forestry is longstanding and well-known. McDougall et al. (2002b) report that the community and private forestry program is the largest component of the 1988 Master Plan for the Forestry Sector and is likely to absorb 47 percent of the forestry sector investments through 2010. "All accessible hill forests" are to be handed over to local communities (New Era Team 2001d).

Nationwide, the Department of Forestry has handed over about 850,000 ha of national forests to around 12,000 forest user groups (FUGs), composed of 1,100,000 household members (McDougall et al. 2002b, 38).

The framework in which the ACM FUGs operate is presented in Chapter 8, Box 8-3. ACM activities on these sites have focused on institutionalizing social learning mechanisms and on enhancing equity in decisionmaking and benefit distribution within the four sites. They have also made significant progress in strengthening the communication flow from the villages to the elites and to the forestry officials, both of whom previously dominated decisionmaking relating to forest management (see, e.g., McDougall et al. 2002b; Dangol 2004).

External Diversity—Low

In Bamdibhirkhoria CFUG, external stakeholders include the following: neighboring communities, other FUG and non-FUG member households, the Range Post; Bamboo Secret, Peace Healing Center, Women's Environment and Development Association, Women's Development Branch, FECOFUN (a FUG network), Village Development Council, District Soil Conservation Office, Machhapuchhre Development Organization, and Mothers' Groups (Dangol et al. 2001a).

In Deurali-Baghedanda CF the main external stakeholders are the Range Post, Health Post, District Forest Office, Village Development Committee, Schools, Range Post Level FECOFUN, AFORDA (a research institution), *Kisan Dekhi Kisan Samma* (KIDEKI, an organization to promote environmentally friendly agricultural practices) and the District Soil Conservation Office (Dangol et al. 2001a). Neighboring communities and FUG and non-FUG member households are also stakeholders. Numerous internal community organizations built on caste, gender, and ethnic lines also divide the community.

In Manakamana CF, the New Era Team (2001c) identified as significant stakeholders neighboring communities, FUG and non-FUG member households, the District Forest Officer, the District Development Committee, the Range Post, the municipality, the Nepal-UK Community Forestry Project, FECOFUN, the Range Post Coordinating Committee, and sawmill operators. They also mentioned as players the *Manakamana Samaj Sewa* Club, dual users (joint members of other FUGs), and the NGO Rural Reconstruction Nepal.

The same group of researchers (New Era Team 2001c) listed the first 10 stakeholders identified in Manakamana as the same for Andheribhajana. The eleventh stakeholder in Andheribhajana is the Women's Development Programme.

Internal Diversity—High

The four sites in Nepal are the most diverse among the case studies in this book. Historically and culturally influenced by the hierarchical worldview of Hinduism, each community has its own hierarchy of castes, each with its own occupation. The 11 major castes and ethnic groups in Bamdibhir include Brahmins, Chettri, Magars, and "untouchables" (Kami, Damai, Sunar, Sarki). In Deurali-Baghedanda, the six major groups include Brahmin, Gurung, Magar, Bhujel, Newar, and "untouchables" such as Kami, Sarki, Sunar and Chandara (Dangol et al. 2001b). Manakamana has eight ethnic and caste groups, plus five households of "other." The Tamang, Newar, Majhi, Chettri, and Brahmin caste and ethnic groups are the major forest users. Andheribhajana has nine ethnic and caste groups (New Era Team 2001c). Chettri, Magar, Tamang, and Brahmin are the major caste and ethnic groups living in the community. Some castes and ethnic groups have also organized themselves into formal clubs and organizations (McDougall et al. 2002a). Refer to Box 9-2 for more information on measures the Nepali communities have taken to address internal diversity as part of their ACM efforts.

Level of Conflict—High

Andheribhajana was characterized by more conflict than our other sites in Nepal. Major conflicts revolved around community forest encroachment by other users, boundaries with neighboring FUGs, forest product distribution, illegal extraction of forest products, and interethnic conflicts between Brahmins and Chettri versus the indigenous ethnic groups (McDougall et al. 2002b).

Internal capacity for conflict resolution was so minimal that the District Forest Office and the municipality were called in to help resolve the internal and external conflicts of this FUG. Additionally, an action committee was formed in the Range Post Coordination Committee (RPCC) to resolve conflicts between Andheribhajana and another FUG (McDougall et al. 2002b). Our ACM facilitator in that community reported being purposely poisoned when she first began working there and would have died had she not received an antidote from her co-worker (Sharma 2001); such an attack seems likely to be correlated with high levels of conflict, although a local belief holds that good things come to a poisoner.

Another instance of conflict and its resolution in Andheribhajana has been further detailed by Laya Prasad Uprety, Kalpana Sharma, and Narayan Sitaula of the New Era Team. Invited to attend an FUG committee meeting on illegal tree cutting in the CF, Sharma discovered intense conflict already under way. Participants wielding sticks and in a belligerent mood exhibited a very low level of trust and respect, shouting, interrupting, and disregarding others. One CFUG member who had cut a pole-sized tree from the CF without consulting the committee turned out to have been encouraged by the committee secretary, who was paid Rs 1,000 in return (78.6 Nepali rupees = 1$U.S.). During the meeting, the secretary supported the culprit, suggesting a penalty of only Rs 5 and insisting that the tree not be confiscated. Some CFUG members responded aggressively, while others sided with the secretary. Sharma's objections were ignored, and the dispute resulted in a fistfight between two sexagenarians.

Sharma and the other ACM researchers subsequently organized training on "leadership development and empowerment" to improve understanding of how poor leadership affected the entire community, including women and marginalized user groups. A total of 55 users including the committee members took part in this training, of which 51 percent were women and 49 percent were men. Interestingly, women from all caste, ethnic, and wealth groups took part. After the training, the CFUG members became more interested in social mobilization to increase their awareness and to disseminate their new knowledge and skills among other users (see McDougall et al. 2002b, 104–105, for a fuller discussion).

Level of Conflict—Medium

Other more general conflicts, prevalent in all the Nepal sites, existed. McDougall et al. (2002b) note that forest user group membership, the distribution and sale of forest products, the selection of FUG committees, fund mobilization, and the delineation of forest boundaries are important issues everywhere. Relations between FUGs and other stakeholders are complex, involving conflict, negotiation, resistance, and debate. The ACM emphasis on encouraging equitable access to decisionmaking and forest benefits may have caused some discomfort, initially at least, to the more powerful members of the FUGs, since increasing equity involves some redistribution of power (McDougall et al. 2002b). In all the FUGs, tensions related to the committees' income, expenditures, and a lack of transparency were common. Routine disclosure in general assembly and *tole* meetings as a part of the ACM approach significantly reduced conflict levels on all sites (McDougall et al. 2002b). None of these conflicts was as severe as in Andheribhajana.

Common problems in Bamdibhir included encroachment, demarcation problems, irregular support from government organizations, breaking of FUG rules, and the weakness of FECOFUN (McDougall et al. 2002b). The FUG committee attempted to address these issues rather unsuccessfully in the past. A latent conflict between the FUG and the chair of its committee was a factor. A conflict between the committee and the members over use of a bamboo nursery fund was successfully addressed after an action group was convened, and the loss of a boundary marker was successfully resolved through an investigation committee (McDougall et al. 2002b). The district-level FECOFUN was considering involving the Village Development Council to investigate and mediate differences and boundary disputes between FUGs (McDougall et al. 2002b).

Deurali-Baghedanda exhibited some conflicts about boundaries within the FUG and between neighboring FUGs. These were satisfactorily resolved during the research period through the FUG committee or with the support of external stakeholders such as the Village Development Committee (VDC) (McDougall et al. 2002b).

Most local conflicts in Manakamana were effectively managed by the FUG committee, though occasionally the assistance of the DFO, FECOFUN, or neighboring FUG representatives was required (McDougall et al. 2002b). During the course of the ACM research, a Forestland Encroachment Monitoring Committee was formed to reduce conflict between the FUG and individual encroachers. Another action group, called the Forest Product Distribution Monitoring Committee, was formed to deal with conflict relating to timber distribution. This group was to help *tole* coordination committees to assess actual needs in their *tole* and to ensure equitable distribution of timber (McDougall et al. 2002b).

Social Capital—Low

McDougall et al. (2002b) report that FUG members and the research team viewed improvements in social capital as critical. Members sought to bring about a "sense of well-being" directly but also to lay the groundwork for community development and effective natural resource management. Recognizing the difficulty of measurement, McDougall et al. identify several means by which social capital may be built: "effective communication; history of positive interactions; joint actions; resolution of conflicts; and social learning processes" (2002b, 132).

Social capital within local formal management entities—FUGs, forest user group committees (FUGC), general assemblies—has been low. Formed as part of the legislation authorizing community forests, these entities are externally defined institutions.

No specific investigation of intra*tole* social capital exists, but levels appear much higher within the more homogeneous social groupings. The *tole*s in our ACM sites, though quite homogeneous in ethnicity and caste, are not necessarily typical for Nepal, and such comparative homogeneity is, in our view, one reason why management at this level worked well. Previously marginalized individuals were able to speak up in smaller, more homogenous social settings. The decisions on all sites to develop or strengthen *tole* involvement in management as part of participatory action research appears to be a recognition of the better intra*tole* social capital (bonding). A representative from each *tole* has the potential to play a role in fulfilling a bridging function to the larger FUGC and general assembly. Informal action groups formed around specific topics also strengthened local social capital. Some of these were existing groups, like the mothers' club in Bamdibhir.

In Bamdibhir, the general assembly for forest management—scheduled to meet every six months—had not met in three years when the ACM team arrived. The FUGC met irregularly, often canceling a planned monthly meeting due to lack of a quorum; *tole*-based committees did not exist. FUG membership knew little about the rules for community forest use and benefit sharing and participated in decisionmaking forums even less. The FUG chairperson dominated decisionmaking and took advantage of interesting training opportunities himself. Women and some caste and ethnic groups were marginalized (McDougall et al. 2002b, 70–73).

Midlevel social capital was lower in Kaski District than in Sangkhuwasaba District when ACM activities began. In Bamdibhir, the important other stakeholders included only the local government (VDC), the Range Post FECOFUN, and the Range Post, none of which met or acted together regularly.

In Deurali-Baghedanda, meetings were somewhat more regular than in Bamdibhir, but the chair, secretary and treasurer of the FUG committee still dominated planning and decisionmaking. Although the community did exhibit a moderate amount of social capital in other spheres, working together to accomplish some joint goals, cooperation did not extend to community forest management. Community members had little say about how the community forest benefits would be distributed or other management concerns (McDougall et al. 2002b). The VDC, Range Post FECOFUN, Range Post, and the NGO Kisan Dekhi Kisan Samma were potentially relevant actors, but as in Bamdibhir none of these stakeholders met or collaborated routinely.

In Manakamana, although initial estimates of social capital were "medium," more information and comparison with other sites led to its recategorization as "low." General assemblies, FUGC, and *tole*-level management committees met irregularly, and interaction between FUGC members and the community was minimal. Decisions were made within a small, elite, wealthy group. Timber distribution determined by *tole* leaders was met with considerable community dissatisfaction; other forest product use was determined by the FUGC. Although training opportunities were usually reserved for FUGC members, one or two women managed to receive training (McDougall et al. 2002b, 70–73). Stakeholders included the FUG, FECOFUN, the Range Post Coordination Committee, the Range Post, and District

Forest Office, with some involvement by Nepal UK Community Forest Project (NUKCFP). Little sharing or joint planning took place in their meetings.

Adheribhajana was also characterized by low levels of social capital, despite the formal forums designed to enhance collaborative effort that did already exist. Like Manakamana, Adheribhajana's social capital was initially identified as "medium" and subsequently put in the "low" category. The general assembly met rarely and the FUGC more often but irregularly. As elsewhere, the FUGC was dominated by the chair and a small, elite group, decisionmaking was generally top-down, and the community had little input. As was noted in the Conflict—High and Conflict—Medium sections, intense, often interethnic strife resulted in disputes over boundaries, poaching, and distribution of benefits. Midlevel collaboration among the FUG, the DFO, and the Range Post existed, with some involvement of the District Soil Conservation Office and the local government (VDC). As in Manakamana, quarterly meetings were used for reports from the FUGs, with little or no sharing or joint planning.

CASE 12

Philippines

Contributing Authors
Herlina Hartanto, Cristina Maria Lorenzo, Linda M. Burton,
Cecil Valmores, and Lani Arda-Minas

Sites in the Philippines
San Rafael-Tanabag-Concepción, Palawan, and Basac, Lantapan, Mindanao

Devolution Status—High

The Philippines, like Nepal, is well-known for its innovative and community-oriented approach to forest management. Malayang (2001, 8) summarizes the situation since 1986. Although not all the instruments listed here are central to this case, their proliferation is an interesting reflection of Philippine policy discourse (see acronym list):

> National government control of forests continues to be high but higher emphasis has been placed on community participation in forest governance with the institution of new tenurial instruments to govern access to forests (e.g., CADCs, CADTs, CBFM). The national government has become more open to the influence of civil society in forest policy and management (e.g., CPPAP, CBFMP, PCSD). More powers have been given to local government units (LGUs) to protect and develop forests. With the crisis in Philippine forestry now more acutely recognized, the debate to stress (as policy) either production or protection forestry has intensified as well. The government seeks to stress both production and protection forestry and civil society seeks to stress protection over production.

Guerrero and Pinto (2001, 427) date a significant shift in Philippine forest policy to 1982, with the Integrated Social Forestry Program (described in Gibbs et al. 1990). The program was designed both to address the massive deforestation under way at that time and to inculcate more people-oriented strategies in forest management. Guerrero and Pinto also briefly describe a number of previously ignored regulations, dating as far back as 1909, which provide some rights to indigenous groups.

Philippines Key ACM Characteristics	
Devolution status	High
Forest type	Humid tropical
Population pressure	Medium
Management goals	Small-scale timber: San Rafael-Tanabag-Concepción
	Large-scale conservation: Basac
External diversity	High
Internal diversity	Medium: San Rafael-Tanabag-Concepción
	Low: Basac
National conflict	High
Local conflict	High: Basac
	Medium: San Rafael-Tanabag-Concepción
Social capital	Medium

The national regulations that have been of particular relevance in our research include the following (see Luna 2001; Magno 2001):

1. *National Integrated Protected Area Systems Act* (1992, NIPAS). NIPAS recognized the rights of indigenous peoples within protected areas. This act provides for a site-based Protected Area Management Board (PAMB), composed of NGOs, local government units, local communities, and representatives of the Department of Environment and Natural Resources (DENR), who together consider land use planning, zoning, and other resource activities in protected areas.

2. *Delineation of Ancestral Lands and Domains Claims* (1993, DENR Administrative Order No. 2). Provincial Special Task Forces on Ancestral Domains review indigenous groups' claims to land and, if approved, provide them with a Certificate of Ancestral Domain Claim (CADC).

3. *Community-Based Forest Management* (1995, CBFM). This program was signed into law with Executive Order No. 263 as the national strategy to ensure the sustainability of the nation's forest resources. It encourages watershed-based management, involving local communities, local government, and NGOs. It provides long-term tenure, so long as management is environmentally friendly, ecologically sustainable, and labor intensive.

4. *Indigenous Peoples' Rights Act* (1997, IPRA). IPRA grants indigenous peoples the rights to their traditional territories and recognizes the applicability of traditional rules governing property. In cases where such rights have been recognized, the group receives a Certificate of Ancestral Domain Title (CADT) from the National Commission on Indigenous Peoples. The constitutionality of this act was challenged in the courts in 2001 and found to be constitutional.

Located some 67 km from Puerto Princesa City in Palawan Province, the ACM site of San Rafael-Tanabag-Concepción is a CBFM area, which means that its management was devolved by the DENR to a People's Organization (PO) called San Rafael, Tanabag, and Concepción Multi-Purpose Cooperative, Inc. (STCMPC). STCMPC is composed of people from the three villages (*barangays*) of San Rafael, Tanabag, and Concepción. STCMPC initially received tenure rights over 1,000 ha in the early 1990s under DENR's Community Reforestation Program as one of nine pilot sites in the Philippines. In 1996, the tenure was expanded to 5,006 ha. The area covered a series of watersheds within the territories of the three *barangays*. The communities have a total population of 3,597 (June 1999 *barangay* census) and include three major groups: the indigenous Batak and Tagbanua; the Cuyunon; and migrants (Lorenzo 2001c, 2001e, 2002; Hartanto 2001).

The second ACM site, Basac, is one of 14 *barangays* in the municipality of Lantapan. Basac is located within the buffer zone of the 47,000 ha Mt. Kitanglad Natural Park, a declared protected area in Bukidnon Province on Mindanao. The core of the park remains pristine forest. Basac's territory includes about 2,800 ha of agricultural land and about 4,000 inhabitants in 700 households (mostly farmers). Ninety-five percent of the inhabitants are Binukid speakers belonging to the Talaandig ethnic group. Basac is accessible by national road and is located some 35 km from Malaybalay, Bukidnon. One of the three POs in Basac with whom the ACM team works is the Bukidnon Upland Farmers Association Incorporated (BUFAI). BUFAI was granted a CBFM Agreement by the DENR in 1999 (Burton 2001).

Forest Type—Humid Tropical

In the San Rafael-Tanabag-Concepción site, rattan, *almaciga* (*Agathis damarra*), and honey are important nontimber forest products (NTFPs), though timber remains important. Felled and damaged trees, amounting to 663 m³ in 2000–2001, can be harvested (Lorenzo 2001c, 2001e). The CBFM area consists of a strip of disturbed forest, surrounded on three sides by closed canopy forest. Of the 18,111 ha in the three *barangays*, 67 percent of the area remains covered in primary forest, with the remainder in secondary forest, brush, and cropland. Human disturbances have been due to logging by concessionaires and past shifting cultivation that increased dramatically with the inflow of migrants.

The most abundant tree species found in the old-growth forest was *guijo* (*Shorea guiso*), followed by *narig* (*Vatica manggachapui*), and *Dipterocarpus sp.* The most dominant tree species was *dao* (*Dracontamelon dao*), followed by *kalantas* (*Toona calantas*) and *amugis* (*Koordersiodendron pinnatum*). *Almaciga* forest was dominated by *bagtik* (*Agathis damarra*). However, residual forest was the most dominant type in the CBFM area. Most common tree species included *taluto* (*Pterocymbium tinctorium*), *agosip* (*Symplocos ahernii*), *sanglai* (*Ahernia glandulosa*), *alupag* (*Litchi chinensis*), *anilau* (*Colona serratifolia*), and *pagsahingin* (*Canarium asperum*). The most dominant species were *Pterocymbium tinctorium*, *Canarium asperum*, and *balakat* (*Zizyphus talanai*) (Hartanto et al. 2000b, 2002c). Palawan was badly affected by the 1998 El Niño, losing at least 15,000 ha of forest. The island is home to 232 wildlife species—23 percent of the fauna of the Philippines. Of these species, 11 are nonflying mammal species and 14 are bird species endemic to the province (Guerrero and Pinto 2001).

In Lantapan, ACM works in a government-sponsored CBFM area, granted to BUFAI and covering an area of 2,800 ha of agricultural land. Mt. Kitanglad, considered one of the major critical watersheds in Northern Mindanao, is the water source for agricultural, domestic, and industrial purposes downstream. Problems in the watershed include high soil erosion (61 percent of the area has slopes steeper than 40 percent), illegal logging, shifting cultivation, fire, and high annual population growth at 4.18 percent (Hartanto 2001). In 1983, 500 ha burned in El Niño related fires. Very little forest is left in the ACM site. Local community members estimated that by 2000 only 4 percent of the tree cover that existed before 1940 still remained (Valmores 2002). Hartanto et al. (2003) summarize as follows: "Around 60 percent of the BUFAI CBFM area is cultivated land, which is planted with food crops. Around 15 percent … is grassland, 15 percent is open canopy forests, and only around 10 percent is closed canopy forests" (p. 37).

Population Pressure—Medium

In the STCMPC 433 people reside, making up about 12 percent of the overall community (Hartanto et al. 2002c). The population in the three communities has grown almost tenfold (see Table A-5) since the 1950s, when there were 25 Batak, 15 Tagbanua, and 30 in-migrants (Lorenzo 2001c, 2001e). Of the total population, 55 percent (1,993) are female (Hartanto et al. 2003).

In Basac, the community is divided into six *puroks* (districts), with 750 households and a population of around 4,000 people, 95 percent of whom are indigenous Talaandig. The remainder are migrant settlers from the nearby provinces of Cebu, Bohol, and Leyte (Hartanto et al. 2003). One hundred eight people, or 4.5 percent of the village population, are members of BUFAI (Hartanto et al. 2002c). The population growth rate is high at 4.18 percent (Hartanto 2001).

Management Goals—Small-Scale Timber

The main activity of STCMPC, when the ACM team began working with them, was small-scale timber production. The CBFM agreement made between STC-MPC and DENR requires that the STCMPC prepare a 25-year management plan and annual work plans for DENR approval. Palawan requires additional approvals by the Palawan Council for Sustainable Development (PCSD), the Community

Table A-5. *Population Trends in San Rafael, Tanabag, and Concepción from 1970 to 2000*

	1970	1980	1990	2000
San Rafael	320	497	849	1,384
Tanabag	136	245	307	412
Concepción	242	375	968	1,093

Source: Lorenzo (2001c, 2001e).

Environment and Natural Resources Office, and the local *barangay* governments. The process of getting these approvals took four years, due to the conflict between STCMPC and the neighboring *almaciga* concessionaire and to the lack of support from the *barangay* captains for CBFM. Another crucial issue delaying PCSD's endorsement was the council's preference for communal forestry, with forest management in the hands of the local government units and with the main thrust on conservation and protection rather than use. The council also feared that income-generating activities such as the use of fallen logs could be abused by the people and could encourage illegal activities in the forest.

Palawan, indeed the entire Philippines, has a log ban, but the STCMPC was granted permission to do salvage logging in their CBFM area. Cyclone damage and the presence of felled logs left by previous concessionaires left room for harvesting according to a management plan. Lumber was produced on the cooperative grounds to generate capital for other CBFM activities.

STCMPC found 863 trees suitable for harvesting in their CBFM area and had permission to extract in one year a total of 667 m³ (or 60 percent of the estimated total volume of 1,112 m³). Although their enterprise estimated a 17 percent profit margin, high fees owed to DENR were a serious problem (Devanadera et al. 2002). Administrative Order No. 98-42 stipulates a production-sharing agreement between POs and the government. STCMPC had to pay 3,000 pesos/m³ (computed as 60 percent recovery rate), which seriously compromised their ability to remain profitable. One ACM activity STCMPC undertook was to initiate a dialogue with various stakeholders to try to solve this problem. As a result, the group successfully petitioned DENR for lower forest charges. STCMPC also diversified its forest product collection activities to include *almaciga* resins, rattan, and honey (Hartanto et al. 2002c).

Management Goals—Large-Scale Conservation

Mt. Kitanglad in Mindanao was declared a park in 1996 (Valmores 2002) after considerable degradation. In Basac, where barely 10 percent of the site is forested, ACM work began in mid-2001 (Valmores 2001).

External Diversity—High

Lorenzo (2001c, 2001e) describes a complex set of external actors in San Rafael-Tanabag-Concepción: two Indigenous Peoples' Organizations (for the Tagbanua and the Batak, respectively); three levels of local government (*barangay*, city, and province); various levels of DENR; the PCSD; six NGOs; private concessionaires; and other business interests.

After a series of complex focus group discussions, Arda-Minas managed to reduce the amazing number of entities operating in Basac to 15 on-site stakeholders, namely the farmers; the *barangay* Council; the Council of Elders; a Mother's Club; a religious group; BUFAI; BUFAI Women's Association; *Bantay bayan* (local security patrol); *Jeep ni Erap* (a government credit scheme); *Lupon* (a local dispute settlement mechanism); Forest Rangers; Department of Agriculture; Basac Tribal Farmers Association; Kitanglad Guard Volunteers; and the Tribal force. Off-site stakeholders include DENR; Protected Area Management Board; Kitanglad Integrated Network; Barangay Integrated Development Assistance for Nutrition Improvement; the World Agroforestry Center; local government units; National Commission for Indigenous People; and Heifer International (Arda-Minas 2002b).

Internal Diversity—Medium to Low

All three Philippine *barangays* of San Rafael, Tanabag, and Concepción consist of two major groups: the indigenous Batak and Tagbanua; and migrants (Cuyunon, Tagalogs, Ilokanos, Bikolanos, and Visayans) (Lorenzo 2001c, 2001e). Basac is 95 percent Talaandig (Arda-Minas 2002b).

Level of Conflict—High

In Basac, the interactions among indigenous peoples and in-migrants have led to two major conflicts in recent times. The first conflict was between the Talaandig tribal leaders and the PO that holds the CBFM agreement (see Box 10-5). That conflict led to the second incident. In March 2001, in preparation for a festival, some timber was felled from land belonging to the father-in-law of Datu Migkitay. Seeing it piled in a conspicuous place, a DENR Forest Officer asked its origin and to see a DENR permit. Datu Migkitay responded that because the timber came from an ancestral domain no permission from DENR was necessary. The disagreement nearly resulted in a clash between Talaandig warriors and the police. At Datu Migkitay's request, the Talaandig Council of Elders met and imposed a *sala* (a cultural penalty based on customary law) on the DENR officers. The *sala* fine—10 water buffalo, 3 pigs, 6 chickens, and 3 meters of cloth—came with a warning that the fine could be multiplied up to sevenfold and would be doubled immediately if DENR did not pay by the deadline of April 28, 2001 (Burton 2001). The controversy, which affected many activities in the Mt. Kitanglad area for some time, was finally resolved with the revocation of the *sala* (Burton 2001; Hartanto et al. 2002c, 19–20).

Level of Conflict—Medium

Quite a few conflicts exist in San Rafael, Tanabag, and Concepción, but they seem to be under control. First, a variety of conflicts among administrative groups are evident. PCSD and DENR, for instance, have been struggling with overlapping functions relating to natural resources for a number of years. Two important kinds of disputes between them have involved the issuance of timber and nontimber forest product extraction permits and other licenses (Esguerra and Hartanto 2002). Second, a history of conflicts between private and public groups in the area have complicated the picture. STCMPC clashed with the neighboring *almaciga* resin concessionaire, who was supported by the *barangay* captains, contributing to a lack of support for community-based forest management in Tanabag and Concepción. DENR had refused to extend the concessionaire's extraction permit at a time when the CBFM area was being expanded to include some of the concessionaire's previously allocated area. With the support of the *barangay* captains, the concessionaire took the matter to court, resulting in considerable lingering antagonism (Hartanto et al. 2002c).

Another significant conflict was over the boundary between the STCMPC's forest and that of the neighboring Bataks, both of which were covered under CBFM agreements. The Batak, supported by a local NGO, had applied for a CADC in 1997. When the IPRA was cancelled, making the CADC unavailable, the Batak applied to DENR for a CBFM area. Meanwhile, the STCMPC forest area was expanded from 1,000 to 5,006 ha, covering an area not fully surveyed. The overlap between the two areas had to be resolved because the Batak depend on *almaciga* resin as an important part of their livelihoods (Lorenzo et al. 2000; Hartanto et al. 2002, 18; Box 5-9).

Social Capital—Medium

The Palawan grouping of three villages—San Rafael, Tanabag, and Concepción—was initially estimated as having a medium level of social capital. The board of directors of the STCMPC had a medium to high level of social capital, shown by their consistent involvement in STCMPC activities, their ability to work together, and their commitment to STCMPC goals. Their relationship with the STCMPC general membership, however, was not characterized by any significant joint action. Indeed, the general members actually expressed little interest initially, because of the group's inability to provide the anticipated economic benefits (Hartanto et al. 2002). Although the board of directors was fairly elected and enjoyed general trust by members, there were some speculations about inequitable benefit sharing (in that the management staff and the members involved in logging received regular pay). The group solved any internal conflicts informally; for external problems, they sometimes relied on external mediators like DENR or the city government (Lorenzo 2002).

Trust and feelings of partnership between the STCMPC and the local government (the *barangays*) were at a low level due to some of the conflicts described earlier (relating to the desirability of CBFM agreements and earlier conflicts with the almaciga concessionaire; Lorenzo 2002, 9–10). Levels of trust between the indigenous Bataks and the STCMPC were quite good, despite occasional conflicts. Around 33 Bataks were members of the STCMPC and some worked in the STCMPC lumber operation (Hartanto et al. 2002; see also Box 5-9).

There appears to be a strong foundation of social capital among the indigenous Talaandig group in Basac, Lantapan. A somewhat charismatic leader was able to bring together a large group of indigenous people to apply for a CADC (before this option was withdrawn by the government). When that idea fell through, the Basac group was able to mobilize to form BUFAI and negotiate a CBFM agreement for itself; nor were the Talaandig shy to take on the DENR apparatus when they felt their rights were being abridged (see Box 10-5). Hartanto et al. (2002) concludes that, despite conflicts between the Council of Elders (who supported the CADC) and BUFAI (which has a CBFM agreement), mutual respect remains high within the indigenous group, based on shared culture and family connections.

Relations between the community members and *barangay* officials, however, were unsatisfactory. Arda-Minas (2002) says that community members did not express their opinions for fear of exclusion from possible benefits from the *barangay*; nor did community members have routine access to information or opportunities from *barangay* leaders who tended to share such benefits with family or political supporters. Hartanto et al. (2002) reports on a number of *barangay* issues relating to governance, brought up in a group discussion among community members, that also have some bearing on social capital. These are as follows:

1. Weak enforcement of rules and regulations
2. Lack of mechanisms for information sharing
3. Lack of transparency in *barangay* financial management
4. Lack of skills and capacity of the *barangay* councils that contributed to the lack of a *barangay* development plan

As in many local contexts, decisions are often made prior to a meeting by the leader, who then simply convinces members to rubber-stamp his opinion.

CASE 13

Zimbabwe

Contributing Authors
Frank Matose, Tendayi Mutimukuru, Richard Nyirenda, Ravi Prabhu, Bev Sithole, and Wavell Standa-Gunda

Site in Zimbabwe
Mafungautsi Forest Reserve

Devolution Status—No Formal Devolution

Statistics about Zimbabwean land use, which set the stage for the serious natural resource-related conflicts that are currently pulling that country apart, are provided by Nhira et al. (1998). See also Chapter 10. Over a quarter of Zimbabwe's woodland area is state land: national parks and wildlife and forest reserves. Before the land reform program, most of the best farmland and more than 30 percent of the woodland was found on commercial farms, where 19 percent of the rural population live. However, 74 percent of the rural population live in the 42 percent of the land defined as communal areas—typically lands with the lowest farming potential.

The land reform program has changed the face of agriculture, with over 90 percent of the former white-owned farms being placed under compulsory acquisition by the government. The resulting shift in population distribution is still a

<table>
<tr><td colspan="2" align="center">Zimbabwe
Key ACM Characteristics</td></tr>
</table>

Devolution status	No formal devolution
Forest type	Dry tropical
Population pressure	High
Management goals	Multipurpose reserve
External diversity	Low
Internal diversity	Medium
National conflict	High
Local conflict	Medium
Social capital	High

process under documentation. A decline in the forested area in the former commercial farms is anticipated as new farmers open lands for agriculture. In the mid-1990s, about 4,000 whites still retained 35 percent of the country's arable and ranch land. The Land Tenure Commission proposed radical changes in tenure and rural governance systems in communal and commercial areas in 1993, without much government response or implementation.

Despite the fame of Zimbabwe's CAMPFIRE program, which is based on local control of the benefits from sport hunting, forest policy in general is much less oriented toward devolution. The National Parks and Wildlife Act was alone in providing any legal mechanism for local woodland management when ACM began (Nhira et al. 1998). Today various versions of the 1992 Land Acquisition Act (amended after 2000) and provisions of the Water Act strengthen devolution. The Rural District Councils Act of 1988 would seem to have the potential to decentralize significant natural resource use and management, but lack of funds and political space to tailor bylaws to local conditions seem to have prevented much progress from being made to date.

Matovu (2002) writes of the Thirteen Principles to Guide the Decentralization Process that were adopted in 1998, the first of which relates to strengthening democracy and citizen participation. However, all authors agree that rhetoric about devolution is significantly more abundant than its implementation. Mandondo and Mapedza (2002) offer a detailed account of current shortcomings.

Mafungautsi Forest Reserve—also called Mafungabusi Forest Reserve—is located in the Gokwe South District of the Midlands Province in central-western Zimbabwe. Originally, the forest reserve of miombo woodlands comprised 105,000 ha. It was reduced to 82,000 ha, with the remainder being ceded to neighboring communities in 1978. Although Mafungautsi was selected by the Forest Commission (FC) as an experiment in comanagement, the state remains legal custodian of the forest via the FC. Any arrangements made with local communities can be revoked if the FC perceives a decline in forest conditions (Matose 2001). Shackleton et al. (2001, 57) suggest that the management committees in Gokwe have been little more than a proxy for the FC, which still controls critical elements such as distribution of benefits.

The FC initiated a "resource sharing project" in 1993. By September 1999, 11 Resource Management Committees (RMCs) had been formed around the forest. In six of the RMCs, members were trained in beekeeping and honey processing. A resource education center for communities was completed. The RMCs are legal entities functioning within local governance structures, under the Rural Development Councils, the local arm of the state government, through local village leadership and chieftancy.

According to Matose (2001, 31), the project's objectives were twofold:

a. Guarantee the conservation and management of Mafungautsi Forest by involving the neighboring communities in the decisionmaking and profit-sharing derived from sustainable forest resource development and utilization programs.
b. Develop appropriate models for forest grazing, access and utilization of other forest resources leading to the conservation of Mafungautsi as a water catchment and state forest, without compromising the environmental agenda.

When ACM research began, the comanagement effort was largely a failure. Antagonism between local communities and the FC was palpable, persistent, and ubiquitous. Community members were not complying with regulations or requests, and FC staff showed marked distrust of community involvement (Sithole 2001). This relationship has clearly improved over the past few years (Matose et al. 2002).

Forests Type—Dry Tropical

Mafungautsi Forest is characterized by miombo woodlands on Kalahari Sand or deciduous dry miombo. The vegetation is dominated by *Brachystegia spiciformis* and *Julbernadia globiflora*. In some areas *Brachystegia spiciformis* is in association with *Brachystegia boehmii*. A sizable quantity of the commercial timber species *Baikiaea plurijuga* and *Pterocarpus angolensis* is present. Vegetation is multistoried, fairly open, with grassy ground cover. Other common tree species in this vegetation type are *Grewia spp.*, *Ximenia caffra*, *Bauhinia tementosa*, *Bauhinia petersiana*, *Ochna spp.*, *Terminalia sericea*, *Peltophorum africana*, and *Burkea africana*. The herbaceous layer is dominated by grasses such as *Digitaria eriantha* (Finger grass), *Schmidta pappophoroides* (Kalahari Sand Quick), *Stipagrostis uniphumis*, *Eragrostis pallens*, and *Perotis patens*, some of which provide grazing for livestock. Plains game inhabit the forest: zebra, kudu, bush pig, warthog, reedbuck, bushbuck, duiker, hyena, genet species, hare species, and buffalo are present, along with a wide variety of bird species. The forest is also susceptible to winter frost, particularly in wetlands.

The Mafungautsi Plateau lies between 1,100 and 1,390 m above sea level. These Kalahari Sands are scored by four rivers—Sengwa, Mbumbusi, Lutope, and Ngodoma—that drain to Lake Kariba in the north. Rainfall varies from 600 to 800 mm/year. Temperatures vary between 5°C and 35°C, with an average of 22°C. The soils consist of deep sands of aeolian origin overlying partially weathered rock. The ridges are interspersed by rivers where the sands have been eroded to give way to heavier textured soils of basalt and sedimentary deposits—wetlands once favored for cultivation (Matose 2001).

Population Pressure—High

The Mafungautsi State Forest covers 82.1 km^2 and has a population density of about 35 people per km^2 (Shackleton et al. 2001). Matose et al. report "a 32% increase in the number of cattle grazing in the forest (11,800 to 15,600 head) between 1993 and 1999"; their impact is estimated to be low (2002).

The ACM team has been working with three RMCs in Gababe; Ndarire; and Batanai. The RMC in Ndarire serves a population that was about 1,000 in 1992. Standa-Gunda et al. note "Those who formed the villages in the 1950s and their lineage remain members of that village. Recent entries into the existing villages outside the 'old families lineage' are not very common" (Standa-Gunda et al. 2003a, 11) Ndarire, ethnically Shona and Ndebele, is the most remote of the three Mafungautsi sites.

The 10 villages of Gababe hold about 2,000 people, mostly Ndebele, with a few Shona and in-migrants from Malawi. Road conditions make access difficult. Mbumbusi-Batanai RMC, one of the most active in Mafungautsi, served a population of about 1,800 in 1992. In-migration from those seeking their fortunes in cotton farming in Gokwe South is said to be greater than the out-migration of youths seeking urban employment (Matose et al. 2002, 29; Nyirenda 2000).

Management Goals—Multipurpose Reserve

While acknowledging that both production and conservation functions are important in the Mafungautsi Forest Reserve, Matose (Matose 2001, 70) writes that "Mafungautsi was set aside primarily as a protective forest with the objective of conserving watersheds for various rivers [e.g., the Zambezi] that flow to Lake Kariba in the north." Established in 1954, Mafungautsi is also considered a "remnant" of the plateau's original vegetation type, which should be conserved for national interests.

The FC, a statewide organization created in 1948, passed through several bureaucratic shifts and is now part of the Ministry of Environment and Tourism. It includes two main arms: the State Authority and the Commercial Forestry sections. FC responsibilities include state forest management, conservation and regulation of timber resources, afforestation, woodland management, and the provision of support services in forestry research, education, and extension (Matose 2001, 10–11).

Mafungautsi Forest Reserve actually has a long history of human settlement, and although the last inhabitants were forcibly evicted in 1986 at the peak of the civil war, the area is surrounded by densely populated communal areas. The FC maintains that these forested environments are stable ecosystems not to be disturbed by human settlement and farming. Local people counter that they have been farming inside forests since long before the forests were reserved and that their farming patches regenerate and return to forest (Matose 2001).

Such disagreements led the FC to initiate a resource-sharing project in Mafungautsi in 1993 within its Forestry Extension Services division (FES). Management responsibility for the forest was transferred from the Indigenous Resources Division to FES in 1994 (Matose 2001, 47). Ownership remains firmly in the hands of the state, with the FC openly expressing its legal custodianship of the forest. As recently as 1997, the FC argued with the U.K. Department for International Development over its right to relocate people if they were interfering with sustainable forest management.

ACM's activities have focused very closely on relations between local communities and the FC. Nyirenda et al. (2001) document the usefulness of "Training for Transformation"—an approach used by the Catholic Development Commission in Zimbabwe to assist communities in taking independent collective action for their own development. Although the team began with a somewhat unenthusiastic relationship with the FC, their relationship developed until they became real partners in joint endeavors, particularly with three kinds of community user groups (broom grass, thatch grass, and honey production).

External Diversity—Low

According to Vermeulen (2000), the only significant stakeholders in Mafungautsi are neighboring small-scale commercial farmers, other communal area residents, the local and central FC, the Rural Development Councils, NGOs, loggers, and thatch cutters. On-site external diversity may be low, but as Standa-Gunda (2003a) points out, numerous more distant stakeholders benefit significantly and directly from conservation efforts in and around Mafungautsi. They too are thus stakeholders. Because a central purpose of the Mafungautsi Forest Reserve is its function as a water catchment, important stakeholders that must not be neglected are the Zimbabwe National Water Authority and the Zimbabwe Electricity Supply Authority (ZESA). Poor management practices of local communities around Mafungautsi could harm ZESA, which produces hydroelectric power at Kariba Dam and would have to import electricity from outside Zimbabwe if water flows changed.

Age

Relations between the ages in Zimbabwe seem traditionally less conflict-ridden than in Cameroon. Common issues of dependency on the part of the young and responsibility on the part of the old are combining with the dynamics of change (Matose 2001; Chitiga and Nemarundwe 2003). The young are drawn to new approaches and question the viewpoints and perspectives of their elders. However, AIDS has decimated the young population and changed such relationships, rendering the young adults dependent on parents when they anticipate more freedom. The recruitment of young men to carry out government policy in this time of political unrest has also created an unusual situation in which the young are widely feared as enforcers (Sithole 2002a).

Internal Diversity—Medium

Mafungautsi is a large forested area, with numerous communities, typically inhabited by Shona, Ndebele, and the indigenous Shangwe people, surrounding it (Matose 2001).

Level of Conflict—Medium

Mafungautsi has been subject to quite serious conflicts in the past, when haphazard cutting of trees and uncontrolled grazing in the forest led to siltation of the rivers that drain into the Zambezi. At that time forest encroachment by neighboring communities was severely penalized, people were evicted from the forest, and other problems—illegal grazing, illegal hunting, unauthorized collection of nontimber forest products—were considered serious. Evictions resulted in deep mistrust between local communities and the FC (Standa-Gunda et al. 2003a, 6). Local community members preferred to convert the forest into agricultural lands, while the FC would not even discuss extraction of resources like timber and poles that were much needed by local communities. Communities responded by extracting even more wood from the forest and refusing to help with fire fighting (Matose et al. 2002, 30).

Partially in response to these conflicts, the government initiated an experimental resource-sharing project in 1994, which involved the creation of RMCs. Several kinds of conflicts revolve around these comparatively new institutions. The FC has dealt directly with the RMCs, sidelining and thus alienating traditional leaders (Matose et al. 2002, 63). Forest Protection Units and local monitoring subcommittees were confused about monitoring responsibilities. Recently, the Forest Protection Units agreed to focus on forest areas with no RMC and to seek information through RMCs where they are available (Mutimukuru et al. 2003, 10).

There has also been a lack of coordination among the RMCs. Members of some RMCs were accused of stealing resources from other RMCs. Mutimukuru et al. (2003, 17) report

> In one instance, an RMC member saw one woman with dug brooms at her homestead and wanted to fine her since this [digging rather than cutting brooms] was not allowed in that RMC. The woman however boasted that she had poached the grass from another RMC and hence the RMC member had no right to punish her.

Matose et al. (2002, 28) report that locals consider conflicts within communities to be too mild to result in violence. All people have equal access to resources from communal grazing areas and woodlands, according to focus groups comprised of all ethnic groups, recent immigrants, and long-term residents. Immigrants of any ethnic group, however, have less access to arable land and may be excluded from other resources by long-term residents. Local communities also expressed dissatisfaction with the distribution of forest benefits, with a disproportionate share going to the wealthy and members of the RMCs.

Social Capital—High

Within Mafungautsi, all three RMCs—Gababe, Ndarire, and Batanai—were reported as having high levels of internal social capital. Gababe was said to have strong traditional and ethnic (Ndebele) ties, including some distrust of outsiders based on beatings sustained during the liberation struggle. In Batanai, strong traditional ties were also reported. In Ndarire, where Standa-Gunda et al. conducted a systematic study of communication networks, the existence of a number of social clubs and cooperatives was seen as a major indicator of high social capital. These researchers found mutual reliance among households for draught power in exchange for labor, and group formation in order to gain access to resources not locally available (Standa-Gunda et al. 2003, 70).

Proceeding on the assumption that building on existing social capital made more sense than starting afresh in resource management, Standa-Gunda et al. (2003, 86) studied and described the communication networks in the five villages of Ndarire RMC as an early step in the ACM process:

> A total of 23 networks were identified in all the five villages. The maximum number of networks households participated in was eleven. Most households were actively involved in four to six networks. The concentration of most households into only six kinship (Akwah) and three churches and all coming from the five villages tended to concentrate information in a few individuals …. These included groups like political, church, soccer, school, the [village head's] security personnel, kinship, and agricultural commodity groups (cotton and cattle dip).

The researchers also discovered traditions that specified which information could be conveyed within a kinship group, stipulating nephews as messengers. Age, gender, and kin relations determined the order in which messages were conveyed. Such rules were typically followed even when two recipients were in the same homestead and the nephews were far away. Following this procedure was particularly important in traditional matters like marriage and appeasing spirits. In Ndarire, such patterns were most pronounced in networks that evolved within African traditions, like kraalhead meetings, kinship networks, and initial announcement of funerals, where specific individuals had to be informed before the general public (Standa-Gunda et al. 2003).

Bridging social capital to outside agencies like the FC was less evident. Indeed, various researchers have documented the suspicion and mistrust, including fear of arrest, that local people felt toward other local stakeholders, including the ACM team, initially (Matose et al. 2002, 36; Mutimukuru et al. 2003; Standa-Gunda et al. 2003). Nyirenda et al. write, "This passiveness or inertia manifests itself as a form of mistrust towards outsiders, a sense of powerlessness to overcome problems, and a feeling of lack of knowledge on what to do to overcome problems and act collectively" (Nyirenda et al. 2001). Effective communication is an important precursor in the formation or maintenance of social capital. Both Matose et al. (2002) and Mutimukuru et al. (2003) note the one-way flow of information from the FC to the local communities, with little or no involvement of community members or even the RMCs in management planning. Other government agencies such as agricultural technical and extension services, the Natural Resource Board, Agricultural Finance Corporation, and Rural District Council, were also characterized by weak information exchange with the communities (Mutimukuru et al. 2003, 15).

Notes

Chapter 1: Introduction

[1] I am not using the term *system* in any deterministic way. Determinism is "a theory or doctrine that acts of the will, occurrences in nature, or social or psychological phenomena are causally *determined* by preceding events or natural laws" (Merriam-Webster Online Dictionary 2004). My use of the term *system* instead focuses on the importance of the interrelations among parts (sometimes called networks). I refer to "soft" systems, where there are many actors with different objectives and no intrinsic purpose to the system as a whole. I am not thinking of a designed system or a system that is necessarily in equilibrium. I agree with Fisher (2003b) that "the concept of system is a heuristic device. Interconnectedness is real, but the system is an abstraction."

[2] This definition was initially crafted by Ravi Prabhu (Prabhu et al. 2001) and modified by the ACM team at an October 2001 meeting of the program's International Steering Committee in Manila.

[3] The site selection process was complicated by the research requirements for the second stream (discussed below). Normally, this process would simply require obtaining the permission of those involved. See Figure 1-3 for the ACM sites.

[4] I use the term *ACM facilitator* to refer to CIFOR's ACM program team members. Team members were often both facilitators and researchers—difficult roles to juggle. They were primarily facilitators in their attempt to catalyze the adaptive and collaborative management process in the field; they were primarily researchers in their attempt to assess and learn from what was happening. I hope that this approach will be more widely used, outside formal research organizations, in contexts where facilitators may be more available than researchers. Neither term is perfect, and the combination is unwieldy.

[5] The "powerless" are not in fact without power, but I use the word to reflect the formal, structural discrepancies that have real impacts in the world (cf. Foucault 1980; Scott 1985).

[6] Several tools, described elsewhere, have already been developed or found useful: future scenarios (Wollenberg et al. 2000); multi-criteria analysis (Mendoza et al. 1999); various pebble distribution methods (Colfer et al. 1999a, b); the who counts matrix (Colfer et al. 1999c); guidelines for use of C&I (Tiani et al. 2002); CORMAS (Purnomo et al. 2001; Purnomo et al.

2003b); CIMAT (Purnomo et al. 2000; Purnomo et al. 2001); participatory mapping (Anau et al. 2004); coview (Haggith et al. 2002); heterogeneity analysis (Pokorny, Cayres, and Nuñes 2001a); micropolitics (Sithole 2002b); and on the ACM process itself (Hartanto et al. 2003; Kusumanto et al. 2005).

[7]The International Steering Committee was led by Don Gilmour (Austrailia), supported by Peter Frost (Zimbabwe), Irene Guijt (Netherlands), K.B. Shrestha (Nepal), Rene de Rueda (Philippines), and Erwidodo and Yunita Winarto (both Indonesia). Their role was to advise the ACM team on direction and critique our efforts in a meeting at least once a year. They have also provided less formal assistance throughout our work.

[8]The Zimbabwe ACM team, now in the process of scaling up and out to the national extension service, instituted a similar, Zimbabwe-focused newsletter in January 2004.

[9]Like other writers who argue along these lines (cf. Scott 1998; Uphoff 1996; and many advocates of farming systems research), I am not arguing against the scientific method or human capacity to predict, given the right circumstances. Box 1-1 defines simple and complicated systems, both potentially predictable, as well as complex systems—those with which we are concerned in this book. I am arguing for a broadening of our perspective when we deal with complex adaptive systems, not for a replacement of the scientific method.

[10]Thanks to Norman Uphoff for pointing out that this observation was also made by Heraclites.

[11]The international research centers are guided by the Consultative Group on International Agricultural Research, in Washington, DC—now called "Future Harvest"—under whose authority CIFOR falls.

[12]In addition to the six sites mentioned above, the Cameroon team has begun collaborating with the Worldwide Fund for Nature (WWF) and the Ministry of Environment and Forests (MINEF) on a project called Local Communities Resettlement in Korup; and with IRM on a project called Community-Based Natural Resource Management in Ngambe Tikar/Djoum. Because these are recent additions, I have not dealt with them in this book.

[13]After Indonesia's decentralization process became effective, the Bulungan District was split into three parts. Our research efforts focus on the area now called Malinau District.

Chapter 2: ACM's Intellectual Underpinnings: A Personal Journey

[1]This is not to argue for the preeminence of rationality in all things. Cardenas (2002) has shown clear examples of human irrationality in daily life, using the techniques of experimental economics; and most students of human behavior consider rational response to be only one of many options for all of us.

[2]Cf. Vanda Radzik's recent comment that "Development that is not engendered is endangered" (Insanally 2003, 44).

[3]Perhaps we have been rediscovering the wheel: Don Reisman (2004) reminds us that Fredrich Engels wrote, some 120 years ago, that the relationship between men and women is the model for all other exploitation in society.

[4]Of course, in areas where political corruption is rife (as in many forested areas of the world), other factors enter that have even more deleterious effects on local people.

[5]The ubiquity of fads in these approaches (and related funding) is probably well known to the readers (cf. Senge 1990).

[6]Many farming systems practitioners made a self-conscious differentiation between interdisciplinary approaches, which try to integrate the different disciplines around a common issue, problem, or opportunity, and multidisciplinary approaches, which involve parallel but not necessarily

integrated work by several disciplines (Gilbert et al. 1980; Colfer 1983a). In these terms, integrated rural development had tended to be multidisciplinary.

[7]Diaw and Kusumanto (2005) document a large number of similar approaches that have evolved in recent years.

[8]Hulme and Murphree (2001b, 292) comment that "Cycles of enchantment and disenchantment are characteristic of programmes that attract aid funds when donors pile onto an apparent panacea and then pull out when it does not deliver miracles ..."

[9]Collinson and Lightfoot (2001, 406–407) propose five key elements in current and future farming systems research, suggesting that the field is addressing some of the problems identified above: characterization of the farming system, multistakeholder visioning, partnerships for concerted action, action learning and tracking change; and building knowledge networks.

[10]Reisman (2004) reminds us of the works of Pinchot early in the last century, which included many of these same ideas. Similarly, many current ideas were also expressed in the early days of the U.S. extension system, when the knowledge and experience of the "common man" (sic) were routinely lauded.

[11]Forestry scientist Robert Nasi (2003) recently suggested a pertinent title for a review of sustainable forest management: "Sustainable tropical forest management, everything is in order but the patient is dying."

[12]A term I dislike because I do not accept many of the assumptions that tend to go along with its use: that more money will solve all problems, that the financially poor are also lacking in other human traits, that the poor themselves are ipso facto incompetent to solve their own problems, and so on.

[13]Ravi Prabhu was team leader on the project, Testing Criteria and Indicators for Sustainable Forest Management. There was considerable experience developing international-level C&I at that time (Tarapoto Process 1995; Montreal Process 1997; FSC 1998), and Indonesia had produced a national set (LEI 1997) (see Prabhu et al. 1999 for listing). The unique features of the CIFOR project were that it focused on the forest management unit level and involved interdisciplinary field testing of the C&I. The project was funded by GTZ, the European Union, U.S. AID, and a variety of smaller donors between 1994 and 1998.

[14]These examples of principles, criteria, and indicators come from the CIFOR (1999a) set.

Chapter 4: Seven Analytical Dimensions

[1]Community management of tropical forests typically revolves around swidden agriculture, which includes elements of such diverse technical spheres as agriculture, agroforestry, forestry, fisheries, and wildlife. It is often technically illegal.

[2]I have been among those arguing for decentralization for two decades, but that does not make me any happier with the results to date. See Resosudarmo and Dermawan (2002) on Indonesia; Pacheco (2002) and Kaimowitz et al. (1999) on Bolivia; Wittman (2002) on Guatemala; Ribot (2001) and Shackleton et al. (2001) on Africa; Anderson (2001) on Asia and Colfer and Capistrano (forthcoming) for a global view. However, Ribot's oft-repeated comment that devolution has not yet truly been tried is fair.

[3]But early evidence from Indonesia suggests that initial attempts to decentralize authority for forest management to local levels has led to disastrous outcomes, at least in the short term, as local power brokers captured most forest benefits and environmental implications and concerns were ignored (see also Barr et al. 2001; Casson 2001).

Chapter 5: Creativity, Learning, and Equity: The Impacts

[1]The emphasis on the forest users constantly enhancing their capacity emphasizes intentional

learning, especially shared learning (i.e., 'social', 'communicative,' or 'transformative' learning). But this is not to play down the role of reproductive learning as well (e.g., knowledge or skills building through training).

[2]The increase in understanding, knowledge, and skills refers to several kinds of 'learning loops,' i.e., the forest manager may be learning about a specific aspect of forests; cause and effect relationships between a policy/management activity and forest or social outcome (thus about the systems); and/or learning how to learn and manage more effectively.

[3]Uncertainties can range from information that can be accessed from somewhere/someone else, to knowledge that has to be generated, to information that can only be speculated about, for example: market prices; understanding of tree species' growing needs; awareness of condition and amounts of Non Timber Forest Products (NTFPs); likelihood of drought; relationship between certain species and other environmental functions.

[4]See de Geus (2000), Persoon and van Est (2000), and Gallopín (2002), for congenial views on the value of thinking about possible futures.

[5]Although Chivi is not technically an ACM site, there was considerable cooperation between the ACM team and researchers at Chivi, who were also interested in involving local communities more meaningfully in their research.

[6]The titles and contents of two of our major, funded projects reflect this interest: The DFID funds were for "developing indicators-based collaborative monitoring arrangements to promote adaptive community-based forest management"; the E.U. project was entitled "Developing Collaborative Monitoring for Adaptive Comanagement of Tropical African Forests."

[7]The Palawan Provincial Planning Development Office had, by October 2001, already included the C&I set produced in our C&I workshop into its Community Based Monitoring System (Lorenzo 2001a).

[8]See the special issue of *Small Scale Forest Management, Economics and Policy*, which is devoted to FLORES (Vol. 2, No. 2, 2003).

[9]This differentiation is artificial and of analytical use primarily, since many local actors are in fact part of institutions that stretch upward. Moreover, what begins locally often expands to incorporate and/or affect events and people at "higher" levels.

[10]This is not to ignore the considerable power in the hands of local communities to affect their destinies, but rather to recognize the potency of the structured and systemic inequities that surround us all.

[11]David Kaimowitz informed me in October 2003, after this manuscript had been sent to the publisher, that the Zimbabwe government had effectively laid off our researcher (Kaimowitz 2003). By February 2004, he had taken up a position in South Africa—another indication of the ubiquity of surprise (and the worrying conditions in Zimbabwe).

[12]This is also considered important to ensure that all the resource management committees comply with regulations (such as the prohibition on digging up broom grass, rather than cutting it; see below).

[13]By February 2004, the Zimbabwe government had also selected six other districts in which to implement ACM methods on an experimental basis, as part of their normal governmental program.

[14]Some of these skills are reported to have been used subsequently by individuals to negotiate more lucrative private deals with timber companies, without the desirable attention to environmental impacts. Individual capture of benefits intended for communities at large remains another "wicked" problem.

[15]Two of the four forest user groups (deemed average or below average on activeness, institutional development, and equity in 2000) earned special recognition for their excellent progress in 2002. Andheribhajana received a second-place prize from the District Forestry Office, in a field of 200 such groups; and Bamdibhir took first place in Kaski District, in a field of 39.

Chapter 6: Devolution

[1]Cf. Gilmour's (2002) four-part division of the extent of decentralization of forest management: well-established, limited, emerging, and almost nonexistent.

[2]Since May 2003, I have been at three international conferences also attended by officials of this grassroots organization.

[3]As will become clear later in this chapter, some logging continues by local communities—salvage logging.

[4]Protected plant and animal species are the subject of regular updates from MINEF (*Ministère de l'Environnement et Forêts* (Ministry of Environment and Forests).

[5]It is important to remember that the federal agencies that influence Pará also have jurisdiction in Acre; however, the effects on the ground are considerably different, because of the stronger state-level support for devolution and the forests in Acre.

[6]All references to statistical tests of these data refer to the analyses made by Dr. Agus Salim, CIFOR's statistician.

Chapter 7: Forest Type and Population Pressure

[1]We followed Ravi Prabhu in defining forest-rich as contexts characterized by "islands of people in a sea of forest" and forest-poor as "islands of forest in a sea of people" (Colfer et al. 2001).

[2]I hesitate to pronounce any ACM sites "unsuccessful" because we are facilitating a long-term process that may not yield immediate improvements. Those sites that have not yet shown dramatic changes may simply be proceeding at a slower pace because of the complexity of interacting systems in which the participants are enmeshed. Moreover, failures represent opportunities for learning.

Chapter 8: Management Goals

[1]Schmink and Wood (1992, 66) mention the custom of referring to the earliest land claim as the first floor, noting that in Pará, "edifices seven to ten 'floors' high were not unknown." Pokorny (2002b) notes that indigenous lands cannot be colonized now.

[2]Pacheco (2001) reports that 1.5 million ha had been assigned to three municipalities as municipal forest reserves by mid-2000, though 2.4 million ha had been requested. Moreover, 41 local associations had requested legal recognition as forest users, but only half had received such recognition because of bureaucratic problems. By March 2000, only 800 ha of the 1.5 million mentioned above had been allocated as forest concessions to local forest groups (BOLFOR 2000, quoted in Pacheco 2001).

[3]Schmink and Wood (1992, 14) differentiate competitive conflicts from resistance:

Competitive conflicts occurred between members of the same stratum, such as the contests between miners and Indians (both subordinate) over access to gold deposits. Resistance, on the other hand, occurred when members of the subordinate group challenged attempts by the dominant stratum to appropriate resources or to impose its will on the manner in which resources were to be exploited.

Although we have not made this differentiation within our research, it may be useful to bear in mind as we examine conflict more generally. Many situations that we describe as conflict would be resistance in Schmink and Wood's terminology.

[4]Peters (1994) also notes that U.S. AID/Madagascar's SAVEM project financed six protected areas, four of which were national parks, for U.S.$29 million. His conclusion for international

conservation NGOs: "If you want money, operate a national park project."

Chapter 9: Human Diversity

[1]Though among the Javanese (from another Indonesian ethnic group), dead ancestors can curse their descendants for failing to honor them properly or observe the appropriate traditions correctly (field observations, 1980s, Sitiung transmigration site, near Baru Pelepat, in Jambi).

Chapter 10: Conflict and Social Capital

[1]In October 2003, large-scale political problems erupted in Bolivia as well.

[2]Another Cameroonian site, Makak, was initially selected and subsequently abandoned because of high levels of conflict, as were sites in Indonesia and the Philippines.

[3]Malinau is part of an area previously called the Bulungan Research Forest (a name also used in this book).

[4]The team's continuing commitment to the original agenda, in the face of serious problems in implementation, is related to the terms under which they were funded and a lack of flexibility to change direction. Such flexibility is crucial to ACM success.

[5]Schafft and Brown (2002) document the phenomenal increase in interest in this concept, from two or three journal articles in 1990 to more than 200 per year by 2000.

[6]This box is an excerpt from an unpublished report, "Sustainable Livelihoods for Livestock Producing Communities, Kyrgyzstan, Inception Phase Report," June 2001. The report was an output from a project funded by the U.K. Department for International Development (DFID) for the benefit of developing countries. The views expressed are not necessarily those of DFID.

[7]In early 2001, when this research was conducted, U.S.$1 equaled 47.50 Kyrgyz som (information provided by Kaspar Schmidt).

[8]Cf. Holling et al.'s (2002c, 91) comment, "Truly transforming changes are panarchical ones that can cascade up a panarchy as a conscious act of wise, purposive design and implementation," as shown in Westley (2002).

Chapter 11: Catalyzing Creativity, Learning, and Equity

[1]Though for this book, I (an anthropologist) have distinguished the two villages, dealing with his site as two.

[2]It is also important to note, however, that the second team leader, Ravi Prabhu (a forester), was very attuned to the importance of social science input and provided at least as much encouragement and support to social scientists as he did to biophysical scientists (and probably as much as I did). In the ACM case, we had developed a truly interdisciplinary team.

[3]In fact, researchers in Malawi worked with a government research institution; and in Bolivia, the BOLFOR project was already intimately intertwined with the *Superintendencia Forestale* (a government agency). These ties were simply one step removed from those linking ACM to government agencies in the other countries.

[4]See Janssen (2002, 250) for an interesting panarchy-based discussion of the dominance of different roles coinciding with the stages of policy development. He places the activist as most important in the change from K (conservation or implementation) to W (release or crisis); the catalyst, in the change from Ω to α (reorganization or alternatives); the decisionmaker, from α to r (exploitation or policy); and the bureaucrat, from r to K. These are interesting ideas, in light of

the likely, normal roles of incumbents in the institutions represented by our various partners.

[5]Cf. Brown and Asman's (1996, 1473) conclusions that "When parties of unequal power have different interests, we would expect some disagreements in the formulation and implementation of programs that affect those interests. The absence of conflict raises questions about the extent to which mutual influence exists in such relationships."

[6]Specific constraints included the fact that the participatory action research component was conceptualized as a relatively small part of a larger research effort, the limited presence of the primary researcher in the field, the lack of receptivity from the Kyrgyz institutional partners, changes in requirements from the Ph.D. committee, and a felt need by the researcher (a forester) for greater expertise and support in qualitative methods.

[7]The Ghana final report, for instance, estimated that there were five ACM sites, globally (Blay 2002, 3).

[8]Prabhu et al. (2002a) provide a useful, and perhaps more easily accessible summary of Kamoto's work in Malawi and adjacent work in Zimbabwe over the same time period. See also Prabhu 2003.

[9]Blay notes that the Ghana team plans to strengthen its attention to the issue of equitable dealings with illegal settlers in the community forest of Namtee. Net forest benefits are now divided 1/3 to the chief, 1/3 to the elders, and 1/3 for community development—none, for instance, for the volunteers who look after the forest (Blay 2002c).

[10]The exchange rate is 78.6 Nepali Rupees (NR) to the U.S. dollar (May 2003).

Chapter 12: Commentary and Conclusions

[1]Agus Salim, a CIFOR statistician, analyzed these variables and their relationship to ACM success: "We defined the level of success into three categories: low, medium, and high. A logistic regression analysis using SPSS ver. 10.0 shows that there are three variables that can affect the level of success. They are the level of national conflicts, the level of population pressure, and the level of external diversity. Highly successful programs are typically found in areas with high national conflict. The higher the population pressure in an area, the less likely it will be successful. Meanwhile, areas with high external diversity are less likely to be successful. There is no strong difference across country and none of the other variables significantly affect the level of success at the 1% level of significance" (e-mail communication, 3 September 2003).

[2]It is also possible that the success of our overall endeavor owes much to the difficulties we had getting the approach accepted. Perhaps the initial antagonism we encountered provided some of the "chaos" that contributed to ACM success in the field.

[3]This is a more fundamental bifurcation than the one mentioned in Chapter 1, in which we worked with communities and evaluated what we did. The philosophical basis of the evaluation is actually what I'm discussing here.

[4]A qualitative researcher develops and routinely tests multiple minihypotheses about the patterns being observed. This kind of hypothesis testing—a part of day-to-day thinking and problem solving—occurred continually among field-based researchers and community members in the process of participatory action research, which by definition requires ongoing hypothesis testing by participants.

[5]Dove (1988) and his contributors have written compellingly about this process in the collection, *The Real and Imagined Role of Culture in Development: Case Studies from Indonesia.* See also Dove 1993.

[6]Though the woman still has to be the secretary!

[7]And in fact, governance issues wound up being addressed in Nepal as well, since addressing the equity issues had implications for local governance! Similarly, addressing the governance issue in Indonesia led researchers and community members to begin to address some equity issues.

[8]Gunderson et al. (2002), for instance, while noting a fundamental similarity between adaptive ecological and adaptive human systems, propose "that the human ones have much greater powers for both rigidity and novelty. The ability of the bureaucracy of a government agency to control information and resist change seems to show a level of individual and group ingenuity and persistence that reflects conscious control by dedicated and intelligent individuals."

[9]In a previous study, we found that in six interdisciplinary teams assessing criteria and indicators in Cameroon, the most effective groups were composed of both national and international scientists. This combination seemed to provide synergies between the in-depth knowledge the local scientists could provide and the alternative perspectives that a foreigner could bring (Prabhu et al. 1998). Senge (1990) also argues persuasively for the value of bringing a variety of mental models to bear on problems.

References

Abdymomunov, R.A., ed. 2001. *Conclusions of the First National Census Conducted in 1999.* Vol. Oblast Jalal-Abad, Book III, Series R. Bishkek, Kyrgyzstan: National Statistical Committee of the Kyrgyz Republic.

Academy of Science of the Kyrgyz Republic, Institute of Biology, Section of Forests, ed. 1992. *Walnut-Fruit Forests of Southern Kyrgyzstan, Part I (Physical-Geographical Conditions) (Russian language).* Bishkek, Kyrgyzstan: Ilim.

Academy of Science of the Kyrgyz SSR, Section of Geography, ed. 1965. *Climate of the Kyrgyz SSR (Russian language).* Frunze: Ilim.

ACM-Acre Team. 2002. Presentation, University of Florida, Gainesville, 7 May 2002.

ACM PAR Team, Indonesia. 2002. Planning for Sustainability of Forests through Adaptive Co-Management: Indonesia Country Report. Bogor, Indonesia: CIFOR.

ACM Team, Bolivia. 2002. ACM Researcher Seminar, University of Florida, Gainesville, 7 May.

ACM Team, Cameroon. 2002. Les Communautés Riveraines du Parc National de Campo Ma'an: Maîtrise de l'Espace Forestier et Vécu Quotidien des Mutations Socio-Ecologiques. [The Riverine Communities of the Campo Ma'an National Park: Mastery of Forest Spaces and Daily Life Experience of Socio-Ecological Change]. Yaoundé, Cameroon: CIFOR.

Acworth, J., H. Ekwoge, J-M. Mbani, and G. Ntube. 2001. Towards Participatory Biodiversity Conservation in the Onge-Mokoko Forests of Cameroon. *Rural Development Forestry Network Paper* 25d (July): 1–20.

Adario, P. 2002. Brazil and IBAMA—A Case Study. New Haven, CT: Yale University.

Adnan, H., S. Hakim, A. Jafar, and Suprihatin. 2002. The Whispering Game: Study of Information Flow and Dynamics of Stakeholders in Forest Management in Pasir District, East Kalimantan Province. Bogor, Indonesia: CIFOR.

Akwah, G. 2003a. Re: Reports. Email communication with Carol Colfer. 29 April.

———. 2003b. Social Capital in Campo Ma'an, Email communication with Carol Colfer. 28 February.

———. 2002. RE: Age in Cameroon. Email communication with Carol Colfer. 26 November.

Alvira, D. 2002. Logging Crews' Perspectives of Sustainable Forest Management in Bolivia. Gainesville: University of Florida.

Amnesty International. 2002. Political Violence in Madagascar. *The Wire.* July.

Anau, N., R. Iwan, M. van Heist, G. Limberg, M. Sudana, and E. Wollenberg. 2005. Negotiating More Than Boundaries in Indonesia. In *The Equitable Forest: Diversity, Community, and Resource Management*, edited by C.J.P. Colfer. Washington, DC: Resources for the Future and CIFOR.

Anderson, J. 2001. On the Edge of Chaos—Crafting Adaptive Collaborative Management for Biological Diversity Conservation in a Pluralistic World. In *Biological Diversity: Balancing Interests through Adaptive Collaborative Management*, edited by L. Buck, C.C. Geisler, J. Schelhas, and E.W. Wollenberg. Boca Raton, FL: CRC Press.

Anon. 2002. Kyrgyzstan's Political Crisis: An Exit Strategy. Osh/Brussels: International Conflict Group Report No. 37.

———. 2003. *East Kalimantan Statistic.* http://www.jakweb.com/id/kaltim/stat/ (accessed 24 January 2003).

Antona, M., and D. Babin. 2001. Multiple Interest Accommodation in African Forest Management Projects: Between Pragmatism and Theoretical Coherence. *International Journal of Agricultural Resources, Governance and Ecology* 1 (3/4): 286–305.

Applegate, G., R. Smith, J.J. Fox, A. Mitchell, D. Packham, N. Tapper, and G. Baines. 2002. Forest Fires in Indonesia: Impacts and Solutions. In *Which Way Forward? People, Forests and Policymaking in Indonesia*, edited by C.J.P. Colfer and I.A.P. Resosudarmo. Washington, DC: Resources for the Future and CIFOR.

Arda-Minas, L. 2001. Women in ACM-Lantapan. *ACM News* 2 (3): 3–4.

———. 2002a. Cut Flowers in Lantapan. *ACM News* 3 (2): 5–6.

———. 2002b. Stakeholder Analysis: Basak, Lantapan, Province of Bukidnon, Philippines. Bogor, Indonesia: CIFOR.

Ardener, S. (ed.). 1975. *Perceiving Women.* London: Malaby Press.

Arnold, M. 1998. Trees as Out-Grower Crops for Forest Industries: Experience from the Philippines and South Africa. *Rural Development Forestry Network Paper* 22a.

Assembe, S. 2003. Re: Population. Email communication with Carol Colfer. 14 January.

———. 2004. Justice Environnementale, Gestion Durable et Manipulation des Revenus: Evalution des Impacts dans le Sud-Cameroun. Yaounde, Cameroon: CIFOR.

Assembe, S., and P.R. Oyono. Forthcoming. An Assessment of Social Negotiation as a Tool of Local Management. A Case Study of the Dimako Council Forest, Cameroon. *Scandinavian Journal of Forest Research.*

ATD (Association Terre et Développement). 2001. La Terre, un Trésor. Ngoumou, Cameroon: ATD.

———. 2002. Rapport de la Tournée d'Information et de Sensibilisation dan les Villages de la Reserve Forestiere d'Ottotomo à Ngoumou. [Report of the Information Tour and Sensitizing of Villages in the Ottotomo Forest Reserve at Ngoumou]. Ngoumou, Cameroon: ATD.

Atkinson, J.M., and S. Errington. 1990. *Power and Difference: Gender in Island Southeast Asia.* Stanford, CA: Stanford University Press.

AtKisson, A. 1999. *Believing Cassandra: An Optimist Looks at a Pessimist's World.* White River Junction, VT: Chelsea Green.

Auzel, Ph., G.M. Nguenang, R. Feteké, and W. Delvingt. 2001. Small-Scale Logging in Community Forests in Cameroon: Towards Ecologically More Sustainable and Socially More Acceptable Compromises. *Rural Development Forestry Network Paper* 25f(i) (July): 1–11.

Axelrod, R., and M.D. Cohen. 1999. *Harnessing Complexity: Organizational Implications of a Scientific Frontier.* New York: The Free Press.

Banuri, T., and F. Apffel-Marglin. 1993. *Who Will Save the Forests? Knowledge, Power and Environmental Destruction.* London/Atlantic Highlands, NJ: Zed Books.

Barr, C., E. Wollenberg, G. Limberg, N. Anau, R. Iwan, I. Made Sudana, M. Moeliono, and T. Djogo. 2001. *The Impacts of Decentralisation on Forests and Forest-Dependent Communities in Malinau District, East Kalimantan.* Bogor, Indonesia: CIFOR.

Barraclough, S.L., and K. Ghimire. 2000. *Agricultural Expansion and Tropical Deforestation.* London: Earthscan Publications Ltd.

Bateson, G. 1972. *Steps to an Ecology of Mind.* New York: Ballantine Books.

Bateson, M.C. 1989. *Composing a Life.* New York: Atlantic Monthly Press.

Baviskar, A. 2001. Forest Management as Political Practice: Indian Experiences with the Accommodation of Multiple Interests. *International Journal of Agricultural Resources, Governance and Ecology* 1 (3/4): 243–63.

Bennett, C.P.A. 2002. Responsibility, Accountability, and National Unity in Village Governance. In *Which Way Forward? People, Forests and Policymaking in Indonesia*, edited by C.J.P. Colfer and I.A.P. Resosudarmo. Washington, DC: Resources for the Future and CIFOR.

Berlin, B., and P. Kay. 1969. *Basic Color Terms: Their Universality and Evolution.* Berkeley: University of California Press.

Blaser, J., J. Carter, and D. Gilmour (eds.). 1998. *Biodiversity and Sustainable Use of Kyrgyzstan's Walnut-Fruit Forests.* Gland, Switzerland, Cambridge, UK, and Bern, Switzerland: IUCN/Intercooperation.

Blay, D.. 2002a. Ghana Country Report: On the Adaptive and Collaborative Management (ACM) of Community Forests Research Project (Ghana). Harare, Zimbabwe: CIFOR.

———. 2002b. The Potential of Community Forests in Ghana to Contribute to Sustainable Livelihood: The Case of Adwenase and Namtie Community Forests. Harare, Zimbabwe: CIFOR.

———. 2002c. A Question of Diversity. Email communication with Carol Colfer. 17 October.

———. 2003a. CIFOR. Email communication with Carol Colfer. 19 February.

———. 2003b. Ghana Country Report: On the Adaptive and Collaborative Management (ACM) of Community Forests Research Project (Ghana). Harare, Zimbabwe: CIFOR.

———. 2003c. Re: Population Data. Email communication with Carol Colfer. 21 January.

Blay, D., A. Asare, E. Mante, C. Adu-Anning, and K. Diaw. 2000. Report on Site Selection for Adaptive Co-Management Project in Ghana. Yaoundé, Cameroon: CIFOR.

Blay, D., K. Diaw, and C. Adu Anning. 2002. Forest Policy for Collaborative Management in Ghana: How Effective Is It? Harare, Zimbabwe: CIFOR.

Bolaños, O. 2001. Forest Management Policies and Human Well-Being. Gainesville: University of Florida.

———. 2003. Diversity and Conflict in Forest Management Projects: The Case of a Colonist Community in the Lowlands of Bolivia. Master's thesis. Anthropology Department, University of Florida, Gainesville.

Bolaños, O., and M. Schmink. 2005. Women's Place Is Not in the Forest: Gender Issues in a Timber Management Project in Bolivia. In *The Equitable Forest: Diversity, Community, and Resource Management*, edited by C.J.P. Colfer. Washington, DC: Resources for the Future and CIFOR.

Borrini-Feyerabend, G., and D. Buchan. 1997. *Beyond Fences: Seeking Social Sustainability in Conservation.* Gland, Switzerland: IUCN.

Braidotti, R., E. Charkiewicz, S. Hausler, and S. Wieringa. 1994. *Women, the Environment, and Sustainable Development: Towards a Theoretical Synthesis.* London: Zed Books/INSTRAW.

Brechin, S.R., P.R. Wilshusen, C.L. Fortwangler, and P.C. West (eds.). 2003. *Contested Nature: Promoting International Biodiversity with Social Justice in the Twenty-first Century.* Albany: State University of New York.

Brockington, D. 2002. *Fortress Conservation: The Preservation of the Mkomazi Game Reserve Tanzania.* Oxford, UK, Bloomington, Indiana, Dar es Salaam, Tanzania: The International African Institute, Indiana University Press, and Mkuku Na Nyota.

Brown, D.L., and D. Asman. 1996. Participation, Social Capital, and Intersectoral Problem Solving: African and Asian Cases. *World Development* 24(9): 1467–79.

Brown, D., and K. Schreckenberg. 2001. Community Forestry: Facing Up to the Challenge in Cameroon. *Rural Development Forestry Network* 25a (July): 19.

Brown, K., and S. Lapuyade. 2001. Changing Gender Relationships and Forest Use: A Case Study from Komassi, Cameroon. In *People Managing Forests: The Links between Human Well Being and Sustainability*, edited by C. J. P. Colfer and Y. Byron. Washington, DC: Resources for the Future and CIFOR.

Brundtland Report. 1992. http://www.brundtlandnet.com/brundtlandreport.htm. Earth Summit, Rio de Janeiro.

Bryant, D., D. Nielsen, and L. Tangley. 1997. *The Last Frontier Forests: Ecosystems and Economies on the Edge.* Washington, DC: World Resources Institute.

Buck, L. 1998. Capacity Building for Protected Area Planning and Management. Ithaca, NY: CIIFAD/Cornell University.

———. 2002. Personal communication. Ithaca, NY, 19 June.

———. 2003. Personal communication. Ithaca, NY, 30 April.

Buck, L., C.C. Geisler, J. Schelhas, and E. Wollenberg (eds.). 2001. *Biological Diversity: Balancing Interests through Adaptive Collaborative Management.* Boca Raton, FL: CRC Press.

Bulychev, A.S., and B.I. Venglovsky. 1978. *Soil Conditions in the Walnut Forest Belt of Southern Kyrgyzstan.* Frunze: Kyrgyzstan.

Burford de Oliveira, N., with C. McDougall, B. Ritchie, H. Hartanto, and T. Setyawati. 2000a. *Developing Criteria and Indicators for Community Managed Forests as Assessment and Learning Tools: Objectives, Methodologies, Results.* Bogor, Indonesia: CIFOR.

————. 2000b. *Developing Criteria and Indicators for Community Managed Forests.* Bogor, Indonesia: CIFOR.

Burton, E. 2001a. Research Plan for ACM Lantapan, Philippines. Bogor, Indonesia: CIFOR.

————. 2001b. State Policies vs. Tribal Hegemony: Conflict in Forest Management Implementation. *ACM News* 2(4): 6–7.

Campbell, B., A. Mandondo, C. Lovell, W. Kozanayi, O. Mabhachi, T. Makamure, F. Mugabe, M. Mutamba, and S. Sibiza. 2000. *Forging New Institutional Arrangements for Common Property Resource Management: A Case Study from Southern Zimbabwe.* Bogor, Indonesia, and Harare, Zimbabwe: CIFOR and Institute for Environmental Studies.

Campbell, B. 2000. Case 8: Governance of Common Property Resources. In *Equity, Well-Being and Ecosystem Health*, edited by C. P. Program. Cali, Colombia: CGIAR.

————. 2003. *Empowering Forest Dwellers and Managing Forests more Sustainably in the Landscapes of Borneo.* Bogor, Indonesia: CIFOR.

Campbell, C., A. Chicchón, M. Schmink, and R. Piland. 2005. Intra-household Differences in Natural Resource Management in Peru and Brazil. In *The Equitable Forest: Diversity, Community, and Resource Mangement*, edited by C.J.P. Colfer. Washington, DC: Resources for the Future and CIFOR.

Cardenas, J.C. 2002. Bringing the Experimental Lab to the Field: Methods and Lessons. Nyeri, Kenya: CAPRI (Collective Action and Property Rights).

Carter, J. 1996. *Recent Approaches to Participatory Forest Resource Assessment, Rural Development Forestry Study Guide 2.* London: Rural Development Forestry Network, ODI.

————. 1997. Collaborative Forest Management in Kyrgyzstan. An Exploratory Approach in Two Model Leskhozes: Background Document for the Preparation of the Kyrgyz-Swiss Forestry Programme (KIRFOR) Project 05. Bern, Switzerland and Bishkek, Kyrgyzstan: Intercooperation and Goslesagentstvo.

————. 2000. Recent Experience in Collaborative Forest Management Approaches: A Review Paper. Bern, Switzerland: Intercooperation.

Carter, J., B. Steenhof, E. Haldimann, and N. Akenshaev. 2003. Collaborative Forest Management in Kyrgyzstan: From Top-Down to Bottom-Up Decision-Making. *IIED Gatekeeper Series* 108:18.

Casson, A. 2001. *Decentralisation of Policies Affecting Forests and Estate Crops in Kutai Barat District, East Kalimantan.* Bogor, Indonesia: CIFOR.

————. 2002. The Political Economy of Indonesia's Oil Palm Subsector. In *Which Way Forward? People, Forests, and Policymaking in Indonesia*, edited by C.J.P. Colfer and I.A.P. Resosudarmo. Washington, DC: Resources for the Future/CIFOR.

Cayres, G., and B. Pokorny. 2000. Projecto Piloto: Manejo Colaborativo Adaptativo no Pará. As Comunidades São João Batista e Nova Jericó no Município de Tailândia – PA. Case Study Report. Translated and summarized by B. Pokorny. Belem: CIFOR.

Center for Statistics Bureau of Bungo District. 2000. Kabupaten Bungo dalam Angka. Center for Statistics Bureau of Bungo District, Jambi Province, Indonesia. Muara Bungo, Sumatra: Center for Statistics.

Cernea, M. 1990. Beyond Community Woodlots: Programmes with Participation. *Social Forestry Network Paper* 11a.

Chambers, R. 1997. *Whose Reality Counts? Putting the First Last.* London: Intermediate Technology Publications.

Chambers, R., A. Pacey, and L.A. Thrupp (eds.). 1993. *Farmer First: Farmer Innovation and Agricultural Research.* London: Intermediate Technology Press.

Chitiga, M., and N. Nemarundwe. 2003. Policies and Gender Relationships and Roles in the Miombo Woodland Region of Southern Africa. In *Policies and Governance Structures in Woodlands of Southern Africa*, edited by Kowero, G., B.M. Campbell, and U.R. Sumaila. Bogor, Indonesia: CIFOR.

CIA World Fact Book. 2003. *Population Density* (1999). CIA. http://www.photius.com/wfb1999/rankings/population_density_0.html (accessed 25 January 2003).

CIFOR. 1999a. *C&I Toolbox.* 9 vols. Bogor, Indonesia: CIFOR.

————. 1999b. *Generic C&I Template*. Edited by R. Prabhu. *CIFOR C&I Toolbox,* Vol. 2. Bogor, Indonesia: CIFOR.

————. 2000. Managing Common Property Resources in Catchments: Are There Any Ways Forward? CIFOR Policy Brief. Bogor, Indonesia.

————. 2001. Creating Space for Local Forest Management: An Evaluation of the Impacts of Forest Devolution Policies in Asia. Bogor, Indonesia: International Fund for Agricultural Development (IFAD).

————. 2002a. *Bulungan Research Forest* [web]. CIFOR, November 25. http://www.cifor.cgiar.org/ scripts/default.asp?ref=research/bulungan/index.htm (accessed 21 January 2003).

————. 2002b. Forest, Science and Sustainability: The Bulungan Model Forest. Bogor, Indonesia: CIFOR/ITTO.

CIFOR Bulungan Research Forest ACM Team. 2001. Pemerintah Daerah Kabupaten Malinau, Villages of Long Loreh, Sengayan, Bila Bekayuk, Pelancau, Long Rat, Nunuk Tanah Kibang, and Langap. Decentralization that Benefits Communities: Involving Local Forest Communities in Multi-Stakeholder Policy Development at the Kabupaten Level. Submitted to U.K. Department for International Development (DFID).

————. 2002. Forest, Science, and Sustainability: The Bulungan Model Forest. Bogor, Indonesia: CIFOR/International Tropical Timber Organization (ITTO).

CIIFAD. 1997. CIIFAD Aannual Report, 1996–97. Ithaca, NY: Cornell International Institute for Food, Agriculture and Development, Cornell University.

CIMAT 2 (C&I modification and adaptation tool). Bogor, Indonesia: CIFOR.

Cincotta, R.P., and R. Engelman. 2000. *Nature's Place: Human Population and the Future of Biological Diversity.* Washington, DC: Population Action International.

Claridge, G., and B. O'Callaghan (eds.). 1997. *Community Participation in Wetland Management: Lessons from the Field.* Kuala Lumpur, Malaysia: Wetlands International.

Clay, J. W. 1988. *Indigenous Peoples and Tropical Forests: Models of Land Use and Management from Latin America, Cultural Survival Report 27.* Cambridge, MA.: Cultural Survival Inc.

Cleaver, F. 2002. Institutions, Agency and the Limitations of Participatory Approaches to Development. In *Participation: The New Tyranny?* edited by B. Cooke and U. Kothari. London: Zed Books.

Cohen, Y. A. 1968a. *Man in Adaptation: The Biosocial Background.* Chicago: Aldine Publishing Co.

————. 1968b. *Man in Adaptation: The Cultural Present.* Chicago: Aldine Publishing Co.

Colchester, M., T. Apte, M. Laforge, A. Mandondo, and N. Pathak. 2003. Learning Lessons from International Community Forestry Networks: Synthesis Report. Bogor, Indonesia: CIFOR.

Colfer, C.J.P., R.L. Wadley, J. Woelfel, and E. Harwell. 1997. From Heartwood to Bark in Indonesia: Gender and Sustainable Forest Management. *Women in Natural Resources* 18(4): 7–14.

Colfer, Carol J. Pierce. 1977. *Women's Communication and Family Planning in Rural America: The Case of Bushler Bay.* Vol. 4, *Case Studies in Family Planning.* Edited by D.L. Kincaid. Honolulu, HI, USA: EWCI (Case Study 4; reprinted 1978).

————. 1979. In Defense of Many Paths. MPH, International Public Health, University of Hawaii, School of Public Health, Honolulu, HI.

————. 1981. Women of the Forest: An Indonesian Example. *Proceedings, Women in Natural Resources: An International Perspective,* University of Idaho, Moscow.

————. 1983a. Integrating Participants: Farming Systems Lessons from Hawaii. *Proceedings of Kansas State University's 1982 Farming Systems Research Symposium.* Edited by C.B. Flora, Manhattan, KS: International Agricultural Programs, Kansas State University.

————. 1983b. On Communication among 'Unequals'. *International Journal of Intercultural Communication* 7: 263–83.

————. 1991a. Indigenous Rice Production and the Subtleties of Culture Change. *Agriculture and Human Values* VIII (1,2): 67–84.

————. 1991b. *Toward Sustainable Agriculture in the Humid Tropics: Building on the Tropsoils Experience in Indonesia.* Vol. No. 91/02, *Tropsoils Technical Bulletin.* Raleigh, NC: North Carolina State University.

————. 2000. First Attempt to Standardize Dimensions on Site Selection Matrix. Bogor, Indonesia: CIFOR.

————. 2001a. *ACM Teamwork Philosophy.* Yaoundé, Cameroon. Powerpoint presentation.

———. 2001b. Fire in East Kalimantan: A Panoply of Practices, Views and [Discouraging] Effects. *Borneo Research Bulletin* 32: 24–56.

——— (ed.). 2002a. News Flashes. In *ACM News*.

———. 2002b. Social Capital and Empowerment: A Comment. *ACM News* 3 (3): 4–5.

———.(ed.). 2005. *The Equitable Forest: Diversity, Community, and Resource Management.* Washington, DC: Resources for the Future and CIFOR.

Colfer, C.J.P., with A.M. Colfer. 1978. Inside Bushler Bay: Lifeways in Counterpoint. *Rural Sociology* 42(2): 204–20.

Colfer, C.J.P., R. Yost, F. Agus, and S. Evensen. 1989. Expert Systems: A Possible Link from Field Work to Policy in Farming Systems. *AI Applications in Natural Resource Management* 3(2): 31–40.

Colfer, C.J.P., with R.G. Dudley. 1993. *Shifting Cultivators of Indonesia: Managers or Marauders of the Forest? Rice Production and Forest Use among the Uma' Jalan of East Kalimantan, Community Forestry Case Study Series 6.* Rome: Food and Agriculture Organization of the United Nations.

Colfer, C.J.P., R.L. Wadley, and E. Widjanarti. 1996. Using Indigenous Organizations from West Kalimantan. In *Indigenous Organisations and Development*, edited by D. M. Warren, P. Blunt and D. Marsden. London: Kegan Paul International.

Colfer, C.J.P., with N.L. Peluso, and S.C. Chin. 1997. *Beyond Slash and Burn: Building on Indigenous Management of Borneo's Tropical Rain Forests, Advances in Economic Botany.* Vol. 11. Bronx: New York Botanical Garden.

Colfer, C.J.P., and others. 1999a. *The BAG (Basic Assessment Guide for Human Well-Being).* Vol. 5, *C&I Toolbox.* Edited by R. Prabhu. Bogor, Indonesia: CIFOR.

———. 1999b. *The Grab Bag: Supplementary Methods for Assessing Human Well-Being.* Vol. 6, *C&I Toolbox.* Edited by R. Prabhu. Bogor, Indonesia: CIFOR.

Colfer, C.J.P., with R. Prabhu, M. Gunter, C. McDougall, N. Miyasaka Porro, and R. Porro. 1999c. *Who Counts Most? Assessing Human Well-Being in Sustainable Forest Management.* Vol. 8, *C&I Toolbox.* Edited by R. Prabhu. Bogor, Indonesia: CIFOR.

Colfer, C.J.P., R.L. Wadley, and P. Venkateswarlu. 1999d. Understanding Local People's Use of Time: A Precondition for Good Co-Management. *Environmental Conservation* 26: 41–52.

Colfer, C.J.P., R.L. Wadley, A. Salim, and R.G. Dudley. 2000. Understanding Patterns of Resource Use and Consumption: A Prelude to Co-Management. *Borneo Research Bulletin* 31: 29–88.

Colfer, C.J.P., and Y. Byron. 2001. *People Managing Forests: The Links between Human Well-Being and Sustainability.* Washington, DC: Resources for the Future and CIFOR.

Colfer, C.J.P., Y. Byron, with R. Prabhu, and E. Wollenberg. 2001. Introduction: History and Conceptual Framework. In *People Managing Forests: The Links between Human Well Being and Sustainability*, edited by C.J.P. Colfer and Y. Byron. Washington, DC: Resources for the Future and CIFOR.

Colfer, C.J.P., and I.A.P. Resosudarmo (eds.). 2002. *Which Way Forward? People, Forests, and Policymaking in Indonesia.* Washington, DC: Resources for the Future and CIFOR.

Colfer, C.J.P., R.G. Dudley, and R. Gardner. 2002. *Women, Forests and Population.* Ithaca, NY.

Colfer, C.J.P. and D. Capistrano (eds.). Forthcoming. *The Politics of Decentralization: Forests, People, and Power.* London: Earthscan.

Collinson, M. (ed.). 2000. *A History of Farming Systems Research.* New York: CABI.

Collinson, M., and C. Lightfoot. 2001. The Future of Farming Systems Research. In *A History of Farming Systems Research*, edited by M. Collinson. New York: CABI Publishing.

Conklin, H.C. 1957. *Hanunóo Agriculture: A Report on an Integral System of Shifting Cultivation in the Philippines.* Rome: Food and Agriculture Organization of the United Nations.

Contreras-Hermosilla, A., and M.T.V. Ríos. 2002. Social, Environmental and Economic Dimensions of Forest Policy Reforms in Bolivia. Washington, DC: Forest Trends.

Cooke, B., and U. Kothari. 2002. The Case for Participation as Tyranny. In *Participation: The New Tyranny?*, edited by B. Cooke and U. Kothari. London: Zed Books.

Cornet, J.-G., and M. Rajapbaev. 2004. *Criteria and Indicators for Sustainable Management of Juniper Forests in Southern-Kyrgyzstan.* Nancy, France: Laboratoire de Politique Forestière de l'Ecole Nationale du Génie Rurale, des Eaux et des Forêts, Commission of the European Communities.

Cowles, P.D., S. Rakotoarisoa, H. Rasolonirinamanana, and V. Rasoaromanana. 2001. Facilitation, Participation, and Learning in an Ecoregion-Based Planning Process: The Case of AGERAS in Toliara,

Madagascar. In *Biological Diversity: Balancing Interests through Adaptive Collaborative Management*, edited by L.E. Buck, C.C. Geisler, J. Schelhas and E. Wollenberg. Boca Raton, LA: CRC Press.

Croll, E, and D. Parkin (eds.). 1992. *Bush Base, Forest Farm: Culture, Environment and Development*. London: Routledge.

Cronkleton, Peter. 2000. Will Rural Communities Adopt Community Forest Management? Reflections from Acre, Brazil. Bogor, Indonesia: CIFOR.

———. 2001a. ACM Workshop in Bolivia. *ACM News* 2 (3): 3–4.

———. 2001b. Bolivia Site Selection Report and Context Analysis. Santa Cruz, Bolivia: CIFOR.

———. 2001c. Collaboration and Adaptation in the Marketing of Timber by Indigenous People in Lowland Bolivia. Bogor, Indonesia: CIFOR.

———. 2001d. Community Forestry and Gender in Bolivia. *ACM News* 2(2): 6–7.

———. 2003a. Age at ACM Sites. Email communication with Carol Colfer. 21 January.

———. 2003b. Re: Conflict in Guarayos. Email communication with Carol Colfer. 26 February.

———. 2003c. Re: External Stakeholders and Conflict. Email communication with Carol Colfer. 3 March.

———. 2003d. Re: Social Capital. Email communication with Carol Colfer. 26 February.

———. 2005. Gender, Participation and the Strengthening of Indigenous Forest Management in Bolivia. In *The Equitable Forest: Diversity, Community, and Resource Management*, edited by C.J.P. Colfer. Washington, DC: Resources for the Future/CIFOR.

Cunha dos Santos, M. 2001a. Background Information: PAE Porto Dias. Gainesville, FL: CIFOR.

———. 2001b. Presentation and Background Material on Porto Dias. Gainesville, FL: University of Florida.

———. 2002a. Adaptive Co-Management, a Case Study: PAE Porto Dias, Acre, Brazil. Gainesville, FL: PESACRE/University of Florida/CIFOR.

———. 2002b. Porto Dias ACM Project: Final Report. Gainesville, Florida: PESACRE/University of Florida/CIFOR.

———. 2002c. Presentation on Porto Dias. Gainesville: University of Florida.

Cunha dos Santos, M., M. Schmink, and N. Porro. 2002. Strengthening Local Communities and Supportive Public Policies. *ACM News* 3(2): 2–4.

Cwebe Working Group. 2001. Using Participatory Systems Analysis: A Case Study of Migrancy, Gender and Local Forest Management on the Wild Coast of South Africa. Harare, Zimbabwe: Faculty of Forestry, University of Stellenbosch, University of Transkei, Institute of Environmental Studies, University of Zimbabwe Center for International Forestry Research (CIFOR), Department of Rural Economy, University of Alberta.

Dangol, S. 2005. Participation and Decisionmaking in Nepal. In *The Equitable Forest: Diversity, Community, and Resource Management*, edited by C.J.P. Colfer. Washington, DC: Resources for the Future and CIFOR.

Dangol, S., M.R. Banjade, and N. Tumbahangphe. 2000. Adaptive Co-Management Program (ACM), Participatory Action Research (PAR) (Kaski Team, Nepal). Bogor, Indonesia: CIFOR.

———. 2001a. A Historical Trend of Conflict, Collaboration and Adaptation in Community Forestry of Nepal (Cases of Bamdibhirkhoria and Deurali-Baghedanda Forest User Groups), Kaski District, Nepal. Bogor, Indonesia: CIFOR.

———. 2001b. Stakeholders in Bamdibhirkhoria and Deurali-Baghedanda Forest User Groups, Kaski District, Nepal. Bogor, Indonesia: CIFOR.

de Geus, M. 2000. Ecological Utopianism and Sustainability. *Focaal: The Study of the Future in Anthropology* 35: 113–25.

Department for International Development (DFID). 2001a. Salam Alik Ayil Okmotu, Uzgen Raion, Osh Oblast: Ak Terek Village Profile. Bishkek, Kyrgyzstan: DFID.

———. 2001b. Sustainable Livelihoods for Livestock Producing Communities, Kyrgyzstan, Inception Phase Report. Bishkek, Kyrgyzstan: DFID, U.K.

Dermawan, A., and I.A.P. Resosudarmo. 2002. Forests and Regional Autonomy: The Challenge of Sharing the Profits and Pains. In *Which Way Forward? People, Forests, and Policymaking in Indonesia*, edited by C.J.P. Colfer and I.A.P. Resosudarmo. Washington, DC: Resources for the Future and CIFOR.

Devanadera, M. D., N. Devanadera, E. Canete, D. Robles, and M. Rodriquez. 2002. People's Initiatives to Influence Change: The Experience of a Community in Palawan, Philippines. Bali,

Indonesia: Paper presented at Rebuilding Indonesia, a Nation of "Unity in Diversity": Towards a Multicultural Society Workshop.

Dewi, S. 2003. Re: A question [on Population Statistics]. Email communication with Carol Colfer. 27 January.

Diamond, J. 1999. *Guns, Germs and Steel: The Fates of Human Societies.* New York: W.W. Norton and Co.

DFID. 2001. Sustainable Livelihoods for Livestock Producing Communities, Kyrgyzstan, Inception Phase Report. Bishkek, Kyrgyzstan: U.K. Department for International Development.

Diaw, C. 1997. Si, Nda Bot, and Ayong: Shifting Cultivation, Land Use, and Property Rights in Southern Cameroon. *Rural Development Forestry Network Paper* 21e (summer).

———. 2002. Re: Classification by Diversity. Email communication with Carol Colfer. 17 October.

Diaw, C., R. Oyono, A.M. Tiani, C. Kouna, C. Jum, G. Akwah, J. Nguiébouri, P. Bigombe, P. Etoungou, S. Assembe, S. Efoua, W. Mala, and Y. Eben. 2002. Developing Collaborative Monitoring for Adaptive Co-Management of Tropical African Forests: Interim Report for Cameroon 2001. Yaoundé, Cameroon: CIFOR.

Diaw, K. 2001. Developing Collaborative Monitoring for Adaptive Co-Management of Tropical African Forests: Case Study Sites in Ghana. Harare, Zimbabwe: CIFOR.

Diaw, C., and T. Kusumanto. 2005. Scientists in Social Encounters: The Case for an Engaged Practice of Science. In *The Equitable Forest: Diversity, Community, and Resource Management*, edited by C.J.P. Colfer. Washington, DC: Resources for the Future and CIFOR.

Diaw, C., F. Tchala Abina, and P.R. Oyono (eds.). Forthcoming. *Adaptive Collaborative Management of Forests and Decentralization Policies in Cameroon.* Bogor, Indonesia: CIFOR.

Djeumo, A. 2001. The Development of Community Forests in Cameroon: Origins, Current Situation and Constraints. *Rural Development Forestry Network Paper* 25b (July): 1–16.

Dove, M. 1988. *The Real and Imagined Role of Culture in Development: Case Studies from Indonesia.* Honolulu: University of Hawaii Press.

———. 1993. A Revisionist View of Tropical Deforestation and Development. *Environmental Conservation* 20(1): 17–55.

Dubois, O., and J. Lowore. 2000. The "Journey" towards Collaborative Forest Management in Africa: Lessons Learned and Some "Navigational Aids." An Overview. London: International Institute for Environment and Development (IIED).

Dudley, R.G. 2002. Dynamics of Illegal Logging in Indonesia. In *Which Way Forward? People, Forests and Policymaking in Indonesia*, edited by C.J.P. Colfer and I.A.P. Resosudarmo. Washington, DC: Resources for the Future and CIFOR.

———. 2000. The Rotten Mango: The Effect of Corruption on International Development Projects Part 1: Building a System Dynamics Basis for Examining Corruption. Paper presented at Systems Dynamics Society meeting, Bergen, Norway (July).

Dupar, M., and Nathan Badenoch. 2002. *Environment, Livelihoods and Local Institutions: Decentralisation in Mainland Southeast Asia.* Washington, DC: World Resources Institute.

Durrenberger, E.P. 2002. Why the Idea of Social Capital Is a Bad Idea. *Anthropology News* 43(9): 5.

Edmunds, D., and E. Wollenberg. 2001. A Strategic Approach to Multistakeholder Negotiations. *Development and Change* 32: 231–53.

Efoua, Samuel. 2000. Rapport Partiel de Travail de Terrain Baseline Study Ngoumou (Ottotomo). [Partial Report of Field Work for the Baseline Study of Ngoumou]. Yaoundé, Cameroon: CIFOR.

———. 2001. Issues for ACM in the Site of Lomié. *ACM News* 2(4): 12–13.

———. no date. La Problematique de la Gestion de la Reserve Forestiére de Ottotomo. Yaoundé, Cameroon: CIFOR.

Ekoko, F. 1997. The Political Economy of the 1994 Cameroon Forestry Law. Yaoundé, Cameroon: CIFOR.

Ekspres, J. 2001. Pemilihan BPD Terdemokratif di Desa Baru Pelepat [The most democratic village council election in Baru Pelepat Village], page 6.

Emmerson, D.K. (ed.). 1999. *Indonesia beyond Suharto: Polity, Economy, Society, Transition.* Armonk, NY: The Asia Society.

Engel, Paul G.H., A. Hoeberichts, and L. Umans. 2001. Accommodating Multiple Interests in Local Forest Management: A Focus on Facilitation, Actors and Practices. *International Journal of Agricultural Resources, Governance and Ecology* 1(3/4): 306–26.

Engelman, R. 1998. *Plan and Conserve: A Source Book on Linking Population and Environmental Services in Communities.* Washington, DC: Population Action International.

EnviroDev. 2003. *East Kalimantan Pasir: Rattan gardens.* EnviroDev. http://www.envirodev.org/forresasia/sites/site2uk.htm (accessed 24 January 2003).

Environmental Investigation Agency (EIA) and Telapak. 2002a. *Above the Law: Corruption, Collusion and Nepotism and the Fate of Indonesia's Forests.* Jakarta, Indonesia: EIA/Telapak.

———. 2002b. *Timber Trafficking: Illegal Logging in Indonesia, Southeast Asia, and International Consumption of Illegally Sourced Timber.* Jakarta, Indonesia: EIA/Telapak.

Escobar, A. 1995. *Encountering Development: The Making and Unmaking of the Third World, Princeton Studies in Culture/Power/History.* Princeton, NJ: Princeton University Press.

Esguerra, E.M., and H. Hartanto. 2002. Stakeholders' Collaboration and Learning: The Implications of Devolution and Decentralisation in Palawan. Bogor, Indonesia: CIFOR.

Etoungou, P. 2002. Decentralization Viewed from Inside: The Implementation of Community Forests in East Cameroon. Yaoundé, Cameroon: World Resources Institute.

Fairhead, J., and M. Leach. 1996. *Misreading the African Landscape: Society and Ecology in a Forest-Savanna Mosaic, African Studies Series 90.* Cambridge, NY: Cambridge University Press.

Feldstein, H.S., and J. Jiggins (eds.). 1994. *Tools for the Field: Methodologies Handbook for Gender Analysis in Agriculture.* West Hartford, CT: Kumarian Press.

Feldstein, H.S., and S.V. Poats (eds.). 1990. *Working Together: Gender Analysis in Agriculture.* West Hartford, CT: Kumarian Press.

Fernandes, E. 1997. Evolving a Strategy for Sustainable Alternatives to Slash-and-Burn Agriculture. In *CIIFAD Annual Report, 1996–97.* Ithaca, NY: Cornell International Institute for Food, Agriculture and Development (CIIFAD).

Ferraro, P.J. 2002. The Local Costs of Establishing Protected Areas in Low-Income Nations: Ranomafana National Park, Madagascar. *Ecological Economics* 43: 261–75.

Fisher, R.J. 1999. Collaborative Forest Management in Kyrgyzstan: Exploring a New Approach. Bern, Switzerland: Swiss Agency for Development and Cooperation.

———. 2003a. Personal communication, review of Chapters 1–4, 21 May 2003.

———. 2003b. Collaborative Forest Management in Kyrgyzstan: Consolidating a New Approach (CFM Report 9/03). Bishkek, Kyrgyzstan: Natural Resources and Environment Sector of the Swiss Agency for Development and Cooperation.

Fisher, R.J., R. Prabhu, and C. McDougall (eds.). Forthcoming. *Adaptive Collaborative Management of Community Forests in Asia: Experience from Nepal, Indonesia and the Philippines.* Bogor, Indonesia: CIFOR.

Fomété, T. 2001. The Forestry Taxation System and the Involvement of Local Communities in Forest Management in Cameroon. *Rural Development Forestry Network Paper* 25b: 17–28.

Fomété, T., and J. Vermaat. 2001. Community Forestry and Poverty Alleviation in Cameroon. *Rural Development Forestry Network Paper* 25h(i) (July): 1–8.

Ford, R., and W.J. McConnell. 2001. Linking Geomatics and Participation to Manage Natural Resources in Madagascar. In *Biological Diversity: Balancing Interests through Adaptive Collaborative Management,* edited by L. Buck, C. Geisler, J. Schelhas, and E. Wollenberg. Boca Raton, FL: CRC Press.

Forest Stewardship Council (FSC). 1998. Ten Principles. Reprinted in *Guidelines for Developing, Testing and Selecting Criteria and Indicators for Sustainable Forest Management: Criteria and Indicators Toolbox.* No. 1, edited by R. Prabhu, C.J.P. Colfer, and R.G. Dudley. Bogor, Indonesia: CIFOR.

Fortmann, L., and J.W. Bruce. 1988. *Whose Trees? Proprietary Dimensions of Forestry.* Boulder: Westview Press.

Foucault, M. 1980. *Power/Knowledge.* New York: Pantheon Books.

Fox, J., with O. Lynch, M. Zimsky, E. Moore, and H. Takeuchi (eds.). 1993. *Legal Frameworks for Forest Management in Asia: Case Studies of Community/State Relations.* Vol. 16. Honolulu, HI: East-West Center Program on Environment Occasional Paper.

Freire, P. 1970. *Pedagogy of the Oppressed.* New York: Herder and Herder.

Freire, P., and A.M.A Freire. 1994. *Pedagogy of Hope: Reliving Pedagogy of the Oppressed.* New York: Continuum.

Freudenberger, K.S. 1999. Flight to the Forest: A Study of Community and Household Resource Management in the Commune of Ikongo, Madagascar. Ithaca, NY: Cornell University.

Frost, P. 2002. Developing Collaborative Monitoring for Adaptive Co-Management of Tropical African Forests: Ghana Case Study. Harare, Zimbabwe: CIFOR.

Gallopín, G. 2002. Planning for Resilience: Scenarios, Surprises, and Branch Points. In *Panarchy: Understanding Transformations in Human and Natural Systems*, edited by L.H. Gunderson and C.S. Holling. Washington, DC: Island Press.

Gardner-Outlaw, T., and R. Engelman. 1999. *Forest Futures: Population, Consumption and Wood Resources*. Washington, DC: Population Action International.

Geisler, C. 2003. Your Park, My Poverty: Using Impact Assessment to Counter the Displacement Effects of Environmental Greenlining. In *Contested Nature: Promoting International Biodiversity with Social Justice in the Twenty-first Century*, edited by S.R. Brechin, P.R. Wilshusen, C.L. Fortwangler and P.C. West. Albany: State University of New York.

Gibbs, C., E. Payuan, and R. del Castillo. 1990. The Growth of the Philippine Social Forestry Program. In *Keepers of the Forest*, edited by M. Poffenberger. West Hartford, CT: Kumarian Press.

Giesen, W. (ed.). 2000. Special Issue on Danau Sentarum National Park. *Borneo Research Bulletin* 31.

Gilbert, E.H., D.W. Norman, and F.E. Winch. 1980. *Farming Systems Research: A Critical Appraisal*. Vol. No. 6, *MSU Rural Development Paper*. East Lansing, MI: Department of Agricultural Economics.

Gilmour, D. 2002. The Future of Collaborative Management of Forests in Asia. Bangkok: CIFOR: Regional ACM Strategy Meeting.

Gilmour, D., and R.J. Fisher. 1991. *Villagers, Forests and Foresters*. Kathmandu, Nepal: Sahayogi Press.

Gilmour, D., G.C. King, and M. Hobley. 1992. Management of Forests for Local Use in the Hills of Nepal: Part 1, Changing Forest Management Paradigms. In *Readings in Social Forestry and Natural Resource Management for Nepal*, edited by D.A. Messerschmidt and N.K. Rai. Kathmandu, Nepal: HMG Ministry of Agriculture–Winrock International.

Gomes, C.V.A., and H.S. Sassagua. 2002. Deforestation Patterns and Household Determinants of Land Use Choices by Rubber Tappers in Amazonia: The Case of the Chico Mendes Extractive Reserve in Acre, Brazil. Gainesville: University of Florida.

Goodman, M.J. 1985. *Women in Asia and the Pacific: Towards an East-West Dialogue*. Honolulu, HI: Distributed for the Women's Studies Program by the University of Hawaii Press.

Goslesagentsvo, and LES-IC. 1997. Short description of the Kyrgyz Leshozes. Bishkek, Kyrgyzstan: Kyrgyz-Swiss Forestry Sector Support Programme.

Greenwood, D.J. 2002. Naturalizing Social Capital: Social Darwinism—Again and Again and Again. Paper read at Workshop on Social Capital and Civic Involvement, 13–14 September, Cornell University, Ithaca, NY.

Grundy, I.M., B.M. Campbell, R.M. White, R. Prabhu, S. Jensen, and T.N. Ngamile. 2002. Participatory Forest Management in Conservation Areas: The Case of Cwebe, South Africa. Bogor, Indonesia: CIFOR.

Guerrero, M.C.S., and E.F. Pinto. 2001. Reclaiming Ancestral Domains in Pala'wan, Philippines: Community-Based Strategies and Perspectives on Adaptive Collaborative Management. In *Biological Diversity: Balancing Interests through Adaptive Collaborative Management*, edited by L.E. Buck, C.C. Geisler, J. Schelhas, and E. Wollenberg. Boca Raton, FL: CRC Press.

Guha, R. 2001. The Prehistory of Community Forestry in India. *Environmental History (Forest History in Asia)* 6(2): 213–238.

Guijt, I. Forthcoming. *Learning from Collaborative Monitoring in ACM*. Bogor, Indonesia: CIFOR.

Guijt, I., and J. Proost. 2002. Monitoring for Social Learning: Insights from Brazilian NGOs and Dutch Farmer Study Clubs. In *Wheelbarrows Full of Frogs: Social Learning in Rural Resource Management*, edited by C. Leeuwis and R. Pyburn. Assen, Netherlands: Koninklijke van Gorcum.

Guijt, I., and M. Shah. 1998a. *The Myth of Community: Gender Issues in Participatory Development*. London: Intermediate Technology Publications.

Guijt, I., and M.K. Shah. 1998b. Waking Up to Power, Conflict and Process. In *The Myth of Community: Gender Issues in Participatory Development*, edited by I. Guijt and M.K. Shah. London: Intermediate Technology Publications.

Gunderson, L.H., and C.S. Holling (eds.). 2002. *Panarchy: Understanding Transformations in Human and Natural Systems*. Washington, DC: Island Press.

Gunderson, L.H., C.S. Holling, and G.D. Peterson. 2002. Surprise and Sustainability: Cycles of Renewal in the Everglades. In *Panarchy: Understanding Transformation in Human and Natural Systems*, edited by L.H. Gunderson and C.S. Holling. Washington, DC: Island Press.

Haggith, M. 2001. First Thoughts on the Role of Modelling in Action Research. *ACM News* 2(4): 13–14.

Haggith, M., and C.J.P. Colfer. 1999. Artificial Intelligence Meets Traditional Wisdom. *Agroforestry Forum* 9(3): 19.

Haggith, M., R. Prabhu, C.J.P. Colfer, B. Ritchie, A. Thomson, and H. Mudavanhu. 2003. Infectious Ideas: Modelling the Diffusion of Ideas Across Social Networks. *Small-Scale Forest Economics, Management and Policy* 2(2): 225–40.

Hakim, S. 2001a. ACM PAR Lumut Mountain Forest Paser District East Kalimantan, Building Common Perception and Research Agenda (Field Note Summary until March 2001). Bogor, Indonesia: CIFOR.

———. 2001b. Kriteria dan Indikator untuk Pengelolaan hutan lestari Hutan Gunung Lumut (Rantau Layung dan Rantau Buta), Paser, Kalimantan Timur [Criteria and Indicators for the Sustainable Forest Management of Lumut Mountain (Rantau Layung and Rantau Buta), Pasir, East Kalimantan]. Bogor, Indonesia: CIFOR.

———. 2001c. Report ACM PAR Paser, May–July 2001. Bogor, Indonesia: CIFOR.

———. 2002. District Level Workshop, Paser District, East Kalimantan, Indonesia. *ACM News* 3(3): 6–8.

———. 2005. Dealing with Overlapping Access Rights in Indonesia. In *The Equitable Forest: Diversity, Community, and Resource Management*, edited by C.J.P. Colfer. Washington, DC: Resources for the Future and CIFOR.

Hakim, S., A. Jafar, and Suprihatin. 2001a. Context study report, adaptive collaborative management (ACM) research in Lumut Mountain Forest (Rantau Layung and Rantau Buta): Paser District East Kalimantan, Indonesia: Stakeholder Analysis. Bogor, Indonesia: CIFOR.

Hakim, S., Suprihatin, and A. Jafar. 2001b. Context Study Report, Adaptive Collaborative Management (ACM) Research in Lumut Mountain Forest (Rantau Layung and Rantau Buta), Paser District, East Kalimantan, Indonesia: Historical Trends Report. Bogor, Indonesia: CIFOR.

Hall, A. 1997. *Sustaining Amazonia: Grassroots Action for Productive Conservation*. New York: Manchester University Press.

Hanson, P.W. 1995. The Politics of Need Interpretation within the Ranomafana National Park Project. Philadelphia: University of Pennsylvania.

Hardner, J.J., and R. Rice. 1999. Rethinking Forest Concession Policies. In *Forest Resource Policy in Latin America*, edited by K. Keipi. Washington, DC: Inter-American Development Bank.

Harris, M. 1968. *The Rise of Anthropological Theory: A History of Theories of Culture*. New York: Crowell.

Hartanto, H. 2001. Site Selection Report: Lantapan, Mindanao, Philippines. Bogor, Indonesia: CIFOR.

Hartanto, H., T. Villanueva, and M.C. Lorenzo. 2000a. Criteria and Indicator-Based Policy Assessment Report: Palawan. Bogor, Indonesia: CIFOR.

Hartanto, H., T. Villanueva, P. Malbrigo, and N. Sapin. 2000b. Criteria and Indicator-Based Biophysical Assessment Report: Palawan. Bogor, Indonesia: CIFOR.

Hartanto, H., A. Paiman, M. Boestami, Anggana, Suryamin, and L. Yuliani. 2001. Context Study Report: Biophysical Assessment (Jambi). Bogor, Indonesia: CIFOR.

Hartanto, H., L. Arda-Minas, L.M. Burton, A. Estanol, M.C. Lorenzo, and C. Valmores. 2002a. ACM in the Philippines: Summary of Outcomes, Lessons Learnt, and their Policy Implications. Bogor, Indonesia: CIFOR.

Hartanto, H., L. Arda-Minas, L.M. Burton, A. Estanol, M.C. Lorenzo, and Cecil Valmores. 2002b. Philippines Country Report: Planning for Sustainability of Forests through Adaptive Co-management (component 2 report). Bogor, Indonesia: CIFOR.

Hartanto, H., with L. Arda-Minas, L.M. Burton, A. Estanol, M.C. Lorenzo, and C. Valmores. 2002c. Planning for Sustainability of Forests through Adaptive Co-management: Philippines Country Report. Bogor, Indonesia: CIFOR.

Hartanto, H., C. Lorenzo, C. Valmores, L. Arda-Minas, E.M. Burton, and A. Frio. 2003. *Learning Together: Responding to Change and Complexity to Improve Community Forests in the Philippines*. Bogor, Indonesia: CIFOR.

Harwood, R. 2000. The Evolution of FSR-E in Asia through the Mid 1970s: A View from IRRI. In *A History of Farming Systems Research*, edited by M. Collinson. Rome: Food and Agriculture Organization.

Hecht, S., and A. Cockburn. 1989. *The Fate of the Forest: Developers, Destroyers, and Defenders of the Amazon.* New York: Verso.

Hemery, G.E., and S.I. Popov. 1998. The Walnut (*Juglans regia* L.) Forests of Kyrgyzstan and Their Importance as a Genetic Resource. *Commonwealth Forestry Review* 77 (4): 272-276.

Hilborn, R., and C.J. Walters. 1992. Designing Adaptive Management Policies. In *Quantitative Fisheries Stock Assessment: Choice, Dynamics, and Uncertainty*, edited by Chapman and Hill. New York: Chapman and Hall.

Hildebrand, P.E. (ed.). 1986a. *Perspectives on Farming Systems Research and Extension.* Boulder, CO: Lynne Reiner Publishers.

———. 1986b. The *Sondeo*: A Team Rapid Survey Approach. In *Perspectives on Farming Systems Research and Extension*, edited by P.E. Hildebrand. Boulder, CO: Lynne Reiner Publishers.

Hobley, M. 1996a. Institutional Change with the Forestry Sector: Centralised Decentralisation. London: Overseas Development Institute.

———. 1996b. *Participatory Forestry: The Process of Change in India and Nepal.* Vol. 3, *Rural Development Forestry Study Guide.* London: Overseas Development Institute.

Holling, C.S. 1978. *Adaptive Environmental Assessment and Management.* New York: John Wiley and Sons.

Holling, C.S., and L.H. Gunderson. 2002. Resilience and Adaptive Cycles. In *Panarchy: Understanding Transformations in Human and Natural Systems*, edited by L.H. Gunderson and C.S. Holling. Washington, DC: Island Press.

Holling, C. S., S.R. Carpenter, W.A. Brock, and L.H. Gunderson. 2002a. Discoveries for Sustainable Futures. In *Panarchy: Understanding Transformations in Human and Natural Systems*, edited by L.H. Gunderson and C.S. Holling. Washington, DC: Island Press.

Holling, C.S., L.H. Gunderson, and D. Ludwig. 2002b. In Quest of a Theory of Adaptive Change. In *Panarchy: Understanding Transformations in Human and Natural Systems*, edited by L.H. Gunderson and C.S. Holling. Washington, DC: Island Press.

Holling, C.S., L.H. Gunderson, and G.D. Peterson. 2002c. Sustainability and Panarchies. In *Panarchy: Understanding Transformations in Human and Natural Systems*, edited by L.H. Gunderson and C.S. Holling. Washington, DC: Island Press.

Holt, T. Van. 2002. Hunting Practices of the Community of Salvatierra and Recommendations for Future Wildlife Use in the Region. Santa Cruz, Bolivia: University of Florida.

Howard, P.L. (ed.). 2003. *Women and Plants: Gender Relations in Biodiversity Management & Conservation.* London: Zed Books.

Hulme, D., and M. Murphree (eds.). 2001a. *African Wildlife and Livelihoods: The Promise and Performance of Community Conservation.* Oxford, UK: James Currey Ltd.

———. 2001b. Conclusions: Community Conservation as Policy: Promise and Performance. In *African Wildlife and Livelihoods: The Promise and Performance of Community Conservation*, edited by D. Hulme and M. Murphree. Oxford: James Currey Ltd.

IDPAS. 2003. *IDPAS Research Facilities.* Interdepartmental Doctoral Program in Anthropological Sciences, State University of New York, Stony Brook. http://www.uhmc.sunysb.edu/anatomy/IDPAS/field.htm (accessed 31 March 2003).

Indriatmoko, Y. 2002. Local Tenure Systems and their Dynamics: A Case Study of Baru Pelepat Village, Muara Bungo, Jambi. Bogor, Indonesia: CIFOR.

———. 2003a. Re: Population in Baru Pelepat. Email communication with Carol Colfer. 26 January.

———. 2003b. Re: Transmigration. Email communication with Carol Colfer. 13 February.

Indriatmoko, Y., and Y. Kusumanto. 2001. Assessment of the Socio-Cultural and Economic Well-Being of Stakeholders in Baru Pelepat Village, Jambi, Indonesia. Context Study for Adaptive Collaborative Management Research. Bogor, Indonesia: CIFOR.

INPE. 2001. *Atlas do Brasil.* São José dos Campos, Brazil: INPE (*Instituto Nacional de Pesquisas Espaciais*, National Institute for Space Research).

Insanally, R. 2003. Development as if Equity Mattered: A Report on the Conference on Public Policy, Natural Resources and Equity. Georgetown, Guyana.

ITTO (International Tropical Timber Organization). 2003. *Tropical Forest Update* (*ITTO Newsletter*): 13(3).

Janssen, M.A. 2002. A Future of Surprises. In *Panarchy: Understanding Transformations in Human and Natural Systems*, edited by L.H. Gunderson and C.S. Holling. Washington, DC: Island Press.

Jordan, B. 1991. *Technology and Social Interaction: Notes on the Achievement of Authoratitive Knowledge in Complex Settings*. Palo Alto, CA: Xerox Palo Alto Research Center, Institute for Research on Learning and Work Practice and Technology System Sciences Laboratory.

———. 1997. Authoritative Knowledge and its Construction. In *Childbirth and Authoritative Knowledge: Cross-Cultural Perspectives*, edited by R. Davis-Floyd and C. Sargent. Berkeley: University of California Press.

Jordan, B., with R. Davis-Floyd. 1993. *Birth in Four Cultures: A Cross-Cultural Investigation of Childbirth in Yucatan, Holland, Sweden, and the United States*. Prospect Heights, IL: Waveland Press.

Joshi, L. 1997. Incorporating Farmers' Knowledge in the Planning of Interdisciplinary Research and Extension. Ph.D. dissertation. University of Wales, Bangor.

Jum, C. 2003a. Ottotomo, Cameroon: Social capital within Communities and between Communities. Email communication with Carol Colfer. 24 February.

———. 2003b. Re: Checking Population Data in Cameroon. Email communication with Carol Colfer. 13 January.

———. 2003c. Social Capital in Ottotomo. Email communication with Carol Colfer. 26 February.

———. no date. Managing Land Use Conflict in Cameroon: Preliminary Lessons from Ottotomo Forest Reserve. Yaoundé, Cameroon: CIFOR.

Jum, C., and M. Abega. 2003. Participatory Action Research for Collaborative Management: Lessons from the Ottotomo Forest Reserve of Cameroon. Yaoundé, Cameroon: CIFOR.

Jum, C., J. Nguiébouri, and W. Mala. 2001. Context Study Report: Campo. Yaoundé, Cameroon: CIFOR.

Jum, C., R. Oyono, and A.M. Tiani. 2001. Assessing the Context of Outside and Inside Management of the Ottotomo Forest Reserve. Yaoundé, Cameroon: CIFOR.

Jum, C., F. Bengono, and Z. Mbia. 2003. Managing Land Use Conflict in Southern Cameroon: Preliminary Observations from Ottotomo Forest Reserve. Yaoundé, Cameroon: CIFOR.

Kaimowitz, D. 2003. FW: Impacts Chapter. Email communication with Carol Colfer. 29 May.

Kaimowitz, D., P. Pacheco, J. Johnson, I. Pávez, C. Vallejo, and R. Vélez. 1999. Local Governments and Forests in the Bolivian Lowlands. *Rural Development Forestry Network Paper* 24b: 16.

Kaimowitz, D., G. Flores, J. Johnson, P. Pacheco, Iciar Pávez, J. Montgomery Roper, C. Vallejos, and R. Vélez. 2000. Local Government and Biodiversity Conservation: A Case from the Bolivian Lowlands. A Case Study for Shifting the Power: Decentralization and Biodiversity Conservation. Washington, DC: Biodiversity Support Program.

Kainer, K., M. Schmink, A.C.P. Leite, and M.J. da Silva Fadell. 2002. Experiments in Forest-Based Development in Western Amazonia. *Submitted to Society and Natural Resources* 25.

Kamoto, J.F.M. 2002a. Adaptive and Collaborative Management of Forest Resources in Malawi: A Socio-Economic Analysis Report of Chimaliro Site in Kasungu District, Malawi. Harare, Zimbabwe: CIFOR.

———. 2002b. Adaptive and Collaborative Management of Forest Resources in Malawi: A Stakeholder Analysis Report of Chimaliro site in Kasungu District, Malawi. Harare, Zimbabwe: CIFOR.

———. 2002c. Adaptive Collaborative Management of Forest Resources in Malawi: A Historical Trend Analysis Report of Chimaliro Site in Kasungu District, Malawi. Lilongwe, Malawi: CIFOR.

———. 2002d. Chimaliro, Malawi Context Study—Ecology and Production C&I. Harare, Zimbabwe: CIFOR.

———. 2002e. Policy C&I. Harare, Zimbabwe: CIFOR.

———. 2002f. Re: A Question of Diversity. Email communication with Carol Colfer. 17 October.

———. 2003a. Assessment of Collaboration and Adaptiveness in Ten Villages in Chimaliro Forest Reserve. Harare, Zimbabwe: CIFOR.

———. 2003b. Chimaliro Specific Insights. Harare, Zimbabwe: CIFOR.

———. 2003c. EU Final Report (part). Harare, Zimbabwe: CIFOR.

———. 2003d. Population. Email communication with Carol Colfer. 11 February.

———. 2003e. Social Capital/ Future Scenarios. Email communication with Carol Colfer. 10 March.

Karsenty, A. 2000. *Economic Instruments for Tropical Forests: The Congo Basin Case*. London: International Institute for Environment and Development, Center for International Forestry Research, Centre de Cooperation Internationale en Recherche Agronomique pour le Développement.

Kayambazinthu, D. 2000. Empowering Communities to Manage Natural Resources: Where Does the Power Lie? (the Case of Malawi). In *Empowering Communities to Manage Natural Resources: Case Studies from Southern Africa*, edited by S. Shackleton and B. Campbell. Bogor, Indonesia: CIFOR (U.S. AID SADC NRM Project No. 690-0251.12 through WWF-SARPO, EU's "Action in Favour of Tropical Forests" through CIFOR and the Common Property STEP Project, CSIR).

Keirsey, D., and M. Bates. 1984. *Please Understand Me: Character and Temperament Types*. Del Mar, CA: Prometheus Nemesis.

Kelly, G.A. 1963. *A Theory of Personality: The Psychology of Personal Constructs*. New York: W. W. Norton.

Klein, M., B. Salla, and J. Kok. 2001. Attempts to Establish Community Forests in Lomié, Cameroon. *Rural Development Forestry Network Paper* 25f (July): 12–28.

Kotey, N.A., J. Francois, J.G.K. Owusu, R.K. Yeboah, S. Amanor, and L. Antwi. 1998. *Falling into Place: Ghana Country Study*. Vol. 4. London: International Institute for Environment and Development (IIED).

Kouna, C. 2001a. Décentralisation de la Gestion Forestière et Développement Local: Performance et Accountabilité dans la Gestion Locale des Revenus Forestiers à l'Est-Cameroun [Decentralization of Forest Management and Local Development: Performance and Accountability in Local Management of Forest Revenues in East Cameroon]. Yaoundé, Cameroon: CIFOR.

Kouna, C. 2001b. Développement Local au Cameroun: Economie Politique de la Performance et de l'Accountabilité dans la Gestion Locale des Revenus Forestiers a l'Est (Arrondissements de Dimako et Lomié) du Cameroun [Local Development in Cameroon: The Political Economy of Performance and Accountability in Local Management of Forest Revenues in the East (Dimako and Lomié regions)]. Yaoundé, Cameroon: CIFOR.

Krishna, A. 2002a. *Active Social Capital*. New York: Columbia University Press.

———. 2002b. How Does an Asset Possessed at the Micro Level Influence Results at the Regional and National Levels? Examining the Fungibility of Social Capital. Paper read at Workshop on Social Capital and Civic Community, 13–14 September, Cornell University, Ithaca, New York.

Krishna, A., and N. Uphoff. 2001. Mapping and Measuring Social Capital through Assessment of Collective Action to Conserve and Develop Watersheds in Rajasthan, India. In *The Role of Social Capital in Development: An Empirical Assessment*, edited by C. Grootaert and T. van Bastelaar. Cambridge, UK: Cambridge University Press.

Kunstadter, P., E.C. Chapman, and S. Sanga. 1978. *Farmers in the Forest: Economic Development and Marginal Agriculture in Northern Thailand*. Honolulu: Published for the East-West Center by the University Press of Hawaii.

Kunstadter, P. 2002. Poverty, Gender and Malaria. Unpublished manuscript. March.

Kusumanto, T. 2000. Site Selection Report: Adaptive Co-management (ACM) Research in Jambi, Indonesia, May 2000. Bogor, Indonesia: CIFOR.

———. 2001a. Context Study Report, Adaptive Collaborative Management (ACM) Research in Jambi, Indonesia: Stakeholder Analysis. Bogor, Indonesia: CIFOR.

———. 2001b. Context Study Report, Adaptive Collaborative Management (ACM) Research in Jambi, Indonesia: Policies. Bogor, Indonesia: CIFOR.

———. 2001c. Criteria and Indicators for Sustainable Community Forest Management: Baru Pelepat, Jambi. Bogor, Indonesia: CIFOR.

———. 2002. Improving learning in policymaking in Jambi. Bogor, Indonesia: CIFOR.

———. 2003. Governance, Democracy and Social Capital: ACM Research in Jambi, Indonesia. *ACM News* 4(1): 6–7.

Kusumanto, T., and Y. Indriatmoko. 2003. RE: Need Your Input—What Has Gone Wrong? Email communication with Carol Colfer. 27 May.

Kusumanto, T., and E.P. Sari. 2001. Up-and-Down the Ladder: A Jambi Case Study of Policy-Related Information Flows. Bogor, Indonesia: CIFOR.

Kusumanto, T., Marzoni, Effi Permata Sari, and Yayan Indriatmoko. 2001. Quarterly Documentation Report No. 4, ACM research, Baru Pelepat Village, Jambi: Indonesia: 1 September–30 November. Bogor, Indonesia: CIFOR.

Kusumanto, T., S. Hakim, L. Yuliani, Y. Indriatmoko, and H. Adnan. 2002a. Final Report to the Asian Development Bank: Indonesia Country Report: Jambi and Pasir districts. Bogor, Indonesia: CIFOR.

————. 2002b. Summary—Adaptive Collaborative Management Research in Indonesia: Expectation, Outcomes, Lessons and Implications. Presentation to Asian Development Bank, at ACM-Asia Concluding Workshop in Bangkok, Thailand.

Kusumanto, T., L. Yuliani, P. Macouri, Y. Indriatmoko, and H. Adnan. 2005. *Learning to Adapt: Managing Forests Together in Indonesia.* Bogor, Indonesia: CIFOR.

Laksono. 1995. C&I Test in East Kalimantan: Social Component. Bogor, Indonesia: CIFOR.

Leach, M. 1994. *Rainforest Relations: Gender and Resource Use among the Mende of Gola, Sierra Leone.* Washington, DC: Smithsonian Institution Press.

Leach, M., and J. Fairhead. 2001. Plural Perspectives and Institutional Dynamics: Challenges for Local Forest Management. *International Journal of Agricultural Resources, Governance and Ecology* 1(3/4): 223–42.

Leach, M., and R. Mearns (eds.). 1996. *Lie of the Land: Challenging Received Wisdom on the African Environment.* Oxford, UK: International African Institute.

Lederer, W., and E. Burdick. 1958. *The Ugly American.* New York: W.W. Norton.

Lee, K.N. 1993. *Compass and Gyroscope: Integrating Science and Politics for the Environment.* Washington, DC: Island Press.

————. 1999. Appraising Adaptive Management. *Conservation Ecology* 3(2): 3.

Legg, C. 2003. CamFlores: A FLORES-Type Model for the Humid Forest Margin in Cameroon. *Small-Scale Forest Economics, Management and Policy* 2(2).

LEI (Lembaga Ekolabel Indonesia). 1997. Draft Set of Criteria and Indicators for Use in Indonesia's Ecolabelling Institution. Bogor, Indonesia: Lembaga Ekolabel Indonesia.

Levang, P., and FPP-Bulungan Team. 2002. People's Dependence on Forests. In *Forest, Science and Sustainability: The Bulungan Model Forest,* edited by CIFOR. Bogor, Indonesia: CIFOR/ITTO.

Limberg, G., R. Iwan, M. Moeliono, M. Sudana, and E. Wollenberg. 2002. Decentralization and Community Participation in Land Use Planning. Bogor, Indonesia: CIFOR.

Lorenzo, M.C. 2001a. Community Based Monitoring System in the Palawan Provincial Planning Development Office (PPDO). *ACM News* 2(4): 6.

————. 2001b. A Cross Visit in the Philippines. *ACM News* 2(4): 4-6.

————. 2001c. Historical Trends Report—Palawan, Philippines. Bogor, Indonesia: CIFOR.

————. 2001d. Oiling the Wheels of Community Based Forest Management (CBFM): Catalyzing Community Participation for Collaborative Actions. Bogor, Indonesia: CIFOR.

————. 2001e. Stakeholder Analysis Report: Palawan, Philippines. Bogor, Indonesia: CIFOR.

————. 2002. Assessment of Adaptiveness and Collaboration. Bogor, Indonesia: CIFOR.

Lorenzo, M.C., and H. Hartanto. 2001. Adaptive and Collaborative Management (ACM) Planning Workshop. Bogor, Indonesia: CIFOR.

Luna, Maria Paz (Ipat) G. 2001. Tenure and Community Management in the Philippines: Policy Change and Implementation Challenges. In *Biological Diversity: Balancing Interests through Adaptive Collaborative Management,* edited by L. Buck, C.C. Geisler, J. Schelhas, and E. Wollenberg. Boca Raton, FL: CRC Press.

Lynch, O.J., and E. Harwell. 2002. *Whose Natural Resources? Whose Common Good? Towards a New Paradigm of Environmental Justice and the National Interest in Indonesia.* Jakarta, Indonesia: Lembaga Studi dan Advokasi Masyarakat (ELSAM).

Lynch, O.J., and K. Talbott. 1995. *Balancing Acts: Community-Based Forest Management and National Law in Asia and the Pacific.* Washington, DC: World Resources Institute.

Machfudh. 2002. General Description of the Bulungan Research Forest. In *Forest, Science and Sustainability: The Bulungan Model Forest,* edited by CIFOR. Bogor, Indonesia: CIFOR/ITTO.

Magno, F. 2001. Forest Devolution and Social Capital, State–Civil Society Relations in the Philippines. *Environmental History (Forest History in Asia)* 6(2): 264–86.

Mahanty, S., and D. Russell. 2002. High Stakes: Lessons from Stakeholder Groups in the Biodiversity Conservation Network. *Society and Natural Resources* 17: 179–88.

Mala, W. 2003a. Population Data for Akok. Email communication with Carol Colfer. 2 May.

————. 2003b. Re: Social Capital. Email communication with Carol Colfer. 25 April.

Mala, W., R. Oyono, C. Diaw, Y. Ebene, and V. Robiglio. 2001. The Context of Adaptive Co-Management at the Forest-Agriculture Interface: Akok and Kaya 2 and the Dynamics of Forest-Agriculture in Southern Cameroon: Scientific Context. Yaoundé, Cameroon: CIFOR.

Mala, W.A., P.R. Oyono, Y. Ebene, and C. Diaw. 2002. The Context of the Interface between Forest and Agriculture in Southern Cameroon: Study Report. Yaoundé, Cameroon: CIFOR.

Malawi Government. 1998. Population and Housing Census: National Statistical Office.

Malayang, B.S. III. 2001. A Model of Environmental Governance and Forest Policy, and Adaptive Collaborative Management in the Philippines. Bogor, Indonesia: CIFOR.

Malla, Y. 2000. Impact of Community Forestry Policy on Rural Livelihoods and Food Security in Nepal. *Unasylva* 51: 37–45.

———. 2001. Changing Policies and the Persistence of Patron-Client Relations in Nepal. *Environmental History* 6(2): 287–307.

Mandondo, A., and E. Mapedza. 2002. *Allocation of Governmental Authority and Responsibility in Tiered Governance Systems: The Case of Environment-Related laws in Zimbabwe*. Edited by J.C. Ribot and P.G. Veit, *Environmental Governance in Africa Working Papers*. Washington, DC: World Resources Institute.

Margoluis, R., S. Myers, J. Allen, J. Roca, M. Melnyk, and J. Swanson. 2001. *An Ounce of Prevention: Making the Link between Health and Conservation*. Washington, DC: Biodiversity Support Program.

Marti, A. 2000. Stakeholders and Local Resource Management in the Walnut-Fruit Forests of Southern Kyrgyzstan: Final Report on Fieldwork. Bishkek and Bern: Kyrgyz Swiss Forestry Support Programme.

Matose, F.M. 2001. Local People and Reserved Forests in Zimbabwe: What Prospects for Co-Management? Ph.D. dissertation. Development Studies, University of Sussex, Brighton, UK.

Matose, F., and R. Prabhu (eds.). Forthcoming. *Adaptive Resource Management in Zimbabwe: What Prospects for Policy Change?* Bogor, Indonesia: CIFOR.

Matose, F., W. Kozanayi, H. Madevu, T. Mutimukuru, R. Nyirenda, and W. Standa-Gunda. 2002. Facilitating Change in Collaborative Forest Management in Zimbabwe: Final report for the E.U. Harare, Zimbabwe: CIFOR.

Matovu, G.W.M. 2002. Africa and Decentralization: Enter the Citizens. *Development Outreach* 4(1): 24–27.

Mauss, M. 2001. *The Gift: The Form and Reason for Exchange in Archaic Societies*. London: Routledge.

McCay, B.J., and J.M. Acheson. 1987. *The Question of the Commons: The Culture and Ecology of Communal Resources*. Tucson: University of Arizona Press.

McCorkle, C.M. 2000. Anthropology, Sociology and FSR. In *A History of Farming Systems Research*, edited by M. Collinson. Rome: Food and Agriculture Organization.

McDougall, C. 2001a. Draft 3: Final Reporting Framework for ACM Main Case Studies. Bogor, Indonesia: CIFOR.

———. 2001b. Gender and Diversity in Assessing Sustainable Forest Management and Human Well Being. In *People Managing Forests: The Links between Human Well Being and Sustainability*, edited by C.J.P. Colfer and Y. Byron. Washington, DC: Resources for the Future and CIFOR.

———. 2002a. Adaptive and Collaborative Management of Community Forests: Nepal Country Report. Bogor, Indonesia: CIFOR.

———. 2002b. Adaptive and Collaborative Management Research Project in Nepal: An Overview of the Project and Preliminary Lessons. Bangkok: CIFOR.

McDougall, C. 2000. Draft Working Model of ACM (#2), Local People, Devolution and Adaptive Co-Management Program, CIFOR (Feb). Internal draft, Bogor, Indonesia.

McDougall, C., S. Dangol, C. Khadka, R.K. Pandey, N. Tumbahangphe, K. Sharma, N.P. Sitaula, L.P. Uprety, C. Lorenzo, H. Hartanto, Y. Indriatmoko, T. Kusumanto, and L. Yuliani. 2002a. Roles, Relations, Access and (In)equity: Insights into Gender and Diversity in Community Forest Management in Nepal, Indonesia and the Philippines. Bogor, Indonesia: CIFOR/ADB.

McDougall, C., Kaski ACM Team, New ERA ACM Team, and Forest Action. 2002b. Planning for the Sustainability of Forests through Adaptive Co-Management: Nepal Country Report. Bogor, Indonesia: ACM Project/MoFSC Internal Research Report, CIFOR.

McIntosh, J.L. 1980. Cropping Systems and Soil Classification for Agrotechnology Development and Transfer. In *Proceedings, Agrotechnology Transfer Workshop* (7-12 July). Soils Research Institute, Bogor, Indonesia, and University of Hawaii, Honolulu.

McShane, T.O., and Wells, M.P. 2004. *Getting Biodiversity Projects To Work: Towards More Effective Conservation and Development.* New York: Columbia University Press.

Meadows, Donella H., Club of Rome, and Potomac Associates. 1974. *The Limits to Growth: A Report for the Club of Rome's Project on the Predicament of Mankind.* 2nd ed. New York: Universe Books.

Melnyk, M. 2001. *Community Health and Conservation: A Review of Projects.* Washington, DC: Biodiversity Support Program.

Mendoza, G.A., P. Macoun, with R. Prabhu, D. Sukadri, H. Purnomo, and H. Hartanto. 1999. *Guidelines for Applying Multi-Criteria Analysis to the Assessment of Criteria and Indicators.* Vol. 9, *CIFOR C&I Toolbox.* Edited by R. Prabhu. Bogor, Indonesia: CIFOR.

Merrill-Sands, D., E. Holvino, with J. Cumming. 2000. Working with Diversity: A Framework for Action. Washington, DC: Gender and Diversity Program of the Consultative Group on International Agricultural Research (CGIAR).

Messerli, M. 2000. Gender Study in Southern Kyrgyzstan: A Preliminary Study of Women's Opinions Concerning Forest Leases in Four Leshozes. Bishkek, Kyrgyzstan, and Berne, Switzerland: Intercooperation: Kyrgyz-Swiss Forestry Support Programme.

Messerschmidt, D.A., and N.K. Rai (eds.). 1992. *Readings in Social Forestry and Natural Resource Management for Nepal.* Kathmandu, Nepal: HMG Ministry of Agriculture–Winrock International.

Metzger, D.G., and G.E. Williams. 1966. Some Procedures and Results in the Study of Native Categories: Tzeltal "Firewood." *American Anthropologist* 68(2): 389–407.

Milner, J. 2002a. Biophysical Assessment of Ntonya Forest Plantation in a Criteria and Indicator Framework. Harare, Zimbabwe: CIFOR.

———. 2002b. Contextualizing the C&I Framework for Adaptive Co-Management in Ntonya Forest, Zomba. Harare, Zimbabwe: CIFOR.

———. 2002c. Initiation of the Collaborative Monitoring Process at Ntonya site: Zomba, Malawi. Harare, Zimbabwe: CIFOR.

MINEFI/MINAT (Cameroon Ministries of Finance and Interior). 1998. Joint Order No. 00122/MINEFI/MINAT of 29 April by *Ministère des Finances* and Minestère de *l'Administration Territoriale.*

Moeliono, M., and T. Djogo. 2001. Policy Reform—ACM CIFOR. *ACM News* 2(4): 10–11.

Momberg, F. 1993. *Indigenous Knowledge Systems: Potentials for Social Forestry Development (Resource Management of Land-Dayaks in West Kalimantan).* Vol. 3, *Berliner Beitrage zu Umwelt und Entwicklung Bd. 3.* Berlin: Technische Universitat Berlin.

Montreal Process. 1997. Progress Report. Ottawa: Liaison Office, Canadian Forest Service, Natural Resources Canada. February. Available at http://www.dpie.gov.agfor/forests/montreal/c-i.html.

Muhtaman, D.R., C.A. Siregar, and P. Hopmans. 2000. *Criteria and Indicators for Sustainable Plantation Forestry in Indonesia.* Bogor, Indonesia: CIFOR.

Muller, U., and B.I. Venglovsky. 1998. L'Economie des Forêts de Montagne dans l'Ex-URSS: l'Exemple du Kirghizistan. *Revue Forestière Française* numéro spécial: 148-60.1

Murphy, Y., and R.F. Murphy. 1985. *Women of the Forest.* 2nd ed. New York: Columbia University Press.

Musuraliev, T.M. 1998. Forest Management and Policy for the Walnut-Fruit Forests of the Kyrgyz Republic. In *Biodiversity and Sustainable Use of Kyrgyzstan's Walnut-Fruit Forests*, edited by D. Gilmour. Gland, Cambridge, and Bern: IUCN and Intercooperation.

Mutimukuru, T., W. Kozanayi, and R. Nyirenda. 2003. Collaborative Monitoring as a Catalyst for Collaboration and Improvement in Community Forest Management in Africa. Harare, Zimbabwe: CIFOR.

Mutimukuru, T., R. Nyirenda, and F. Matose. 2005. Learning Amongst Ourselves: Adaptive Forest Management through Social Learning in Zimbabwe. In *The Equitable Forest: Diversity, Community, and Resource Management*, edited by C.J.P. Colfer. Washington, DC: Resources for the Future and CIFOR.

Mvondo, S.A., and F. Sangkwa. 2002. Dynamics of Council Forests Creation and Management in Cameroon: The Case of the Dimako Council Forest. Yaoundé, Cameroon: CIFOR/WRI.

Myers, I.B., with P.B. Myers. 1990. *Gifts Differing.* Palo Alto, CA: Consulting Psychologists Press.

Nasi, R. 2003. RE: Moving On. Email communication with Carol Colfer. 14 March.

Nasi, R., A-M. Tiani, and J. Nguiébouri. 2001. Tournée dans l'UTO Campo Ma'an, Village de Nkoelon (23–28/06/2001): "Les Eco-Gardes ne Protègent que le Gibier Mort! Nous on Travaille Pour le Gibier Vivant" (un Villageois de Nkoelon) [Visit to the Campo Ma'an Management Unit, Village of Nkoelon (23–28 June 2001): "The Ecoguards Protect Nothing but Dead Wildlife/Bushmeat! We Work for the Living Wildlife/Bushmeat" (A villager of Nkoelon)]. Yaoundé, Cameroon: CIFOR.

Nawir, A.A., L. Santoso, and I. Mudhofar. 2002. Towards Mutually-Beneficial Company-Community Partnerships in Timber Plantation: Lessons Learnt from Indonesia. Bogor, Indonesia: CIFOR.

Nemarundwe, N. 2005. Women, Decisionmaking, and Resource Management in Zimbabwe. In *The Equitable Forest: Diversity, Community, and Resource Management*, edited by C.J.P. Colfer. Washington, DC: Resources for the Future and CIFOR.

Nepstad, D.C., A.G. Moreira, and A.A. Alencar. 1999. *Flames in the Rain Forest: Origins, Impacts and Alternatives to Amazonian Fires.* Brasilia: Pilot Program to Conserve the Brazilian Rain Forest.

Neumann, R.P. 1991. Local Challenges to Global Agendas: Conservation, Economic Liberalization and the Pastoralists Rights Movement in Tanzania. *Antipode* 27: 363–82.

———. 1998. *Imposing Wilderness: Struggles over Livelihood and Nature Preservation in Africa.* Berkeley: University of California Press.

New Era Team. 2000. Adaptive and Collaborative Management in Community Forestry in Nepal: A Site Selection Report of Manakamana and Andheribhajana Forest User Groups. Bogor, Indonesia: CIFOR.

———. 2001a. Criteria and Indicator (C&I) Based Policy Assessment at the Central, District and ACM Field Site Levels of Sankhuwasabha District. Bogor, Indonesia: CIFOR.

———. 2001b. Criteria and Indicator (C&I) Based Socio-Economic Background Study of Manakamana and Andheribhajana CFUGs in Sankhuwasabha District of Eastern Nepal. Bogor, Indonesia: CIFOR.

———. 2001c. Adaptive and Collaborative Management in Community Forestry in Nepal: A Stakeholder Analysis Report of Manakamana and Andheribhajana CFUGs of Sankhuwasabha District. Bogor, Indonesia: CIFOR.

———. 2001d. A Background Report Highlighting the Need for the Use of Adaptive Co-Management (ACM) in the Forestry Sector in Nepal. Bogor, Indonesia: CIFOR.

Nguiébouri, J., A.M. Tiani, and G.A. Neba. 2001. Les Critères et Indicateurs et la Gestion Durable des Forêts: Rapport d l'atelier de Campo Organisé par le CIFOR du 26–28 Fevrier [Criteria and Indicators and Sustainable Forest Management: A Report of the Campo Workshop Organized by CIFOR, 26–28 February]. Yaoundé, Cameroon: CIFOR.

Nguiébouri, J., A.M. Tiani, G.A. Neba, and C. Diaw. 2002. Etude du Contexte de la Gestion Forestière de la Region de Campo Ma'an (draft) [Context study of Forest Management in the Campo Ma'an Region (draft)]. Yaoundé, Cameroon: CIFOR.

Nguinguiri, J-C. 1999. *Les Approches Contractuelles dans la Gestion des Ecosystèmes Forestiers d'Afrique Centrale. Revue des Initiatives Existantes [Contractual Approaches to the Management of Forested Ecosystems in Central Africa. A Review of Existing Initiatives].* Pointe-Noire, Congo: CIFOR/CORAF.

Ngulube, M. 2000. Adaptive Collaborative Management in Malawi: Policy and Initiatives (draft). Harare, Zimbabwe: CIFOR.

Ngulube, M.R., P.W. Chirwa, L. Mwabumba, and D. Kayambazinthu. 2001. Progress Report for the 1995–1998 Forestry Research Programme and Profiles for the Approved 1998–2001 projects. Zomba, Malawi: Forestry Research Institute of Malawi.

Nhira, C., S. Baker, P. Gondo, J.J. Mangono, and C. Marunda. 1998. *Contesting Inequality in Access to Forests: Zimbabwe Country Study.* Vol. 5. London: International Institute for Environment and Development.

Norman, D.W. 1982. *The Farming Systems Approach to Research.* Vol. 3, *Farming Systems Research Paper.* Manhattan, KS: Kansas State University.

Nuñes, W., G. Cayres, and B. Pokorny. 2000. Projecto Piloto: Manejo Colaborativo Adaptativo no Pará. As Comunidades Recreio, Menino Jesus e Jaratuba no Município de Muaná—PA. Case Study Report. Translated and summarized by B. Pokorny. Belém: CIFOR.

Nyirenda, R. 2000. Zimbabwe Adaptive Collaborative Management Research Project: Mafungautsi Forest: Report on District Level Stakeholders Identification. Harare, Zimbabwe: CIFOR.

Nyirenda, R., T. Mutimukuru, and F. Matose. 2001. *Overcoming Inertia: Stimulating Collective Action among Forest Users in Mafungautsi Forest, Zimbabwe.* Bogor, Indonesia: CIFOR.

Ojha, H., B. Pokharel, with K. Paudel, and C. McDougall. 2002. Comparative Case Studies on Adaptive Collaborative Management: A Review of Eight Community Forestry Sites in Nepal. Bogor, Indonesia: CIFOR.

ONADEF. 1999. Management Plan. Yaoundé, Cameroon: National Forestry Development Agency (ONADEF).

Ostrom, E. 1990. *Governing the Commons: The Evolution of Institutions for Collective Action.* Cambridge: Cambridge University Press.

Oyono, P.R. 1997. Ruptures Socio-Economiques et Surrexploitation du Palmier Raphia par les Populations Forestières de Lomié (sud-est Cameroun) [Socioeconomic Ruptures and Over-Exploitation of Raphia Palms by the Forest Populations of Lomié (southeastern Cameroon)]. *Bulletin Arbres, Forêts et Communautés Rurales* 11: 27–33.

———. 1998. Cameroon Rainforest: Economic Crisis, Rural Poverty, Biodiversity. *Ambio* 27: 557–59.

———. 2002a. *Can Devolution Work in Central Africa? Some Emerging Insights from the Cameroonian Model of Forest Management Decentralization.* Read at the meeting of the joint research program WRI-CIFOR on accountability and environment in decentralizations: Local democracy-environment links in Cameroon.

———. 2002b. Re: Classification by Diversity. Email communication with Carol Colfer. 22 October.

———. 2002c. Policy Change, Organizational Choices, "*Infraoutcomes*" and Ecological Uncertainties of the Decentralization Model in Cameroon. Bellagio, Italy: CIFOR.

———. 2003a. External Diversity. Email communication with Carol Colfer. 13 January.

———. 2003b. History, Devolution and Forest Minorities in Cameroon: Introducing Natural Resource Sociology Analysis and Perspectives. Yaoundé, Cameroon: CIFOR.

———. 2003c. Notes on the Foundations of the Conflit de Langage over Land and Forest in Southern Cameroon. Paper read at the International Conference on Competing Jurisdictions: Settling Land Claims in Africa, 24-26 September, Amsterdam, Netherlands.

———. Institutional Deficit, Representation, and Decentralized Forest Management in Cameroon. Forthcoming. Elements of Natural Resource Sociology for Social Theory and Public Policy. Working Paper 15, Environmental Governance in Africa Series, Washington, DC: World Resources Institute.

Oyono, P.R., and S. Assembe. 2003. Explaining the Dilemma of "post-PAR" Conditions: The Case of Dimako Research Site (East Cameroon). *ACM News* 4 (2).

Oyono, P.R., and S. Efoua. 2001. Rapport de l'atelier d'échange sur les critères et indicateurs de gestion durables des forêts (Lomié, 26–28 mars) [Report on the workshop for exchange of ideas on criteria and indicators for sustainable forest management (Lomié, 26–28 March)]. Yaoundé, Cameroon: CIFOR.

Oyono, P.R., C. Kouna, and W. Mala. Benefits of Forests in Cameroon: Global Structure, Issues Involving Access and Decision-Making Hiccoughs. *Forest Policy and Economics* 7(3): 357–68.

Pacheco, P. 2001. The Role of Forestry in Poverty Alleviation: Bolivia. Rome: Forestry Department, FAO.

———. 2002. The Implications of Decentralization in Forest Management: Municipalities and Local Forest Users in Lowland Bolivia. Bellagio, Italy: World Resources Institute Conference on Decentralization and the Environment.

Padoch, C., and N.L. Peluso. 1996. *Borneo in Transition: People, Forests, Conservation, and Development.* Kuala Lumpur and New York: Oxford University Press.

Pandey, R.K. 2002. Self-monitoring as an Effective Tool to Enhance Adaptive and Collaborative Management (ACM) in Community Forestry in Nepal. *ACM News* 3(4): 9–10.

Peluso, N.L. 1992. *Rich Forests, Poor People: Resource Control and Resistance in Java.* Berkeley: University of California Press.

———. 1994. *The Impact of Social and Environmental Change on Forest Management: A Case Study from West Kalimantan, Indonesia.* Vol. 8, *FAO Community Forestry Case Study Series.* Rome: Food and Agriculture Organization.

———. 2002. Some Questions about Violence and Decentralization: A Preliminary Exploration of the Case of Indonesia. Unpublished manuscript, Berkeley, CA.

Peluso, N.L., and E. Harwell. 2002. Territory, Custom, and the Cultural Politics of Ethnic War in West Kalimantan, Indonesia. In *Violent Environments*, edited by N.L. Peluso and M. Watts. Ithaca, NY: Cornell University Press.

Persoon, G.A., and D.M.E. van Est. 2000. The Study of the Future in Anthropology in Relation to the Sustainability Debate. *Focaal: The Study of the Future in Anthropology* 35: 7–28.

Peters, J. 1994. Reconciling Conflicts between People and Parks at Ranomafana, Madagascar: Management Interventions and Legislative Alternatives. Raleigh: North Carolina State University.

———. 1998. Transforming the Integrated Conservation and Development Project (ICDP) Approach: Observations from the Ranomafana National Park Project, Madagascar. *Journal of Agricultural and Environmental Ethics* 11(1): 17–47.

Peters, W.J. 1997. Local Participation in Conservation of the Ranomafana National Park, Madagascar. *Journal of World Forest Resource Management* 8: 109–35.

Peterson, G.D., C.R. Allen, and C.S. Holling. 1998. Ecological Resilience, Biodiversity and Scale. *Ecosystems* 1: 6–18.

Poats, S.V., M. Schmink, and A. Spring (eds.). 1988. *Gender Issues in Farming Systems Research and Extension*. Boulder, CO: Westview Press.

Poffenberger, M. (ed.) 1990. *Keepers of the Forest: Land Management Alternatives in Southeast Asia*. New Hartford, CT: Kumarian Press.

Poffenberger, M., with P. Battacharya, A. Khare, A. Rai, S.B. Roy, N. Singh, and K. Singh. 1996. *Grassroots Forest Protection: Eastern Indian Experiences*. Vol. 7. Berkeley, CA: Asia Forest Research Network.

Pokorny, Benno. 2002a. Re: Diversity on your Sites. Email communication with Carol Colfer. 17 October.

———. 2002b. Timber Communities. Email communication with Carol Colfer. 15 June.

———. 2003a. Devolution in the Amazon. Email communication with Carol Colfer.

———. 2003b. Forest Descriptions. Email communication with Carol Colfer. 13 January.

———. 2003c. Re: Population. Email communication with Carol Colfer. 28 January.

Pokorny, B., R. Prabhu, C. McDougall, and R. Bauch. 1999. Unity in Diversity? Stakeholder Assessment of Sustainable Forest Management in the Eastern Amazon Region. Belem, Brazil: CIFOR.

Pokorny, B., G. Cayres, and W. Nuñes. 2001a. Participatory Analysis of Heterogeneity as a Tool to Consolidate Collaborative Initiatives at Community Level. Bogor, Indonesia: CIFOR.

Pokorny, B., G. Cayres, W. Nuñes, D. Segebart, and R. Drude. 2001b. Final report, pilot project: Adaptive co-management in Pará. Belem, Brazil: CIFOR.

Pokorny, B., G. Cayres, and W. Nuñes. 2005. Improving Collaboration between Outsiders and Communities in the Amazon. In *The Equitable Forest: Diversity, Community, and Resource Management*, edited by C.J.P. Colfer. Washington, DC: Resources for the Future and CIFOR.

Pollini, J. 2000. Land Tenure, Shifting Cultivation and Natural Resources Conservation in Madagascar: The Case of Ranomafana National Park. Ithaca, New York: Cornell University.

Porro, N.M. 2001. Rights and Means to Manage Cooperatively and Equitably: Forest Management among Brazilian Transamazon Colonists. In *People Managing Forests: The Links between Human Well Being and Sustainability*, edited by C.J.P. Colfer and Y. Byron. Washington, DC: Resources for the Future and CIFOR.

———. 2002. Rupture and Resistance: Gender Relations and Life Trajectories in the Babaçu Palm Forests of Brazil. Ph.D. dissertation. Anthropology, University of Florida, Gainesville.

Porro, R., and N. Miyasaka Porro. 1998. Methods for Assessing Social Science Criteria and Indicators for the Sustainable Management of Forest: Brazil Test. Bogor, Indonesia: CIFOR.

Porro, R., A.M. Tiani, B. Tchikangwa, M.A. Sardjono, A. Salim, C.J.P. Colfer, and M.A. Brocklesby. 2001. Access to Resources in Forest-Rich and Forest-Poor Contexts. In *People Managing Forests: The Links between Human Well Being and Sustainability*, edited by C.J.P. Colfer and Y. Byron. Washington, DC: Resources for the Future and CIFOR.

Poschen, P. 1993. Forestry, a Safe and Healthy Profession? *Unasylva* 172(44): 3–12.

Poteete, A., and E. Ostrom. 2003. In Pursuit of Comparable Concepts and Data about Collective Action. *CAPRi Working Paper* 29.

Potter, L., and S. Badcock. 2001. *The Effects of Indonesia's Decentralisation on Forests and Estate Crops in Riau Province: Case Studies of the Original Districts of Kampar and Indragiri Hulu*. Vol. 6 and 7. Bogor, Indonesia: CIFOR.

Prabhu, R. 2003. Developing collaborative monitoring for adaptive co-management of tropical African forests (Contract Number: B7-6201/99-05/FOR). Final report for the period January 1, 2000–December 31, 2002. Harare, Zimbabwe: CIFOR.

Prabhu, R., C.J.P. Colfer, P. Venkateswarlu, L.C. Tan, R. Soekmadi, and E. Wollenberg. 1996. Testing Criteria and Indicators for the Sustainable Management of Forests: Phase 1. Final Report. Bogor, Indonesia: CIFOR.

Prabhu, Ravi, W. Maynard, Richard Eba'a Atyi, C.J.P. Colfer, Gill Shepherd, P. Venkateswarlu, and F. Tiayon. 1998. Testing and developing criteria and indicators for sustainable forest management in Cameroon: The Kribi test, final report. Bogor, Indonesia: CIFOR.

Prabhu, R., C.J.P. Colfer, and R.G. Dudley. 1999. *Guidelines for Developing, Testing, and Selecting Criteria and Indicators for Sustainable Forest Management: A C&I Developer's Reference*. Edited by R. Prabhu, *C&I Toolbox*. Bogor, Indonesia: CIFOR.

Prabhu, R., C.J.P. Colfer, and C. Diaw. 2001. Sharing Benefits from Forest Utilisation: Trojan Horses, Copy Cats, Blind Mice and Busy Bees. Georgetown, Guyana: Cropper Foundation, Iwokrama Centre, Woods Hole Research Center.

Prabhu, R., F. Matose, L. Mwabumba, J. Kamoto, H. Madevu, J. Milner, R. Nyirenda, and W. Standa-Gunda. 2002a. Developing Indicators Based Collaborative Monitoring Arrangements to Promote Adaptive Community Based Forest Management. Harare, Zimbabwe: DFID/CIFOR.

Prabhu, R., C. McDougall, H. Hartanto, L. Yuliani, Y. Yasmi, and C.J.P. Colfer. 2002b. National and Regional Efforts to Adopt and Disseminate Research Findings, vol. 4. Bogor, Indonesia: CIFOR/Asian Development Bank, RETA 5812: Planning for Sustainability of Forests through Adaptive Co-management.

Pretty, J. and N. Uphoff. 2002. Human Dimensions of Agroecological Development. In *Agroecological Innovation: Increasing Food Production with Participatory Development*, edited by N. Uphoff. London: Earthscan Publications.

Prigogine, I. 1997. *The End of Certainty: Time, Chaos and the New Laws of Nature*. New York: The Free Press.

Purnomo, H. H. Priyadi, Y. Yasmi, L. Yuliani, and R. Prabhu. 2001. Development of Multi-Stakeholder Scenarios of Forest Management: A Multi-Agent System Simulation Approach. Bogor, Indonesia: CIFOR.

Purnomo, H., Y. Yasmi, R. Prabhu, C.J.P. Colfer, K. Kartawinata, and A. Salim. 2000. Sustainable Forest Management from the Perspective of Local Stakeholders: A Case Study from Bulungan Research Forest Indonesia. Bogor, Indonesia: CIFOR.

Purnomo, H., Y. Yasmi, R. Prabhu, S. Hakim, A. Jafar, and Suprihatin. 2003. Collaborative Modelling to Support Forest Management: Qualitative Systems Analysis at Lumut Mountain, Indonesia. *Small-Scale Forest Economics, Management and Policy* 2(2).

Purnomo, H., Y. Yasmi, R. Prabhu, L. Yuliani, H. Priyadi, and J.K. Vanclay. 2003. Multi-Agent Simulation of Alternative Scenarios of Collaborative Forest Management. *Small-Scale Forest Economics, Management and Policy* 2(2).

Putnam, R.D. 1995. Bowling Alone: America's Declining Social Capital. *Journal of Democracy, January*: 65–78.

Quist-Arcton, O. 2002. Violence Breaks Out as Rival Camps Clash. *AllAfrica.com*, 28 February.

Rakotondrabe, G. 1999. CBLUPM Helps Villagers Take Charge. In *CIIFAD Annual Report, 1997-98*. Ithaca, NY: CIIFAD (Cornell International Institute for Food, Agriculture and Development).

Rakotoson, L. 1999. Forest Use Agreements in Anjamba. In *CIIFAD Annual Report, 1997-98*. Ithaca, NY: CIIFAD (Cornell International Institute for Food, Agriculture and Development).

Rakotoson, L., D.H. Razaivaovololoniaina, and E. Razafindraboto. 1999. Rapport D'activites sur la Negociation de L'utilisation des Ressources et Essay D'elaboration d'une Convention sur la Gestion Communautaires des Ressources Forestieres dan la Region de Ranomafana-Ifanadiana [Report of Activities on the Negotiations for the use of Resources and the Effort to Elaborate a Convention on Community Management of Forest Resources in the Region of Ranomafana-Ifanadiana]. Villa NTSOA, Mahamanina Fianarantsoa, Madagascar: CIFOR/PAPAMS (Protected and Peripheral Area Management System).

Rakotoson, L., A Randrianarivo, and F. Razafindrakoto. 2000. Atelier sur la Promotion des Plantes Medicinales et Aromatiques, de la Pharmacopée Traditionelle et de L'ethno-eco-tourisme dans la Région de Ranomafana [Workshop on the Promotion of Medicinal and Aromatic Plants, of the Traditional Pharmacopea and of Ethno-eco-tourism in the Ranomafana region]. Ranomafana, Madagascar: CIFOR/LDI/MICET/ University of Michigan.

Ramírez, R. 2001. Understanding the Approaches for Accommodating Multiple Stakeholders' Interests. *International Journal of Agricultural Resources, Governance and Ecology* 1(3/4): 264–85.

Redford, K.H., and C. Padoch. 1992. *Conservation of Neotropical Forests: Working from Traditional Resource Use.* New York: Columbia University Press.

Reisman, Don. Email communication with Carol Colfer. February 2004.

Research & Consulting Group. 2000. National Opinion Poll in the Kyrgyz Republic. SGI CMA, Research & Consulting Group, Kyrgyzstan.

Resosudarmo, I.A.P., and A. Dermawan. 2002. Forests and Regional Autonomy: The Challenge of Sharing the Profits and Pains. In *Which Way Forward? People, Forests and Policymaking in Indonesia*, edited by C.J.P. Colfer and I.A.P. Resosudarmo. Washington, DC: Resources for the Future and CIFOR.

Rhee, S. 2002. Community Management of Natural Resources in a de facto decentralizing Indonesia: A Case Study from East Kalimantan. *Tropical Resources: The Bulletin of the Yale Tropical Resources Institute* 21: 25–30.

Ribot, J.C. 2001. Local Actors, Power and Accountability in African Decentralizations: A Review of Issues. Geneva: U.N. Research Institute for Social Development.

————. 2002. Democratic Decentralization of Natural Resources: Institutional Choice and Discretionary Power Transfers in Sub-Saharan Africa. *Public Administration and Development* 23 (1): 53-65.

Rice, D. 2001. Forest Management by a Forest Community: The Experience of the Ikalahan. Bogor, Indonesia: CIFOR.

Richardson, G.P. 1991. *Feedback Thought in Social Science and Systems Theory.* Philadelphia: University of Pennsylvania Press.

Ritchie, B., C. McDougall, M. Haggith, and N. Burford de Oliveira. 2000. *An Introductory Guide to Criteria and Indicators for Sustainability in Community Managed Forest Landscapes.* Bogor, Indonesia: CIFOR.

Robiglio, V., W. Mala, and C. Diaw. 2003. Mapping Landscapes: Integrating GIS and Social Science Methods to Model Human-Nature Relationships in Southern Cameroon. *Small-Scale Forest Economics, Management and Policy* 2(2): 171-184.

Rocheleau, D., and R. Slocum. 1995. Participation in Context: Key Questions. In *Power, Process and Participation: Tools for Change*, edited by R. Slocum, L. Wichhart, D. Rocheleau and B. Thomas-Slayter. London: Intermediate Technology Publications.

Rocheleau, D., B. Thomas-Slayter, and E. Wangari (eds.). 1996. *Feminist Political Ecology: Global Issues and Local Experiences.* London: Routledge.

Roe, E. 1994. *Narrative Policy Analysis: Theory and Practice.* Durham, NC: Duke University Press.

Rosaldo, M.Z., and L. Lamphere (eds.). 1974. *Woman, Culture and Society.* Stanford, CA: Stanford University Press.

Ruitenbeek, J., and C. Cartier. 1998. *Rational Exploitations: Economic Criteria & Indicators for Sustainable Management of Tropical Forests.* Vol. 17, CIFOR Occasional Papers. Bogor, Indonesia: CIFOR.

————. 2001. The Invisible Wand: Adaptive Co-management as an Emergent Strategy in Complex Bio-Economic Systems. *CIFOR Occasional Paper* No. 34.

Russell, D., and C. Harshbarger. 2003. *Groundwork for Community-Based Conservation: Strategies for Social Research.* New York: Altamira Press.

Russell, D., and N. Tchamou. 2001. Soil Fertility and the Generation Gap: The Bene of Southern Cameroon. In *People Managing Forests: The Links between Human Well-Being and Sustainability*, edited by C.J.P. Colfer and Y. Byron. Washington, DC: Resources for the Future.

Ryan, J. 2003. *Madagascar Biodiversity Monitoring.* Hobart and William Smith Colleges in Geneva, NY http://academic.hws.edu/bio/Pages/RyanResearch.html (accessed 31 March 2003).

Salim, A., C.J.P. Colfer, and C. McDougall. 1999. *The Scoring and Analysis Guide for Assessing Human Well-Being.* In *C&I Toolbox,* vol. 7, edited by R. Prabhu. Bogor, Indonesia: CIFOR.

Salim, A., M.A. Brocklesby, A.M. Tiani, B. Tchikangwa, M.A. Sardjono, R. Porro, J. Woelfel, and C.J.P Colfer. 2001. In Search of a Conservation Ethic. In *People Managing Forests: The Links between Human Well-Being and Sustainability*, edited by C.J.P. Colfer and Y. Byron. Washington, DC: Resources for the Future and CIFOR.

Salim, E. 2002. Afterword. In *Which Way Forward? Forests, People and Policymaking in Indonesia*, edited by C.J.P. Colfer and I.A.P. Resosudarmo. Washington, D.C.: Resources for the Future and CIFOR.

Sanday, P. 1974. Female Status in the Public Domain. In *Woman, Culture and Society*, edited by M.R. Rosaldo and L. Lamphere. Stanford, CA: Stanford University Press.

Sankar, S., P.C. Anil, and M. Amruth. 2000. *Criteria and Indicators for Sustainable Plantation Forestry in India*. Bogor, Indonesia: CIFOR.

Sarin, Madhu. 1993. *From Conflict to Collaboration: Local Institutions in Joint Forest Management*. Vol. 14, *Joint Forest Management Working Paper*. New Delhi, India: The Ford Foundation.

———. 1997. Meeting of the Group on Gender and Equity in Joint Forest Management. New Delhi: Society for the Promotion of Wasteland Development.

———. 1998. Putting Community Institutions under the Microscope: Selective Inclusion, Participation, and Rights under Joint Forest Management (preliminary draft). Anand, India: Prepared for the Workshop on Shared Resources Management, IRMA.

Sarin, M., with N.M. Singh, N. Sundar, and R.K. Bhogal. 2003. Devolution as a Threat to Democratic Decision-making in Forestry? Findings from Three States in India. London: Overseas Development Institute.

Sardjono, M.A. 2004. *Mosaik Sosiologis Kehutanan: Masyarakat Lokal, Politik dan Kelestarian Sumberdaya*. [The Sociological Mosaic of Forestry: Local Communities, Politics, and Sustainable Natural Resources]. Jogjakarta, Indonesia: Debut Press.

Sayer, J., and B. Campbell. 2003. *The Science of Sustainable Development: Local Livelihoods and the Global Environment*. Cambridge: Cambridge University Press.

Sayer, J.A., N. Ishwaran, J. Thorsell, and T. Sigaty. 2000. Tropical Forest Biodiversity and the World Heritage Convention. *Ambio* 29(6): 302–09.

Schafft, K.A., and D.L. Brown. 2002. Social Capital, Social Networks, and Social Power. Paper read at Workshop on Social Capital and Civic Involvement, 13–14 September, Cornell University, Ithaca, NY.

Scheffer, M., F. Westley, W.A. Brock, and M. Holmgren. 2002. Dynamic Interactions of Societies and Ecosystems—Linking Theories from Ecology, Economy, and Sociology. In *Panarchy: Understanding Transformations in Human and Natural Systems*, edited by L.H. Gunderson and C.S. Holling. Washington, DC: Island Press.

Schindler, B., B. Steel, and P. List. 1996. Public Judgments of Adaptive Management: A Response from Forest Communities. *Journal of Forestry* 94(6): 4–12.

Schmidt, K. 2002a. Diversity, Age and Gender in Kyrgyzstan. 10 December.

———. 2002b. Knowledge and Strategies of Local People in Forest Management: A Research Project Contributing to the Development of Collaborative Forest Management in the Walnut-Fruit Forest in Kyrgyzstan. Jalal-Abad, Kyrgyzstan: Research Fellow Partnership Program (RFPP), SDC: ZIL.

———. 2002c. *A Research Project Contributing to the Development of Collaborative Forest Management in a Post Soviet Context: CFM:ACM Research Group Kyrgyzstan*. Bogor, Indonesia: Powerpoint presentation at CIFOR.

———. 2003a. Conflicts in Kyrgyzstan. 24 March.

———. 2003b. Forest, Population and Biophysical Information on the Kyrgyz Sites. Email communication with Carol Colfer. 30 March.

———. 2003c. Kyrgyzstan: External Stakeholders. Email communication with Carol Colfer. 30 May.

———. 2003d. Kyrgyzstan: Devolution Status. Email communication with Carol Colfer. 30 May.

———. 2003e. Knowledge and Strategies of Local People in Forest Management. A Research Project Contributing to the Development of Collaborative Forest Management in the Walnut-Fruit Forest in Kyrgyzstan. Second Annual Progress Report. Unpublished. Jalal-Abad, Kyrgyzstan.

Schmink, M. 2000. Lessons from the Field. *ACM News* 1(1): 4.

Schmink, M., and C. Wood. 1992. *Contested Frontiers in Amazonia*. New York: Columbia University Press.

Sciarrone, R. 2002. The Dark Side of Social Capital: The Case of Mafia. Paper read at Workshop on Social Capital and Civic Involvement, 13–14 September, Cornell University, Ithaca, NY.

Scott, J.C. 1985. *Weapons of the Weak*. New Haven, CT: Yale University Press.

———. 1998. *Seeing Like a State: How Certain Schemes to Improve the Human Condition Have Failed*. New Haven, CT: Yale University Press.

Segebart, D., R. Drude, and B. Pokorny. 2000. Projecto Piloto: Manejo Colaborativo Adaptativo no Pará. A Comunidade Canta-Galo no Município de Gurupá—PA. Case Study Report. Translated and summarized by B. Pokorny. Belém: CIFOR.

Sen, G. 1994. Development, Population and the Environment: A Search for Balance. In *Population Policies Reconsidered: Health, Empowerment and Rights*, edited by G. Sen, A. Germain, and L. Chen. Boston and New York: International Women's Health Coalition.

Senge, P.M. 1990. *The Fifth Discipline: The Art and Practice of the Learning Organization*. New York: Currency Doubleday.

Shackleton, S., B.M. Campbell, with M. Cocks, G. Kajembe, E. Kapungwe, D. Kayambazinthu, B. Jones, S. Matela, G. Monela, A. Mosimane, N. Nemarundwe, N. Ntale, N. Rozemeijer, C. Steenkamp, B. Sithole, J. Urh, and C. van der Jagt. 2001. *Devolution in Natural Resource Management: Institutional Arrangements and Power Shifts (a Synthesis of Case Studies from Southern Africa)*. Bogor, Indonesia: Funded by U.S. AID SADC NRM Project No. 690-0251.12, through WWF-SARPO, EU's "Action in Favour of Tropical Forests" through CIFOR and the Common Property STEP Project, CSIR.

Shackleton, S., B. Campbell, E. Wollenberg, and D. Edmunds. 2002. Devolution and Community-based Natural Resource Management: Creating Space for Local People to Participate and Benefit? *ODI Natural Resource Perspectives* 76: 1-6.

Shaner, W.W., P.F. Philipp, W.R. Schmehl, and Consortium for International Development. 1982. *Farming Systems Research and Development: Guidelines for Developing Countries*. Boulder, CO: Westview Press.

Sharma, K. 2001. The Dangers of Superstition at ACM Site in Nepal. *ACM News* 2(2): 5.

———. 2002. Gender Dynamics of Community Forest Management through Collaborative and Adaptive Strategies: A Case Study from the ACM Sites of Sankhuwasabha District, Nepal. Bogor, Indonesia: CIFOR.

Shiel, D. 2002. Biodiversity Research in Malinau. In *Forest, Science and Sustainability: The Bulungan Model Forest*, edited by CIFOR. Bogor, Indonesia: CIFOR/ITTO.

Shiva, V. 1989. *Staying Alive: Women, Ecology and Development*. London: Zed Books.

Siagian, Y. 2001. The Adaptive and Collaborative Management of Community Forests Research Project: A Participatory Research and Gender Analysis Impact Assessment in Jambi: Indonesia Case Study (Baru Pelepat Village & Batu Kerbau Village, Indonesia). Bogor, Indonesia: CIFOR.

Sigot, A., L.A. Thrupp, and J. Green (eds.). 1995. *Towards Common Ground: Gender and Natural Resource Management in Africa*. Nairobi, Kenya/Washington, DC: African Centre for Technology Studies and World Resources Institute.

Simmonds, N.W. 1985. *Farming Systems Research: A Review*. Washington, DC: World Bank.

Simula, M., and D. Burger. 2003. Brazil's Mission to Achieve SFM: An ITTO Mission to Diagnose Brazil's Efforts for Achieving Sustainable Forest Management Recommends that ITTO Focus its Assistance on Strategic Interventions. *Tropical Forest Update (ITTO Newsletter)* 13 (1/2).

Sinclair, F.L, and D.J. Walker. 1998. Acquiring Qualitative Knowledge about Complex Agroecosystems. Part I: Representation as Natural Language. *Agricultural Systems* 56(3): 341-63.

Singhanetra-Renard, A. 1993. Malaria and Mobility in Thailand. *Social Science and Medicine* 37(9): 1147–54.

Siskind, J. 1973. *To Hunt in the Morning*. New York: Oxford University Press.

Sist, P., D. Sheil, K. Kartawinata, and H. Priyadi. 2002. Reduced-Impact Logging in Indonesian Borneo: Some Results Confirming the Need for New Silvicultural Prescriptions. In *Forest, Science and Sustainability: the Bulungan Model Forest*, edited by CIFOR. Bogor, Indonesia: CIFOR/ITTO.

———. 2003. Reduced-Impact Logging in Indonesian Borneo: Some Results Confirming the Need for New Silvicultural Prescriptions. *Forest Ecology and Management* 179: 415–27.

Sitaula, N.P. 2001. Equity Considerations for Sustainable Forest Management: Experiences from Manakamana, Nepal. Bogor, Indonesia: CIFOR.

Sithole, B. 2001. Uncovering Shrouded Identities: Institutional Considerations and Differentiation in an Adaptive Co-management Project in Zimbabwe. Bogor, Indonesia: CIFOR.

———. 2002a. Re: Age in Zimbabwe. Email communication with Carol Colfer. 21 November.

———. 2002b. *Where the Power Lies: Multiple Stakeholder Politics over Natural Resources: A Participatory Methods Guide.* Bogor, Indonesia: CIFOR.

———. 2005. Becoming Men in Our Dresses! Women's Involvement in a Joint Forestry Management Project in Zimbabwe. In *The Equitable Forest: Diversity, Community, and Resource Management*, edited by C.J.P. Colfer. Bogor, Indonesia: Resources for the Future and CIFOR.

Soemarwoto, O. 2003. Kata pengantar [Introductory remarks]. In *Kemana Harus Melangkah? [Which Way Forward?]*, edited by I.A.P. Resosudarmo and C.J.P. Colfer. Jakarta, Indonesia: Yayasan Obor, Resources for the Future, and CIFOR.

Sonwa, D.J., S.F. Weise, M. Tchatat, B.A. Nkongmeneck, A.A. Adesina, Ousseynou Ndoye, and J. Gockowski. 2001. The Role of Cocoa Agroforests in Rural and Community Forestry in Southern Cameroon. *Rural Development Forestry Network Paper* 25g(i) (July): 1–10.

Stacey, R.D., D. Griffin, and P. Shaw. 2000. *Complexity and Management: Fad or Radical Challenge to Systems Thinking?* London: Routledge.

Stall, S., and R. Stoecker. 1998. Community Organizing or Organizing Community? Gender and the Crafts of Empowerment. *Gender and Society* 12(6): 729–56.

Standa-Gunda, W., R. Prabhu, and S. Dewi. 2003. Spatial Lags Hypothesis: Discerning a Conceptual Framework for Communication Networks. Harare, Zimbabwe: CIFOR.

Standa-Gunda, W., T. Mutimukuru, R. Nyirenda, M. Haggith, and J.K. Vanclay. 2003. Participatory Modelling to Enhance Social Learning, Collective Action and Mobilization among Ssers of the Mafungautsi Forest, Zimbabwe. *Small-Scale Forest Economics, Management and Policy* 2(2): 313–26.

Stankey, G.H., and R.N. Clark. 1998. Adaptive Management Areas: Roles and Opportunities for the PNW Research Station. Portland, OR: Forest Service, Pacific Northwest Research Station.

Stankey, G.H., and B. Shindler. 1997. Adaptive Management Areas: Achieving the Promise, Avoiding the Peril. Portland, OR: Forest Service, Pacific Northwest Research Station.

State Forest Service, Forest Inventory Unit. 2002. Project: Organisation and Development of Forest Management of the Model Leshoz Ortok (Russian language). Bishkek, Kyrgyzstan: State Forest Service.

Staudt, K.A. 1975–76. Women Farmers and Inequities in Agricultural Services. *Rural Africana, African Studies Center, Michigan State University* 29: 14.

———. 1978. Male Preference in Government Agricultural Policy Implementation. *Development and Change* 9: 439-57.

Stone, S. 2002. Community-based Timber Management, Porto Dias, Acre: Participation and Perceived Benefits and Costs. Gainesville, FL, 6 May.

Strehlke, B. 1993. Forest Management in Indonesia: Employment, Working Conditions and Occupational Safety. *Unasylva* 172(44): 25–30.

Styger, E., J.E.M. Rakotoarimanana, R. Rabevohitra, and E.C.M. Fernandes. 1999. Indigenous Fruit Trees of Madagascar: Potential Components of Agroforestry Systems to Improve Human Nutrition and Restore Biological Diversity. *Agroforestry Systems* 46: 289–310.

Summer Institute of Linguistics. 2003. *Part of the Ethnologue* (13). Summer Institute of Linguistics 1996. http://www.christusrex.org/www3/ethno/Came.html (accessed 19 February 2003).

Sunderlin, W., and J. Pokam. 2002. Economic Crisis and Forest Cover Change in Cameroon: The Roles of Migration, Crop Diversification, and Gender Division of Labour. *Economic Development and Cultural Change* 50(3): 581–606.

Tarapoto Process. 1998. In *Guidelines for Developing, Testing and Selecting Criteria and Indicators for Sustainable Forest Management:* Criteria and Indicators Toolbox Series No. 1, by R. Prabhu, C.J.P. Colfer, and R.G. Dudley. Bogor, Indonesia: CIFOR 1999.

Tchikangwa, B., M.A. Brocklesby, A.M. Tiani, M.A. Sardjono, R. Porro, A. Salim, and C.J.P Colfer. 2001. Rights to Manage Cooperatively and Equitably in Forest-Rich and Forest-Poor Contexts. In *People Managing Forests: The Links between Human Well Being and Sustainability*, edited by C.J.P. Colfer and Y. Byron. Washington, DC: Resources for the Future and CIFOR.

Terborgh, J. 1999. *Requiem for Nature.* Washington, DC: Island Press.

Thomas-Slayter, B., E. Wangari, and D. Rocheleau. 1996. Feminist Political Ecology: Crosscutting Themes, Theoretical Insights, Policy Implications. In *Feminist Political Ecology: Global Issues and*

Local Experiences, edited by D. Rocheleau, B. Thomas-Slayter and E. Wangari. New York: Routledge.

Thorburn, C. 2002. Regime Change: Prospects for Community-Based Resource Management in Post-New Order Indonesia. *Society and Natural Resources* 15: 617–28.

Tiani, A.M. 2000. Etude du Contexte: La Gestion du Massif Forestier d'Ottotomo: Tableau des Acteurs et les Premiers Enseignements en vue de L'opérationalisation de la Co-gestion Adaptative (ACM). [Context Study: The Management of the Ottotomo Forest: Table of Actors and First Lessons in View of the Operalization of Adaptive Co-Management]. Yaoundé, Cameroon: CIFOR.

———. 2001. The Place of Rural Women in the Management of Forest Resources. In *People Managing Forests: The Links between Human Well Being and Sustainability*, edited by C.J.P. Colfer and Y. Byron. Washington, DC: Resources for the Future and CIFOR.

———. 2003a. Genre, Diversité et ACM. [Gender, Diversity, and ACM]. Email communication with Carol Colfer. April 17.

———. 2003b. RE: Social Capital in Ottotomo. Email communication with Carol Colfer. 28 February.

Tiani, A.M., J. Nguiébouri, and C. Diaw. 2001. Criteria and Indicators for Collaborative Monitoring and Adaptive Co-management of Tropical Forests. *ACM News* 2(3): 7–8.

———. 2002. Criteria and Indicators as Tools for Adaptative and Collaborative Forest Management: A Guide. Also available in French, Yaoundé, Cameroon: CIFOR.

Tiani, A.M., G. Akwah, J. Nguiébouri, and C. Diaw. 2002. Les Communautés Riveraines du Parc National de Campo Ma'an: Maîtrise de l'Espace Forestier et Vécu Quotidien des Mutations Socio-Ecologiques. [The Riverine Communities of the Campo Ma'an National Park: Mastery of Forest Space and Daily Solutions to Socio-Ecological Change]. Yaoundé, Cameroon: CIFOR.

Tiani, A.M., G. Akwah, and J. Nguiébouri. 2005. Women in Campo Ma'an National Park: Uncertainties and Adaptations in Cameroon. In *The Equitable Forest: Diversity, Community, and Resource Management*, edited by C.J.P. Colfer. Washington, DC: Resources for the Future and CIFOR.

Townsend, J. G., with U. Arrevillaga, J. Bain, S. Cancino, S. Frenk, S. Pacheco, and E. Perez. 1995. *Women's Voices from the Rainforest*. London: Routledge.

Transparency International. 2002. *Corruption Perception Index*. http://www.transparency.org/cpi/2001/cpi2001.html (accessed 5 June 2002).

Tsing, A. L. 1993. *In the Realm of the Diamond Queen*. Princeton, NJ: Princeton University Press.

United Nations. 1995. Women's Education and Fertility Behaviour: Recent Evidence from the Demographic and Health Surveys. New York: U.N. Population Division.

Uphoff, N. 1996. *Learning from Gal Oya: Possibilities for Participatory Development and Post-Newtonian Social Science*. London: Intermediate Technology Publications.

———. 2000. Understanding Social Capital: Learning from the Analysis and Experience of Participation. In *Social Capital: A Multifaceted Perspective*, edited by P. Dasgupta and I. Serageldin. Washington, DC: The World Bank.

Uphoff, N., and C.M. Wijayaratna. 2000. Demonstrated Benefits from Social Capital: The productivity of Farmer Organizations in Gal Oya, Sri Lanka. *World Development* 28(11).

Uprety, L.P., N. P Sitaula, and K. Sharma. 2001a. Role of ACM for Effective Collaboration and Shared Learning. *ACM News* 2(4): 11–12.

———. 2002a. ACM Learning Tour (Nepal): 6 to 9 March 2002. *ACM News* 3(3): 5–6.

———. 2002b. Facilitation Skills Training on Community Forestry. *ACM News* 3(2): 4.

———. 2002c. Inclusion for Policy Refinement. *ACM News* 3(2): 1–2.

Uprety, L., K. Sharma, and N.P. Sitaula. 2001b. ACM Workshops in Nepal. *ACM News* 2(3): 6.

Vabi, M.B., C.N. Ngwasiri, P.T. Galega, and P.R Oyono. 2000. *The Devolution of Forest Management Responsibilities to Local Communities: Context and Implementation Hurdles in Cameroon*. Yaoundé, Cameroon: WWF-World Wide Fund for Nature, Cameroon Programme Office.

Valmores, C. 2001. Shifting Gears for a Genuine Devolution of Local Forest Management: The Case in Talaandig Community, Philippines. Bogor, Indonesia: CIFOR.

———. 2002. Historical Trends Report, ACM Lantapan. Bogor, Indonesia: CIFOR.

———. 2003. ACM Revives Old Traditions in the Philippines. *ACM News* 4(1): 7–8.

van den Berg, Yolanda, and K. Biesbrouck. 2000. *The Social Dimension of Rainforest Management in Cameroon: Issues for Co-Management*. Kribi, Cameroon: The Tropenbos-Cameroon Programme.

van Eijk, T. 2000. Holism and FSR. In *A History of Farming Systems Research*, edited by M. Collinson. Rome: Food and Agriculture Organization.

van Haaften, E.H., and F.J.R. van de Vijver. 1996. Psychological Stress and Marginalization as Indicators of Human Carrying Capacity in Deforesting Areas. *International Journal of Sustainable Development and World Ecology* 3: 32–42.

van Haaften, H. 1995. Final Report/Diary. Bogor, Indonesia: CIFOR.

van Holt, T. 2002. Hunting Practices of the Community of Salvatierra and Recommendations for Future Wildlife Use in the Region. Unpublished Report. Santa Cruz, Bolivia: BOLFOR.

Vanclay, J.K., F.L. Sinclair, and R. Prabhu. 2003. Modelling Interactions Amongst People and Forest Resources at the Landscape Scale. *Small-Scale Forest Economics, Management and Policy* 2(2): 117–20.

Vayda, A.P. 1983. Progressive Contextualization: Methods for Research in Human Ecology. *Human Ecology* 11: 165–81.

Vayda, A.P., C.J.P. Colfer, and M. Brotokusumo. 1980. Interactions between People and Forests in East Kalimantan. *Impact of Science on Society* 30: 179–90.

Venglovsky, B.I. 1998. Potential and Constraints for the Development of the Walnut-Fruit Forests of Kyrgyzstan. In *Biodiversity and Sustainable Use of Kyrgyzstan's Walnut-Fruit Forests*, edited by D. Gilmour. Gland, Cambridge, and Bern: IUCN and Intercooperation.

Venkateswaran, S. 1995. *Environment, Development and the Gender Gap*. New Delhi: Sage Publications.

Vermeulen, S.J. 2000. Setting the Context of the Mafungautsi Forest (Zimbabwe) Project in a Criteria and Indicators Framework. Harare, Zimbabwe: CIFOR.

Vogel, C.G., and R. Engelman. 1999. *Forging the Link: Emerging Accounts of Population and Environment Work in Communities*. Washington, DC: Population Action International.

Waldrop, M.M. 1992. *Complexity: The Emerging Science at the Edge of Order and Chaos*. New York: Touchstone Books, Simon and Schuster.

Walker, B., A. Kinzig, and J. Langridge. 1999. Plant Attribute Diversity, Resilience and Ecosystem Function. *Ecosystems* 2(2): 95–113.

Walker, D.H., and F.L. Sinclair. 1998. Acquiring Qualitative Knowledge about Complex Agroecosystems. Part 2: Formal representation. *Agricultural Systems* 56(3): 365–86.

Warner, K. 1991. *Shifting Cultivators: Local Technical Knowledge and Natural Resource Management in the Humid Tropics*. Vol. 8, Community Forestry Note. Rome: Food and Agriculture Organization.

Warren, D.M., P. Blunt, and D. Marsden (eds.). 1996. *Indigenous Organisations and Development*. London: Kegan Paul International.

WEHAB, Working Group. 2002. A Framework for Action on Health and the Environment. Johannesburg, South Africa: World Summit on Sustainable Development.

Wells, M., S. Guggenheim, A. Khan, W. Wardojo, and P. Jepson. 1999. Investing in Biodiversity: A Review of Indonesia's Integrated Conservation and Development Projects. Jakarta, Indonesia: The World Bank.

Wenger, E. 1999. *Communities of Practice*. Cambridge: Cambridge University Press.

Westley, F. 2002. The Devil in the Dynamics: Adaptive Management on the Front Lines. In *Panarchy: Understanding Transformations in Human and Natural Systems*, edited by L.H. Gunderson and C.S. Holling. Washington, DC: Island Press.

White, A., H. Gregersen, A. Lundgren, and G. Smucker. 2001. Making Public Protected Areas Systems Effective: An Operational Framework. In *Biological Diversity: Balancing Interests through Adaptive Collaborative Management*, edited by L. Buck, C.C. Geisler, J. Schelhas, and E. Wollenberg. Boca Raton, FL: CRC Press.

Whittaker, A., and C. Banwell. 2002. Positioning Policy: The Epistemology of Social Capital and its Application in Applied Rural Research in Australia. *Human Organization* 61(3): 252–61.

Wickham, T. 1997. Community-Based Participation in Wetland Conservation: Activities and Challenges of the Danau Sentarum Wildlife Reserve Conservation Project, Danau Sentarum Wildlife Reserve, West Kalimantan, Indonesia. In *Community Participation in Wetland Management: Lessons from the Field*, edited by G. Claridge and B. O'Callaghan. Kuala Lumpur, Malaysia: Wetlands International.

Wickramasinghe, A. 1994. *Deforestation, Women and Forestry: The Case of Sri Lanka*. Utrecht, the Netherlands: International Books.

Wilde,V.L., and A.Vainio-Mattila. 1995. *Gender Analysis and Forestry: International Training Package*. Rome: Food and Agriculture Organization.

Wittman, H. 2002. Negotiating Locality:The Impact of Forest Sector Decentralization on Communal Forest Management in Guatemala's Western Highlands. Master's thesis. Rural Sociology, Cornell University, Ithaca, NY.

Wollenberg, E. 2001. From the ACM-BRF Team in Malinau. *ACM News* 2(4): 10.

———. 2003a. Chapter 4: Devolution Status. Email communication with Carol Colfer. 14 April.

———. 2003b. Re: Population Data. Email communication with Carol Colfer. 25 January.

Wollenberg, E., D. Edmunds, and L. Buck. 2000. *Anticipating Change: Scenarios as a Tool for Adaptive Forest Management (a Guide)*. Also available in Spanish and Indonesian. Bogor, Indonesia: CIFOR.

Wollenberg, E., J. Anderson, and D. Edmunds. 2001a. Pluralism and the Less Powerful: Accommodating Multiple Interests in Local Forest Management. *International Journal of Agricultural Resources, Governance and Ecology* 1(3/4): 199–222.

Wollenberg, E., D. Edmunds, L. Buck, J. Fox, and S. Broch (eds.). 2001b. *Social Learning in Community Forests*. Bogor, Indonesia: CIFOR.

Wollenberg, E., A.A. Nawir, A. Uluk, and H. Pramono. 2001c. *Income Is Not Enough: The Effect of Economic Incentives on Forest Product Conservation*. Bogor, Indonesia: CIFOR.

Wollenberg, E., R. Iwan, G. Limberg, M. Moeliono, S. Rhee, and I Made Sudana. 2004. Muddling towards cooperation: A CIFOR Case Study of Shared Learning in Malinau District, Indonesia. *Currents* 33: 20–24.

Wrangham, R. 2002. Changing Policy Discourses and Traditional Communities, 1960–1999. In *Which Way Forward? People, Forests and Policymaking in Indonesia*, edited by C.J.P. Colfer and I.A.P. Resosudarmo. Washington, DC: Resources for the Future and CIFOR.

Wright, P.C. 1997. The Future of Biodiversity in Madagascar: A View from Ranomafana National Park. In *Natural Change and Human Impact in Madagascar*, edited by S.M. Goodman and B.D. Patterson. Washington, DC: Smithsonian Institution Press.

Yorque, R., B. Walker, C.S. Holling, L.H. Gunderson, C. Folke, S.R. Carpenter, and W.A. Brock. 2002. Toward an Integrative Synthesis. In *Panarchy: Understanding Transformations in Human and Natural Systems*, edited by L.H. Gunderson and C.S. Holling. Washington, DC: Island Press.

Yuliani, L., T. Kusumanto, Y. Indriatmoko, and S. Hakim. 2002. Indonesia Country Report: ACM PAR Main Case Studies. Bogor, Indonesia: CIFOR.

Yunusova, I. 1999. The Kyrgyz Forestry Concept: A Participatory Process for Forest Policy Formulation in Kyrgyzstan. In *Regional Forest Programmes: A Participatory Approach to Support Forest Based Regional Development*, edited by J. Vayrynen. European Forest Institute (EFI) Proceedings No. 32. Joensuu, Finland: EFI.

Zandstra, H.G., E.C. Price, J.A. Litsinger, and R.A. Morris. 1981. *A Methodology for On-Farm Cropping Systems Research*. Los Baños, Philippines: International Rice Research Institute (IRRI).

Acronyms

ACM	Adaptive Collaborative Management
AFC	Agricultural Finance Corporation, Zimbabwe
AFIS	*Asociación Forestal Indígena Selvatierra* (Indigenous Forestry Association of Selvatierra), Bolivia
AFORDA	[a research institution], Nepal
AGERAS	*Appui à la Gestion Environnementale Régionalisée et à l'Approche Spatiale* (Aid in Regional Environmental Management and Spatial Approaches), Madagascar
AGRITEX	Agricultural Technical and Extension Services, Zimbabwe
AIFUS	*Asociación Indígena Forestal Uribicha - Selvatierra* (Indigenous Forestry Association of Urubichá-Selvatierra), Bolivia
AIMCU	*Asociación Indígena de Manejo Curucú* (Indigenous Management Association of Curucú), Bolivia
AMPROJA	*Associação de Produtores Denominada* (a farmer's association), Brazil
ANAE	National Association for Environmental Action, Madagascar
ANAFOR	*Agence Nationale d'Appui au Développement Forestier* (recently replaced ONADEF) (National Aid Agency for Forestry Development—recently replaced ONADEF), Cameroon
ANGAP	National Association for the Management of Protected Areas, Madagascar
ANRMC	Area Natural Resource Management Committee, Malawi
APFT	*Avenir des Peuples des Forêts Tropiques* (Future of Tropical Forest Peoples, an NGO), Cameroon
APPEC	*Appui à la Promotion des Pygmées dans leur Environement Culturel* (Aid for the Promotion of Pygmies in their Cultural Environment), Cameroon
ARQMG	*Associação dos Remanescentes dos Quilombos de Gurupá* (Association of Quilombos), Brazil
asl	Above sea level
ASLASI	*Associação dos Lavradores do Setor de Ipixuna* (a farmers' association), Brazil

ASLs	*Agrupaciones Sociales de Lugar* (Local Social Associations), Bolivia
ASPOMACRE	*Associação de Produtores do Acre* (Association of Rural Producers of Acre [in Porto Dias]), Brazil
ASPRAA	*Associação dos Pequenos Produtores do Alto Rio Atuá* (Association of Small Producers of the Alto Rio Atuá), Brazil
ATD	*Association Terre et Developpement (*Ottotomo) (Land and Development Association), Cameroon
Bappeda	*Badan Perencanaan Pembangunan Daerah* (Regional Planning Board), Indonesia
BASA	*Banco da Amazônia* (Bank of Amazonia), Brazil
BATRIFA	Basac Tribal Farmers Association, Philippines
BCFP	Blantyre City Fuelwood Project, Malawi
BIDANI	Barangay Integrated Development Assistance for Nutrition Improvement, Philippines
BIOMA	*Yayasan Biosfer-Manusia* (Human-Biosphere-Human Organization, an NGO), Indonesia
BNRMC	Block Natural Resource Committees, Malawi
BOLFOR	Bolivia-USA Sustainable Forestry Project (USAID-funded), Bolivia
BPD	*Badan Perwakilan Desa* (Village Representative Body), Indonesia
BRDFI	Budyong Rural Development Foundation, Inc. , Philippines
BUFAI	Basac Upland Farmers Association, Inc., Philippines
CA	Communal Area, Zimbabwe
CADC	Certificate of Ancestral Domain Claim, Philippines
CADEF	*Centre d'Appui et de Développement des Femmes* (Center for the Aid and Development of Women), Cameroon
CADT	Certificate of Ancestral Domain Title, Philippines
CANFORET	*Cantonnement des Eaux et Forêts* (Forest Service Cantonment), Madagascar
CAPRI	Collective Action and Property Rights (a system-wide initiative among the Future Harvest research centers)
CBFM	Community Based Forest Management, Philippines
CBFMA	Community based forest management agreement, Philippines
CBFMP	Community based forest management program, Philippines
CEADES	*Colectivo para Estudios Aplicados para el Desarrollo Social* (Applied Studies Collective for Social Development), Bolivia
CEFDJA	*Centre d'Etudes Forestières du Dja* (Dja Center for Forestry Studies), Cameroon
CENRO	Community Environment and Natural Resources Office, Philippines
CEPFILD	*Cercle pour la Promotion des Forêts et des Initiatives Locales de Développement* (Circle for the Promotion of Forests and Local Development Initiatives), Cameroon
CF	Community Forest, Nepal
CFCS	Chimaliro Forest Compound Staff, Malawi
CFK	*Compagnie Forestiere Kritikos* [Kritikos Forestry Company, (a logging company)], Cameroon
CFM	Collaborative Forest Management, Kyrgyzstan
CFP	Community Forestry Program, Philippines

CGIAR	Consultative Group on International Agricultural Research
CIAD	*Centre International d'Appui au Développement* (International Center for Aid and Development, an NGO), Cameroon
CIFOR	Center for International Forestry Research
CIRAD	*Centre de Coopération Internationale de Recherche Agronomique pour le Développement* (Center for International Cooperation on Agronomic Research for Development), France
CNRM	Collaborative Natural Resource Management
CNS	*Conselho Nacional dos Seringueiros* (National Rubber Tappers' Council), Brazil
COFA	*La Compagnie Forestiere Assan* (The Assan Forestry Company, a (logging company)), Cameroon
COPNAG	*Central de Organizaciones de los Pueblos Nativos Guarayos* (Confederation of Guarayo Native Peoples), Bolivia
CORMAS	Common Pool Resources and Multi-Agent System
CPPAP	Conservation of Priority Protected Areas Project, Philippines
CTA	*Centro dos Trabalhadores da Amazônia* (Center for Amazonian Workers), Brazil
dbf	Diameter at breast height
DENR	Department of Environment and Natural Resources, Philippine
DFID	Department for International Development, UK
DFO	District Forestry Officer, Malawi
DFO	District Forest Officer, Nepal
DPRD	*Dewan Perwakilan Rakyat Daerah* (Regional House of Representatives), Indonesia
DSCO	District Soil Conservation Office, Nepal
ECOFAC	*Utilisation Rationnelle des ecosystèmes Forestiers d'Afrique Centrale* (Rational Use of Forested Ecosystems of Central Africa), Cameroon
ELAC	Environmental Legal Assistance Center, Philippines
EMATER	*Assistência Técnica e Extensão Rural* (State Extension Agency), Brazil
EMBRAPA	*Empresa Brasileira de Pesquisa Agropecuária* (Brazilian Agricultural Research Corporation), Brazil
EWW	Enterprise Works Worldwide, Philippines
FAFSA	*Fampandrosoana ny Faritra Sahavoemba* [(an association for the economic and socio-cultural development of Sahavoemba)], Madagascar
FASE	*Federação de Orgãos para Assistência Social e Educacional* (Federation of Organizations for Social Assistance and Education), Brazil
FC	Forestry Commission, Zimbabwe
FECOFUN	Federation of Community Forest Users, Nepal
FES	Forestry Extension Services, Zimbabwe
FETAGRI	*Federação dos Trabalhadores na Agricultura* (Federation of Rural Workers), Brazil
FIMARA	*Fitsaboana Malagasy eto Ranomafana* ([planned association for traditional and "'modern'" medical practitioners]), Madagascar
FLMP	Forest Land Management Programme, Philippines
FLORES	Forest Land Oriented Resource Envisioning System

FORCE	*Fédération des Organisations Rurales Pour le Cameroun Economique* (Federation of Rural Organizations for the Cameroonian Economy, [a farmers' organization]), Cameroon
FPU	Forest Protection Unit, Zimbabwe
FRIM	Forestry Research Institute of Malawi, Malawi
FSR&D	Farming Systems Research and Development
FSR&E	Farming Systems Research and Extension
FUG	Forest User Group, Nepal
FUGC	Forest User Group Committee, Nepal
FUNAP	*Fundação de Amparo ao Trabalhador Preso* (Foundation in Support of Imprisoned Workers), Brazil
GELOSE	*Gestion Locale et Securisée des Resources Renouvelables* (Local Management and Security (of tenure) of Renewable Resources), Madagascar
GIC	*Groupe d'initiative Commune* (Common Initiative Group), Cameroon
HEVECAM	rubber plantation agroindustry, Cameroon
HFC	*La Société Forestiere de Campo (*The Campo Forestry Society, a (logging company)), Cameroon
HPHH	*Hak Pemungutan Hasil Hutan* (Forest product harvesting rights), Indonesia
IBAMA	*Instituto Brasileiro de Meio Ambiente e dos Recursos Naturais Renováveis* (the Brazilian Institute for the Environment and Renewable Resources), Brazil
ICDP	Integrated Conservation and Development Project
ICRAF	International Center for Research on Agro-Fforestry
IGA	Income Generating Activity, Malawi
IITA	International Institute of Tropical Agriculture, Cameroon
INCRA	*Instituto Nacional de Colonizaçao e Reforma Agrária* (the federal institute for colonization and agrarian reform), Brazil
INPA	*Instituto Nacional de Pesquisas Amazonica* (National Institute of Amazonian Research), Brazil
INRA	*Instituto Nacional de Reforma Agraria* (National Service of Agrarian Reform), Bolivia
IPPK	*Izin Pemanfaatan dan Pengelolaan Kayu* (Permit for Timber Use and Management), Indonesia
IPRA	Indigenous People's Rights Act, Philippines
IRAD	*Institut de Recherche Agronomique pour le Développement* (Institute for Agronomic Research for Development), Cameroon
IRMP	Integrated Resource Management Program, Philippines
ITERPA	*Instituto de Terras do Pará* (Land Institute of Pará), Brazil
JFM	Joint Forest Management, [India]
KGV	Kitanglad Guard Volunteers, Philippines
KIDEKI	*Kisan Dekhi Kisan Samma* [an organization to promote environmentally friendly agricultural practices], Nepal
KIN	Kitanglad Integrated NGOs, Philippines
KIRFOR	Kyrgyz–Swiss Forestry Sector Support Program
LGU	Local Government Unit, Philippines
MANFU	*Manejo Forestal Urubichá* (Urubichá Forest Management), Bolivia

MICET	Madagascar Institute for the Conservation of Tropical Environments, Madagascar
MINAGRI	*Ministère de l'Agriculture* (Ministry of Agriculture), Cameroon
MINAT(D)	*Ministère de l'administration Territorial et de la Décentralisation* (Ministry of Territorial Administration), Cameroon
MINEF	*Ministère de l'Environnement et Forêts* (Ministry of Environment and Forests), Cameroon
MINEPIA	*Ministère de L'elevage des Pêches et des Industries Animales* (Ministry of Fish Culture and Animal Husbandry Industries), Cameroon
NCIP	National Commission on Indigenous Peoples , Philippines
NGO	Non-Governmental Organization, Philippines
NORMS	Natural and Organizational Resources Management Services, Nepal
NRB	Natural Resource Board, Zimbabwe
NTFP	Non Timber Forest Product
ONADEF	*Office National de Développement des Forêts* (National office of forest development), Cameroon
PAD	*Pendapatan Asli Daerah* (Local Regional Revenue), Indonesia
PAE	*Projeto de Assentamento Agro-Extrativista* (Agroextractive Settlement Project), Brazil
PAMB	Protected Area Management Board, Philippines
PAR	Participatory Action Research
PCSD	Palawan Council for Sustainable Development, Philippines
PDA	*Projetos de Colonização Dirigida* (Directed Settlement Projects), Brazil
PEPIPALM	*Projet d'Appui au Développement des Palmeraies Villageoises* (aid project for village development of palms), Cameroon
PESACRE	*Agroforestry Research and Extension Group of Acre* (Group for Research and Extension in Agroforestry Systems of Acre), Brazil
PFRA	Participatory Forest Resource Assessment, Malawi
PO	People's Organization, Philippines
PPG7	*Programa Piloto para a Proteção das Florestas Tropicais do Brasil* (Program for the protection of the Amazonian forests funded by the G-7 countries and Brazil), Brazil
PPLH	*Pusat Penelitian Lingkungan Hidup* (Environmental Studies Center), Indonesia
PRA	Participatory Rural Appraisal
PRODEX	*Programa de Desenvolvimento do Extrativismo* (A credit program for rubber growers, through BASA), Brazil
PRONAF	*Programa de Nacional de Fortalecimento a Agricultura Familiar* (National Program to Strengthen Small Scale Agriculture), Brazil
PRONAF	*Programa Nacional de Fortalecimento da Agricultura Familiar* (National Program to Support Small Scale Farmers), Brazil
PRONAFINHO	*Programa de Nacional de Fortalecimento a Agricultura Familiar* (National Program to Strengthen Small Scale Agriculture), Brazil
PTFPP	Palawan Tropical Forest Protection Program , Philippines
RDC	Rural District Council, Zimbabwe
RIMCU	Research Institute for Mindanao Culture, Philippine

RMC	Resource Management Committee, Zimbabwe
RNP	Ranomafana National Park, Madagascar
RPCC	Range Post Coordination Committee, Nepal
RRMP	Rain-fed Resource Management Program, Philippines
RRN	Rural Reconstruction Nepal, Nepal
SADC	Southern African Development Community
SDC	Swiss Agency for Development and Co-operation
SDDL	*Soutien au Développement Durable dans la Région de Lomié* (Sustainable Development Support Project in the Lomié Region), Cameroon
SFID	*Société Forestiére et Industrielle de la Doumé* (Forestry and Industrial Company of Doumé), Cameroon
SIMILE	[a system dynamics computer program]
SNV	*Stichting Nederlandse Vijwilligers* (Netherlands Development Organization)
SOCAPALM	*Société Camerounaise de Développement du Palmier à Huile* (Cameroonian Oil Palm Development Company), Cameroon
STCMPC	San Rafael, Tanabag, Concepción Multi Purpose Cooperative, Philippines
STR	*Sindicato dos Trabalhadores Rurais* (Rural Workers' Union), Brazil
SUDAM	*Superintendência para o Desenvolvimento da Amazônia* (Superintendency for Development of the Amazon), Brazil
TCO	*Terras Comunales de Origen* (Communal Indigenous Territory), Bolivia
UFA	*Unité Forestière d'Aménagement* (Forestry Development Unit), Cameroon
UFI	*Unidad Forestal Indígena* (Indigenous Forest Management Group), Bolivia
UNDP	United Nations Development Programme
UnMul	*Universitas Mulawarman* (Mulawarman University), Samarinda, East Kalimantan, Indonesia
UPTD	*Unit Pelaksana Teknis Daerah* (Regional Technical Implementation Unit), Indonesia
UTO	*Unité Technique Opérationnelle* (Technical Operations Unit), Cameroon
VDC	Village Development Committee, Nepal
VIDCO	Village Development Committee, Zimbabwe
VNRMC	Village Natural Resource Committees, Malawi
WADCO	Ward Development Committee, Zimbabwe
WARSI	Warung Informasi Konservasi (Conservation Information Forum), an NGO network, Indonesia
WEDA	Women's Environment and Development Association, Nepal
WID	Women in Development
WWF	World Wide Fund for Nature
YAP	*Yayasan Adat Punan* (Punan Traditional Organization), Indonesia
ZIL	Centre for International Agriculture at the Swiss Federal Institute of Technology, Zurich

Glossary

Acai: kind of palm harvested in Para [Brazil]

Adat: customary law [Indonesia]

ak sakal: village elder (lit. "white beard") [Kyrgyzstan]

alimaong: Talaandig tribal warriors [Philippines]

Arenda: lease [Kyrgyzstan]

Ashar: voluntary group labour within families or a community [Kyrgyzstan]

Ayil okmotu: administrative center [Kyrgyzstan]

Bagyeli: a pygmy, hunter-gatherer group in Campo Ma'an [Cameroon]

Baka: a pygmy, hunter-gatherer group in Dimako/Lomie [Cameroon]

Bakum: important ethnic group in eastern Cameroon [Cameroon]

Banjar: ethnic group from South Kalimantan [Indonesia]

Bantay bayan: local security patrol in Basac [Philippines]

Barangay: village [Philippines]

Batac: indigenous group in Palawan [Philippines]

Batak: ethnic group from northern Sumatra [Indonesia]

Batanga: important ethnic group in southern Cameroon [Cameroon]

Beteseleo: In-migrating settler ethnic group [Madagascar]

Bugis: ethnic group from Sulawesi [Indonesia]

Bulu: important ethnic group in southern Cameroon [Cameroon]

Bupati: district head [Indonesia]

Camat: head of the subdistrict [Indonesia]

Cambas: people from the lowlands [Bolivia]

Campesinos: settlers [Bolivia]

Capoeira: young secondary forests [Brazil]

Capoeirão: older secondary forests [Brazil]

Colocações: family rubber tapping area set aside with an extractive zone [Brazil]

comité de vigilance: vigilance committee [Cameroon]

Commune: a district, municipality, council, an administrative unit [Cameroon]

Dalit: occupational castes (e.g., blacksmiths, tailors, cobblers) [Nepal]

de facto: informally, or in fact [Latin]

de jure: legally, or according to the law [Latin]

ejidos: rural landholding communities [Mexico]

eka'a: rotating work groups [Cameroon]

fihavanana: family [Madagascar]

firaisankina: solidarity [Madagascar]

firaisina: union [Madagascar]

floresta alta: high forests [Brazil]

floresta baixa: low forests [Brazil]

fokonolona: territorial division of a descent group, used by park authorities [Madagascar]

Forets et Terroirs: Forests and Territories, a French-Cameroonian project [Cameroon]

gaharu: an expensive, aromatic heartwood from *Aquilaria* trees [Indonesia]

ganyu: piece work from the wealthy for the poor [Malawi]

gendarmes: police [Cameroon]

GEOVIC: U.S.-Cameroon mining operation in Lomié [Cameroon]

Groupement: administrative grouping of villages [Cameroon

Groupes d'Initiatives Communes: government-sponsored, Common Initiative Groups [Cameroon]

haribon: eagle (shortened from *haring ibon* in Tagalog), NGO name [Philippines]

Hévécam: parastatal rubber plantation [Cameroon]

Inhutani: one of numerous, numbered, parastatal timber companies [Indonesia]

Javanese: ethnic group from Java [Indonesia]

Jeep ni Erap: a government credit scheme, Basac [Philippines]

Jernang: patron of Orang Rimba in Jambi [Indonesia]

Kabupaten: District [Indonesia]

Kenyah: Borneo Dayak ethnic group, swidden cultivators [Indonesia]

kollas: people of the highlands [Bolivia]

kraalhead: village head [Zimbabwe]

Lehibeniala: forest guards [Madagascar]

leshoz: state forest enterprise [Kyrgyzstan]

loi: law [Cameroon (French)]

Lundaye: Borneo Dayak ethnic group, swidden cultivators [Indonesia]

lupon: a local dispute settlement mechanism, Basac [Philippines]

Mabi: important ethnic group in southern Cameroon [Cameroon]

mansa: declaration of possession [Brazil]

mengalir: to go with the stream [Indonesia]

Merap: Borneo Dayak ethnic group, swidden cultivators [Indonesia]

Minangkabau: ethnic group in Sumatra [Indonesia]

Mpanjaka: traditional chief [Madagascar]

Municipio: municipality, district [Brazil]

Mvae: important ethnic group in southern Cameroon [Cameroon]

mvog: lineage [Cameroon]

nda bot: family [Cameroon]

ndo'o: Irvingia gabonensis *[Cameroon]*
Nuevo Ley Forestal: New Forestry Law [Bolivia]
orang rimba: hunter gatherers in Sumatra (previously called *Kubu*) [Indonesia]
Padi: a partner NGO in East Kalimantan, rice in Bahasa Indonesia [Indonesia]
pahina: Talaandig traditional collective action [Philippines]
pangas: a kind of weapon/tool [Malawi]
pataua: an Amazonian crop from Para [Brazil]
PEPIPALM: project to encourage cocoa production [Cameroon]
pirateros: illegal loggers [Bolivia]
Pol: important ethnic group in eastern Cameroon [Cameroon]
Prefeitura: municipal (district) government [Brazil]
Punan: A hunter-gatherer group in Bulungan, East Kalimantan [Indonesia]
quilombos: descendants of slaves [Brazil]
Raiamandreny: notables [Madagascar]
ribeirinhos: people living along the rivers [Brazil]
sala: customary fine among the Talaandig [Philippines]
Sandawa: Talaandig traditional festival [Philippines]
sempolo: rotation work groups, Kalimantan [Indonesia]
shirine: a rotating system of responsibility for hospitality [Kyrgyzstan]
Socapalm: parastatal oil palm plantation [Cameroon]
sondeo: an early form of participatory rural appraisal used in FSR&D [global]
sous-prefet: head of a commune or district, an administrative unit [Cameroon]
stool: a hereditary symbol of chieftainship, site of land ownership [Ghana]
Superintendencia Forestal: Forestry Superintendency [Bolivia]
Tagbanua: indigenous group in Palawan [Philippines]
Tanala: forest people of Ranomafana [Madagascar]
terai: lowland, forested, southern area of Nepal [Nepal]
Terra amarela: yellow soils (probably heavy, high in clay) [Brazil]
terra branca: poor soils (lit. 'white soils') [Brazil]
terra firme: dry land [Brazil]
terra preta: better soils [Brazil]
terra roxa: red soil, probably due to iron oxide [Brazil]
Tidung: Borneo Dayak ethnic group, swidden cultivators [Indonesia]
tole: hamlet [Nepal]
tontines: rotating credit groups [Cameroon]
tontines de travail: rotating work groups [Cameroon]
touke: middleman, person with capital [Indonesia]
Tribunal de première instance: modern court of justice at district level [Cameroon]
Tropenbos: Dutch research institution, NGO [Netherlands]
tulugan: Talaandig tribal meeting place [Philippines]
turun bertahun: a traditional, Minangkabaru yearly planning activity [Indonesia]
ugnayan: relationship, coordination [Philippines (Tagalog)]
ugnayan: relationship, coordination [Philippines (Tagalog)]
varzea: flood plain [Brazil]
ventes de coupe: sales of standing volume (logging contract) [Cameroon]
vleis: wetlands [Zimbabwe]
Yassa: important ethnic group in southern Cameroon [Cameroon]

Index

ACM. *See* Adaptive collaborative management (ACM)

Acre, Brazil. *See* Brazil

Action groups, 72
   See also User groups

Adaptive collaborative management (ACM)
   assessment process, 6–10, 60–62, 163, 186, 193–94, 198–99
   conceptual evolution of, 44
   conclusions, 184–87
   defined, 6, 60
   dimensions, analytical, 51–59
   early stages, 101
   farming systems research and, 38
   future implications, 194–98
   global team integration, 175–77
   pace of change, 192–93
   participatory process, 4–6
   as patchwork quilt, 30
   research partners and, 170–75
   team management, 15–17
   theoretical models, 7, 13–15, 44–50, 186
   variables matrix, 20–21

Adaptive collaborative management (ACM), impact of
   age differentiation and, 139
   assessing, 60–62, 193–94
   chart of, 20–21, 90–91
   community heterogeneity and, 143–44
   conflict levels and, 146–48, 149–51
   in conservation parks, 128
   cumulative effect of, 87–91
   devolution status and, 103–4

diversity and, 133–36
   forest types and, 109
   gender differentiation and, 137–38
   management goals and, 127–30
   population density and, 110
   social capital and, 157

Adaptiveness (diachronic element), 4, 62–69

Age differentiation, 139–40
   See also under specific countries

Agreements, clandestine, 100

Agriculture
   cash crops, 119
   farm settlements in forests, 115
   farmers' organizations, 156
   farming systems research, 37–38
   plots, size of, 119–20
   rubber tapping conflicts, 216
   swidden, 188–89

Agro-extractive reserves, 22, 73, 84, 212

Agroforestry, 39

Akok, Cameroon
   conflict levels, 247–48
   devolution status, 239
   forest types, 240–41
   logging, large-scale, 242–43
   social capital, 156, 249–51

Almaciga extraction, 46, 82–83

Anthropological studies, 31–33

Associations, 74–75, 156, 206

Authoritative knowledge, 32, 115

Bagyeli (pygmies), 246

Bamdibhir, Nepal, 66, 94–95